Information Science and Statistics

Series Editors
M. Jordan, J. Kleinberg, B. Schölkopf

T0180564

C.S. Wallace

Statistical and Inductive Inference by Minimum Message Length

With 22 Figures

 Springer

C.S. Wallace
(deceased)

Series Editors
Michael Jordan
Department of Computer Science
University of California, Berkeley
Berkeley, CA 94720
USA
jordan@stat.berkeley.edu

Professor Jon Kleinberg
Department of Computer Science
Cornell University
Ithaca, NY 14853
USA

Professor Bernhard Schölkopf
Max Planck Institute for Biological
   Cybernetics
Spemannstrasse 38
72076 Tübingen
Germany

Library of Congress Cataloging-in-Publication Data
Wallace, C. S. (Christopher S.), d. 2004.
    Statistical and inductive inference by minimum message length / C.S. Wallace.
      p. cm. — (Information science and statistics)
    Includes bibliographical references and index.

      1. Minimum message length (Information theory)   2. Mathematical
    statistics—Methodology.   3. Induction (Mathematics)   I. Title.   II. Series.
    QA276.9.W35   2005
    519.5—dc22   2004059195

ISBN-13: 978-1-4419-2015-7   Printed on acid-free paper.
e-ISBN-13: 978-0-387-27656-4

Printed in the United States of America.   (EB)

9 8 7 6 5 4 3 2 1

springeronline.com

Preface

My thanks are due to the many people who have assisted in the work reported here and in the preparation of this book. The work is incomplete and this account of it rougher than it might be. Such virtues as it has owe much to others; the faults are all mine.

My work leading to this book began when David Boulton and I attempted to develop a method for intrinsic classification. Given data on a sample from some population, we aimed to discover whether the population should be considered to be a mixture of different types, classes or species of thing, and, if so, how many classes were present, what each class looked like, and which things in the sample belonged to which class. I saw the problem as one of Bayesian inference, but with prior probability densities replaced by discrete probabilities reflecting the precision to which the data would allow parameters to be estimated. Boulton, however, proposed that a classification of the sample was a way of briefly encoding the data: once each class was described and each thing assigned to a class, the data for a thing would be partially implied by the characteristics of its class, and hence require little further description. After some weeks' arguing our cases, we decided on the maths for each approach, and soon discovered they gave essentially the same results. Without Boulton's insight, we may never have made the connection between inference and brief encoding, which is the heart of this work.

Jon Patrick recognized in the classification work a possible means of analysing the geometry of megalithic stone circles and began a PhD on the problem. As it progressed, it became clear that the message-length tools used in the classification method could be generalized to apply to many model-selection and statistical inference problems, leading to our first attempts to formalize the "Minimum Message Length" method. However, these attempts seemed to be incomprehensible or repugnant to the referees of statistical journals. Fortunately, Peter Freeman, a proper statistician who had looked at the stone circle problem, saw some virtue in the approach and very kindly spent a year's sabbatical helping to frame the idea in acceptable statistical terms, leading to the first publication of MML in a statistical journal [55]. Acceptance was probably assisted by the simultaneous publication of the independent but related work of Jorma Rissanen [35].

Over the 35-year gestation of this book, I have benefited greatly from the suggestions, comments and criticisms of many colleagues and anonymous referees. The list includes Mike Georgeff, Peter Cheeseman, Ray Solomonoff, Phil Dawid, David Hand, Paul Vitanyi, Alex Gammerman, Ross Quinlan, Peter Tischer, Lloyd Allison, Trevor Dix, Kevin Korb, Murray Jorgenson, Mike Dale, Charles Twardy, Jon Oliver, Rohan Baxter and especially David Dowe, who has contributed significantly both to the range of applications of MML and to the development of new approximations for message lengths and MML estimators.

I must also thank Julie Austin, who typed and proofread the early chapters, and Steve Gardner and Torsten Seeman, who helped convert the original draft into LaTeX.

Finally, without the constant support of my wife Judith, I would never have managed to complete the work.

Victoria, Australia, August 2004   C.S. Wallace

Disclaimer

The reader should be warned that I make no claim to be an authority on statistical inference, information theory, inductive reasoning or the philosophy of science. I have not read widely in any of these fields, so my discussions of others' work should be treated with some suspicion. The ideas in this book are those of a one-time physicist who drifted into computing via work on computer hardware and arithmetic. In this uncertain progress towards enlightenment, I encountered a succession of problems in analysing and understanding data for which I could find no very satisfactory solution in standard texts. Over the years, the MML approach was developed from rather ad hoc beginnings, but the development was driven mostly by the challenge of new problems and informal argument with colleagues, rather than by a proper study of existing work. This casual, indeed almost accidental, evolution partly excuses my paucity of citations.

Editorial Notes

This book is essentially the manuscript left behind by Christopher Wallace when he died on August 7, 2004.

We wanted to publish a book that was as close as possible to the original manuscript. We have therefore made only minimal changes to the manuscript. We have corrected typing and spelling errors. We have also attempted as best as we could to include all the references that the author intended to include. Where the author made it clear that he wanted to add citations, but did not indicate to what they referred, we have included our best guesses of what these references might be.

Colleagues in the School of Computer Science and Software Engineering at Monash University are in the process of developing a web site that will provide additional material and references on Minimum Message Length to assist the readers of this book.

We acknowledge the contributions of Craig Callender, H. Dieter Zeh, Douglas Kutach and Huw Price for their helpful comments on Chapter 8, the editorial assistance of Sarah George and Yuval Marom in preparing this chapter, and that of Jeanette Niehus in preparing the index. We are also indebted to all the people who helped produce this manuscript after Chris' death. In particular, we thank the following people who assisted us in proofreading the final version of this book: Lloyd Allison, David Dowe, Graham Farr, Steven Gardner, Les Goldschlager, Kevin Korb, Peter Tischer, Charles Twardy and Judy Wallace.

Victoria, Australia, February 2005   D.W. Albrecht and I. Zukerman

Acknowledgment

Judith, Christopher Wallace's wife and executor, gratefully acknowledges the tireless efforts of his colleagues at Computer Science and Software Engineering, Monash University in the final preparation of his manuscript for publication. Thanks are due to Professor Gopal Gupta for his advice and support, and in particular to Dr Albrecht and Associate Professor Zukerman for the very many hours spent editing the final text.

Contents

1. Inductive Inference

1.1 Introduction

The best explanation of the facts is the shortest.

This is scarcely a new idea. In various forms, it has been proposed for centuries, at least from Occam's Razor on. Like many aphorisms, it seems to express a notion which is generally accepted to be more or less true, but so vague and imprecise, so subject to qualifications and exceptions, as to be useless as a rule in serious scientific enquiry. But, beginning around 1965, a small number of workers in different parts of the world and in different disciplines began to examine the consequences of taking the statement seriously and giving it a precise, quantitative meaning. The results have been surprising. At least three related but distinct lines of work have been developed, with somewhat different aims and techniques. This book concentrates on just the one line in which the author has worked, but two other important lines are briefly surveyed in Chapter 10. The major claims of this line refer to several fields.

- Bayesian Inference: The new method unifies *model selection* and *estimation*, usually treated as separate exercises. In many cases, the results obtained by treating both questions at once are superior to previous methods. While closely related to existing Bayesian statistical theory, it provides a sound basis for point estimation of parameters which, unlike "MAP" and "mean of posterior" estimates, do not depend on how the assumed data distribution is parameterized.
- Best Explanation of the Data: For the engineer, scientist or clinician who needs to work with a single, well-defined "best guess" hypothesis, the new result is more useable than methods which provide only "confidence intervals" or posterior densities over a sometimes complex hypothesis space.
- Induction: The work gives a new insight into the nature of inductive reasoning, i.e., reasoning from a body of specific facts and observations to general theories. Hitherto, there has been no accepted logical basis for inductive reasoning despite its great importance.
- Philosophy of Science: The discovery, refinement (and sometimes wholesale replacement) of scientific theories is essentially inductive, and the philosophy of science has been hampered by a lack of a logic for induction. The

new insight is at least a step towards a theory of scientific enquiry which is both normative and descriptive.
- Machine Learning: As a branch of Artificial Intelligence research, machine learning is an attempt to automate the discovery of patterns in data, which amounts to the formation of a theory about the data. One result of the new work has been a sound criterion for assessing what has been "learnt", leading to successful new algorithms for machine learning applications.

These claims may seem rather dry, of interest only to the specialist statistician, machine learning expert, or logician. There are much wider implications.

If the basis of the new approach is sound, it seems to lead to a clearer understanding of the role and methods of science and the validity of its claim to be a search for objective truth about the world. It also places scientific enquiry in the same conceptual basket as the development of human language, traditional techniques of navigation, tool-use, agriculture, hunting, animal husbandry, and all the other skills our species has learnt. They all, including science as we practice it, are based on inductive reasoning from real-world observations to general theories about how the world behaves. In the earliest developments of human culture, these "theories" were possibly not consciously formulated: the emergence of vocal signals for danger, food, enemy, friend, come-here, etc. and the earliest skills for finding food, more likely came from many generations of gradual refinement of simple instinctive actions, but in logical terms they are theories indeed: recognition of similarity or more subtle regularity among many things or happenings. In this light, science and engineering are not wholly revolutionary initiatives of recent centuries, but just the gradual systemization of what humans have always done.

Viewing science as common sense used carefully, we find that the new insight gives strong theoretical support for the belief that, given the same physical environment, any sufficiently long-lived, large and motivated community of intelligent beings will eventually come to the same, or at least equivalent, theories about the world, and use more or less equivalent languages to express them. This conclusion is not unqualified: the development of theories about different aspects of the world may well proceed at different rates and not follow the same paths of refinement and replacement in different communities, but, if our account of induction is right, convergence will occur. The idea that there may be different, correct, but incompatible views of reality seems untenable. If two sets of belief are incompatible but equally valid, it can only be that they are equally wrong. Note that we are not asserting that scientific communities will inevitably converge to a finite set of fundamental theories which then express everything which can be learnt. Our account of inductive reasoning admits the possibility that complete and ultimate "theories of everything" may never be reached in a finite time, and perhaps may not even be expressible, as will be explained later.

It has been said that to a man with a hammer, all problems look like nails. Having perhaps acquired a new and shiny hammer in the shape of a theory of induction, we will of course fall to temptation and swing it at a couple of "nails" which it may miss. These attempts have been left to the end of this work, but may have some value.

This book presents the basic theory of the new approach and shows in numerous examples its application to problems in statistical inference and automated inductive inference, which is usually called "learning" in the Artificial Intelligence literature. I emphasize statistical and machine-learning applications because in these limited arenas, the new approach can be applied with sufficient rigour to allow its performance to be properly assessed. In less well-understood and wider arenas, the approach can arguably be shown to have some merit, but the arguments cannot at this stage be made compelling and must involve some arm-waving. By contrast, statistical inference is relatively simple and its language of probability well defined in operational terms, even if it rests on somewhat ambiguous conceptual foundations. If our approach cannot at least handle problems of some difficulty in this relatively simple field, it cannot be credible. We therefore think it important to show that it performs well in this field, and believe the examples given in this book demonstrate that it does. Moreover, at least some of the examples seem to show its performance to better that of previous general principles of statistical inference, and we have so far found no problems where its performance is notably inferior.

The formal arguments in-principle for the approach, as opposed to specific demonstrations of performance on particular problems, are mainly confined to statistical inference, but are extended to a less-restricted formal treatment of inductive inference. The extensions are based on the theory of Universal Turing Machines, which deals with the capabilities of digital computers, and as far as is currently known, also covers the capabilities of all sufficiently general reasoning systems, including human reasoning. The extensions draw on the work of Turing, Solomonoff, Chaitin and others and provide formal arguments for supposing that our approach is applicable to the inductive inference of any theories whose implications are computable. According to some theorists, this range includes all theories which can be explicitly communicated from one person to another and then applied by the recipient, but we will not pursue that argument. We will, however, argue that the approach provides the basis for a partial account of scientific inference which is both normative and descriptive. It says how science ought to choose its theories, and fairly well describes how it actually does, but has little to contribute to an account of what drives science as a social activity, i.e., what determines the direction of social investment in different areas of enquiry.

This first chapter continues with an informal introduction to inductive inference and our approach to it. It outlines the more obvious problems associated with inductive inference and mentions a couple of well-known approaches to these problems. The critique of classical approaches is neither comprehensive nor fair.

To advance any further with the argument, the informal discussion must be followed by a formal treatment which is inevitably quantitative and mathematical. We assume some familiarity with elementary statistics and simple distributions such as the Normal and Binomial forms. Where less familiar statistical models are used, the models will be briefly introduced and described. The mathematics required is restricted to elementary calculus and matrix algebra.

The chapter concludes with a brief introduction to probability and statistics, essentially just to establish the notations and assumptions used later. The nature of the inferences which can be made using conventional non-Bayesian and Bayesian reasoning are outlined, and certain criticisms made. We do not pretend this critique is comprehensive or impartial. It is intended merely to clarify the distinctions between conventional statistical inference and the method developed in this work. Not all of the workers who have contributed to the new approach would necessarily agree with the critique, and the results obtained with the new approach do not depend on its validity. These sections might well be merely skimmed by readers with a statistical background, but the criticisms may be of interest.

The second chapter introduces the elementary results of Information Theory and Turing Machine theory which will be needed in the sequel. These results are needed to define the notion of the "length" of an explanation and to sharpen the concepts of "pattern" and "randomness". There is nothing really novel in this treatment, and it could well be skipped by readers who are familiar with Shannon information and Kolmogorov-Chaitin complexity. However, at the time of writing (2004) it seems some of this material is still not as well understood as one might expect. In recent years several papers have been published on applications of Minimum Message Length or Rissanen's related Minimum Description Length, which have made significant and in some cases serious errors in estimating the length of "explanation" messages. The commonest errors have arisen from a failure to realize that for the statement of a hypothesis about a given body of data, the shortest useable code need not in general allow the encoding of all hypotheses initially contemplated, and should never state a parameter of the asserted hypothesis more precisely than is warranted by the volume and nature of the data. A couple of examples are discussed in later chapters. Other errors have arisen because the authors have applied approximations described in early MML papers in which the limitations of the approximations were not emphasized.

The third, fourth and fifth chapters formally develop the new approach to statistical inference. In Chapter 3, the development is exact, but leads to a method which is computationally infeasible except in the simplest problems. Chapters 4 and 5 introduce approximations to the treatment, leading to useable results. A number of simple inference problems are used in these chapters to illustrate the nature and limitations of the exact and approximate treatments. Chapter 6 looks in more detail at a variety of fairly simple problems, and introduces a couple of techniques needed for some rather more difficult problems. Chapter 7 gives examples of the use of MML in problems where the possible models include models of different order, number of free parameters or logical structure. In these, MML is shown to perform well in selecting a model of appropriate complexity while simultaneously estimating its parameters. Chapter 8 is speculative, presenting an argument that, while deductive (probabilistic) logic is properly applied in predicting the future state of a system whose present state is partly known, useful assertions about the past state of the system require inductive reasoning for which MML appears well suited. Chapter 9 considers whether scientific enquiry can be seen as conforming to the MML principle, at least over the long term. Chapter 10 briefly discusses two bodies of work, Solomonoff's predictive process and Rissanen's Normalized Maximum Likelihood, both of which embody the same "brief encoding" notion as Minimum Message Length but apply it to different ends.

1.2 Inductive Inference

The term "inductive" is sometimes used in the literature to apply to any reasoning other than deductive, i.e., any reasoning where the conclusions are not provably correct given the premises. We will use the term only in the above narrower (and more common) sense. Deductive reasoning, from general theories and axioms to specific conclusions about particular cases, has been studied and systematized since Aristotle and is now fairly well understood, but induction has been much more difficult to master. The new results give an account of inductive reasoning which avoids many of the difficulties in previous accounts, and which has allowed some limited forms of inductive reasoning to be successfully automated.

With the exception of "knowledge" that we are born with or which comes through extraordinary routes such as divine inspiration, our knowledge of the external world is limited to what our senses tell us and to the inferences we may draw from this data. To the extent that a language like English allows, the information gained about our surroundings can be framed as very specific propositions, for instance "I feel warm", "I hear a loud rhythmic sound of varying pitch", "I see blue up high, green-brown lower down, brown at the bottom". Groups of such elementary sensory propositions can be interpreted by most people to arrive at propositions whose terms involve some abstraction from the immediate sense data. "That motor car has rust in its doors", "Joe is lying down", "Percy said 'Crows are black' ". Let us accept that at least this degree of abstraction may be taken for granted, to save us the trouble of having to treat all our information in purely sensory terms. Attempts such as Carnap's [7] to base all reasoning on natural-language sentences relating to uninterpreted sense data have not been fruitful. Then virtually all of our knowledge comes from simple observational propositions like those above, which I will call directly available. Note that even when we are taught by our mothers or learn from books, the propositions directly available to us are not "crossing the road is dangerous" or "France is a major wine producer", but the observational propositions "I heard Mother say crossing the road is dangerous", "I read in this book maps and tables implying that France is a major wine producer".

Each observation tells us something about a specific object or event at a specific time, and when framed as a proposition has no generality. That is, the propositions do not concern classes of things or events, only single isolated things or events. Yet somehow we work from collections of such specific propositions to very general ones, which make assertions about wide classes of objects and events most of which we have never observed, and never will. "Apples and pears both contain malic acid", "1960 V-8 Roadrats are a bad buy", "All power corrupts", "The universe became transparent to electromagnetic radiation one year after the big bang". The process(es) used to obtain such general propositions from masses of specific ones is called "inductive" reasoning or "induction".

This inductive process is fundamental to our culture, our technology, and our everyday survival. We are perhaps more used to regarding deductive reasoning, and in particular the formalized quantitative deduction of the sciences, as being the hallmark of rational activity, but deduction must be based on premises. The premises of scientific deductions include many general propositions, the "natural laws" of the physical world. Except for deductions based on hypotheses accepted for the sake of argument, deductive reasoning requires the fruits of induction before it can start, and our deduced conclusions are no better than the inductively derived premises on which they are based. Thus, induction is at least as important a mode of reasoning as deduction.

The priority of induction is even stronger than these arguments have shown. When we look at the specific propositions illustrated above, we find that their very expression relies on previous inductive steps. Before I can frame an observation as an assertion like "This car has rust in its doors", I and my cultural forebears must somehow form the belief that there is a class of observable phenomena which share so many features correlating usefully with features of other phenomena that the class deserves a name, say, "rust". This is an inductive conclusion, no doubt based on many thousands of observations of pieces of iron. Inductive conclusions of this type must lie behind the invention of all the common nouns and verbs of natural languages. Indeed, without induction, language could use only proper nouns: the subject of every sentence could be named only as itself, with its own unique grunt. Induction is needed for us to invent and accept the general proposition that all observed phenomena satisfying certain criteria are likely to share certain unobserved but interesting properties, and hence are worth a common name. Thus, we can claim that every generally used common noun and verb in a natural language is the product of inductive inference. We do not claim those inductions were all necessarily sound. The empirical justification for some of them, such as those leading to the nouns "dragon", "miracle" and the compound "free will", may be quite unsound.

With every such word are associated two clusters of propositions. The first cluster one might call the defining propositions — those which allow us to recognize an instance of the class named by the word. For example:

Cows tend to be between 1.5 and 3 m long.
Cows usually have 4 legs.
Cows move against the background.
Cows have a head at one end, often bearing horns, etc., etc.

Then there is a second cluster of propositions which are not needed for recognition, but which allow useful inferences. We may call them "consequential".

Cows are warm.
Cows can (sometimes) be induced to give milk.
Those that can't are often dangerous.
Cows need vegetation to eat.
etc., etc.

The two clusters often overlap: some propositions may be used as defining in some instances, but treated as consequential if not directly observed. The concept of "cow" is accepted because we can recognize an instance of "cow" using a subset of the propositions, and can then infer the probable truth of the remaining propositions in this instance, even though we have not observed them. The induction embodied in the "cow" concept is thus the general proposition that any phenomenon observed to satisfy a certain defining cluster of propositions will also (usually) satisfy the associated consequential propositions.

Some common nouns result from conscious, systematic, "scientific" reasoning. Terms such as "electron", "quark", "cyclonic depression" and "catalyse" label clusters of propositions whose association was unobvious and discovered only after much directed effort. Others, like "man" and "fire" (in whatever language) predate history. We suggest that inductive processes are not necessarily a matter of conscious rational thought, or even of any sort of reasoning.

Any biological organism can be regarded as, at least in a metaphorical sense, embodying such clusters of propositions. Organisms have means, not always neural, of detecting properties of their environment, and many have means to detect with some reliability when several properties are present simultaneously or in a specific sequence. That is, many organisms are equipped to detect when a cluster of propositions is true of their environment. Detection of such a defining cluster instance may trigger behaviour expected to be advantageous to the organism if the environment has other properties not directly detectable by the organism. That is to say, the organism may behave in a way whose benefit depends on certain consequential propositions being true of its environment, although the truth of these propositions cannot at the time be detected by the organism. For instance, seedlings of some species, when grown in a closed dark box with a tiny hole admitting some light, will grow towards the hole even though the light admitted by it is far too weak to support photosynthesis. This behaviour is beneficial to the species because, in the environments naturally encountered by its seedlings, it is indeed usually the case that instances of weak light coming from some direction are associated with useful light being available in a region located in that direction. We do not suggest that such a seedling "knows" or has "inferred" a concept "light source" as a cluster of defining and consequential propositions. However, it is not unreasonable to suggest that the genetic endowment of the species incorporates in some way an association among a cluster of possible properties of its environment, and that other species which grow in environments where such clustering is not evident will not show such behaviour. Further, we suggest that the incorporation of such a cluster of environmental properties differs from a "concept" formed by a reasoning agent only in that the latter is expressible in language-centred terms such as "proposition" and "assertion".

It seems to us proper to regard the genetic makeup of organisms as incorporating many powerful theories about the natural environment. These have not been induced by any reasoning agent. Rather, mutation and other mechanisms result in the creation of many organisms each incorporating a different set of theories. The inductive process which infers good theories is the natural selection of those organisms which carry them. Our aim in this work is to characterize what are good inductions, regardless of how the inductions have been made. The phototropism of a growing seedling, the alarm cry of a seagull, the concept of momentum, and wave equations of quantum mechanics are all the result of the inductive inference of general propositions from hosts of specific "observations". The methods of induction may differ greatly among these examples, but any satisfactory account of how specific data can lead to general assertions should cover them all.

What then are the problems in giving an account of, or logic for, inductive inference? Clearly, one negative requirement which an inferred general proposition must satisfy is that it should not be contradicted by the accepted data. Thus, we cannot conclude that all crows are black if our observations record a large number of black crows, but a few white ones as well. In strictly logical terms, a single counter-example is sufficient to show a general proposition to be false. In practice, the matter is not so clear-cut. Two qualifications are (almost always) implicit in our assertion of a general proposition. First, we are well aware that a general proposition cannot be proved true by any number of conforming observations, so when we assert "All crows are black", most of us mean something a little less. We expect our audience to understand the assertion as an acceptable abbreviation for something like "My best guess is that all crows are black. However, my evidence is incomplete, and I will be prepared to modify my assertion in the light of conflicting evidence". If the assertion is so read, the discovery of one or two white crows among millions of black ones would not cause us to apologize abjectly for misleading. Outside of deductive argument from unquestioned premises, an assertion of universal scope is either meant to be understood in this less than literal way or else is almost certainly unjustified. How can anyone possibly know that all crows, past, present and future, are black?

The second qualification is logically more of a problem. Even if we read the assertion "All crows are black" in its absolute sense, will we really abandon it if we see a single white crow fly past? Not necessarily. The counter-example proposition "The crow that just flew past was white" may itself be suspect. Was the bird indeed a crow? Had it been bleached white by some joker? Was it really white or was it a trick of the light or that last drink over lunch? In the real world, a general assertion which has been long and widely held by people knowledgeable in the field is not rejected on the grounds of a single reported contrary observation. There is always the possibility of error in the observation, perhaps even of malicious misrepresentation or delusion.

Even a steady trickle of contrary reports may not suffice to discredit the proposition. If one observer can make an error, so can others, and perhaps one or two mistaken observations per year must be expected. We also tend to be skeptical of observations, no matter how frequent, which cannot be confirmed by others. Joe may report, frequently and consistently, seeing crows which seem to others black, but which to Joe are distinguished from their fellows by a colour which he cannot otherwise describe. Even if Joe shows he is able consistently to distinguish one crow from another by "colour", we will suspect that the exceptional nature of the observations lies in Joe rather than in what everyone else calls the crows' colour. Similar doubts may arise with respect to observational apparatus: does it really measure what we think it measures?

Before we reject a well-regarded general assertion, it seems we need to be satisfied that, perhaps only under certain specified but achievable conditions, any competent observer can obtain as many counter-examples as are wished. If the bird can be caught, and we find that any competent ornithologist will confirm that it is indeed a crow and indeed white, we will abandon the proposition at least in its absolute sense. If anyone willing to visit lower Slobovia with a pair of binoculars can count a dozen white crows within a day of arrival, we will abandon it completely. But note what needs to be established before we regard the proposition as false: we need to establish that any competent person can consistently accumulate contrary data. This is itself a general proposition. It asserts something about a possibly large class of phenomena: the unlimited observations of any number of competent observers. Note that the large class of observations might all relate to the one white crow.

To summarize, except in special domains, a well-supported general proposition is not regarded as disproven by a single contrary observation or even a limited number of contrary observations. It is usually rejected only when we come to accept a new general proposition, namely that we can get as much credible contrary evidence as we like. The evidence may come from one or many counter-instances (white crows), but the evidential observations are in principle unlimited. However, the requirement that the inductive inference should not conflict with the data, while valid and necessary, does not much advance our understanding of how the inference can be formed from and be supported by the data.

A third qualification may be implicit in a general assertion. It may be true only in an approximate sense, its context implying that it is meant as an approximation. For instance, Boyle's Law that the pressure of a confined gas rises in proportion to its absolute temperature is still taught at school, but is only approximately true of most gasses.

A second necessary requirement of an inductive inference is that the generalization should be *falsifiable*. The importance of this requirement was first clearly stated by Popper (1934), whose writings have influenced much modern discussion of induction. The first requirement we asked was that the inference not be falsified by the data we have. Now we require also that it be possible for future data to falsify the inference. That is, we require it to be at least conceivable, given all we know of the source and nature of the data, that we might find data sufficient to make us reject the proposition. Essentially, this requirement is equivalent to requiring the inferred proposition to have empirical content. If we can deduce, from what is already known about the nature and source of the data, that it is impossible for future data to meet our criteria for falsifying the proposition, then the proposition is telling us nothing of interest. To accept the proposition is to exclude no repeatable observation except those already excluded as impossible.

Popper rightly criticizes theories, advanced as inductive inferences from known data, which are so phrased and hedged with qualifications that no conceivable new data can be considered as damning. He finds most of his bad examples among social, political and economic theories, but examples are not unknown in other domains. His requirement places on any proposed account of inductive inference the duty of showing that any inference regarded as acceptable in the proposed framework must ipso facto be falsifiable. This duty we hope to fulfill in the present work.

The two requirements above, that an inductively derived general proposition be falsifiable but not yet falsified, are far from sufficient to describe inductive inference. They also perhaps place undue emphasis on the notion of disproof. Many useful inductive inferences (let us call them theories for brevity) are known to be false in the form originally inferred, yet are still regarded as useful premises in further reasoning, provided we are careful. A classic example are Newton's laws of motion and gravitation. Reproducible sources of data in apparent conflict with these laws have been known at least since the early 1900s. The reaction to this "falsification" was, first, a series of quite successful attempts to modify the interpretation of these laws, and then the inference of the new theories of Relativity. The new theories were rapidly recognized as superior to the old Newtonian theories, explaining simply all of the results which appeared to falsify the old theories, at least in their original form. Yet the old unreconstructed Newtonian "laws" continue to be used for the great majority of engineering calculations. Although known to be wrong, they in fact fit and explain vast bodies of data with errors that are negligible compared with the measurement errors and uncertainties of the data. In fact, one of the early concerns of the exponents of the Relativistic theories was to show that the new theories did *not* contradict Newtonian theories except under extreme conditions. Similarly, the new quantum theories which replaced Newtonian mechanics under other extreme conditions had to be shown not to contradict the old theory to any measurable degree outside these extreme conditions.

We are forced to conclude that an account of inductive inference must accommodate the fact that theories which have been conclusively falsified can remain acceptable (albeit within a circumscribed domain of phenomena) even though their basic concepts have been shown to be mistaken. The account must also accommodate the fact that two theories can both command general acceptance even though their formulations appear mutually inconsistent. We accept the present situation that Relativistic theory is basically concerned with the relationships among "events" regarded as having precise locations in space and time, while quantum theory denies the possibility of precisely locating an event involving only finite energy. We accept that Relativistic theory describes gravitational effects in geometric terms, while quantum theory, insofar as it can treat gravity at all, must invoke as yet undiscovered particles.

Any satisfactory account of induction must therefore not be overly concerned with absolute notions of truth and falsification. In practice, we do not expect our inferences to be true. We tolerate falsifying data provided it relates to conditions outside our immediate arena, and we tolerate the co-existence (and even the joint application to the one problem) of theories whose conceptual bases seem to belong to different universes.

1.3 The Demise of Theories

The strictly logical requirement that a theory not be falsified cannot be accepted at face value. Merely being shown to be wrong is not sufficient to damn an inference. We can accept the requirement that a theory be falsifiable, i.e., that we can conceive of data which would falsify a theory, as otherwise the theory is empirically vacuous, but we cannot accept that such a falsification will necessarily lead us to reject the inference, because history shows otherwise. How, then, do we ever come to reject theories?

One possible route to rejection is the accumulation of falsifying data. When a theory is falsified, we may not reject it but we must at least qualify it by restricting its application to arenas not including such data, and/or weakening its assertions to approximations. If falsifying data is found in a sufficient range of conditions, the theory may become so qualified and restricted that it ceases to be falsifiable, i.e., becomes empty. The cases of data which do appear to conform to the theory may be found to be no more than might be expected to arise by chance in the absence of the theory, in which case we may decide that the amended theory explains nothing and should be abandoned.

Another route to rejection is that the theory is never decisively falsified, but is supplemented by a theory of greater accuracy or wider applicability. That is, it is found that all the data explained by the theory is explained as well or better by a new theory, which may in addition explain data not covered by the old theory.

A third route is the usual fate of most hypotheses proposed in a scientific investigation. The theory may be compatible with known data, but not regarded as adding much to our understanding of that data compared with other possible theories about the same data. A new experiment or observation is designed such that its expected outcome, if the theory is valid, is one which would not be expected without the theory. The observation is performed, and does not conform with the prediction of the theory. The theory is then rejected as having little explanatory power for the old data, and not fulfilling the hope of explaining the new data.

A fourth, less common, route is that the theory is supplanted by a new theory which is not (at least initially) in better conformity with the known data either in accuracy or scope, but which is in some way simpler or more "elegant" than the old. The criteria of simplicity and elegance are not obviously quantifiable, especially the latter, and people may legitimately disagree in their assessments of theories on these criteria. However, there might be general agreement that, for instance, of two theories otherwise similar in structure, the one needing fewer numeric values to be assumed in order to explain a set of data is the simpler. Similarly, of two theories requiring the same number of assumed quantities, we might assess as the simpler the theory having the shorter mathematical or logical description.

An example may serve to clarify these notions. Observation of the apparent positions of the planets, sun and moon gave rise to the "Ptolemaic" theory, which supposed the motions of the bodies to be composed of simple circular motions with constant radii and speeds. To fit the observations, it was necessary to assume that the motions of most of the heavenly bodies were epicyclic. That is, a body moved round a circle whose centre was moving round another circle whose centre might be moving round yet another circle. This theory could be made to fit the observations quite well, to predict future movements with fair accuracy, and to predict events such as eclipses. It is structurally a simple theory: the circle is one of the simplest geometric shapes by any criterion. However, it required the assumption of a rather large number of numeric values, the radii and speeds of rotation of all the circles, of which there were two, three or more for each body. These quantities had to be assumed: the theory gave no explanation of their values and asserted no useful relationships among them.

The later Keplerian theory was in marked contrast. Structurally, it might be considered more complex or less elegant, since it assumed the motions to be elliptical rather than circular, and to take place with varying rather than constant speed. Each ellipse requires both a major and minor axis to be specified rather than just a radius. In these respects, the new theory seems messier and more complex than the Ptolemaian. However, only one ellipse is needed per body, rather than several circles. The speeds of motion, while not constant, have a fixed and simple relationship to the position of the body round its elliptic path, and the one number for each body required to describe this relationship was shown to have a fixed relation to the size of the ellipse. Thus, the number of values which had to be assumed dropped from half a dozen per body to essentially two. (We are deliberately oversimplifying here: the descriptions of the orbital planes of the bodies involve more numbers but these are essentially the same in both theories.)

The smaller number of arbitrary constants required by Kepler's laws could be held to outweigh his use of more complex geometry, but the issue was not clear-cut on this score. Of course, as observational accuracy increased, it was found that Kepler's theory required only minor refinement to maintain its agreement with observation, whereas more and more circles had to be added to the epicycles of the Ptolemaian model, each with its new inexplicable numbers. The "simplicity" argument in favour of Kepler's model became overwhelming.

This sketch of how theories may be rejected, usually but not always in favour of a new theory, has argued that rejection is not a simple matter of falsification. Rather, it involves factors such as the scope of data explained, the accuracy of explanation, the number of inexplicable or arbitrary values which must be assumed, and some notion of structural simplicity or elegance. Together, we may call these factors the *explanatory power* of the theory. The explanatory power increases with the volume and diversity of the data explained, and with the accuracy of the explanation. It decreases with the structural complexity of the theory, and with the number and precision of the parameters, initial conditions and unobservable quantities which must be assumed in the explanation.

In the above, we have used the terms *explanation* and *accuracy* without elaboration. While it is our intent to use these words in accordance with normal usage, both are sufficiently loosely used in everyday speech as to demand some definition.

1.4 Approximate Theories

The notion of the accuracy of a theory, as applied to some data, rests on belief that a theory is rarely intended or taken in an absolute sense. If we assert the theory that a floating ship displaces its own weight of water, we do not intend to claim that careful measurement of the ship and the displaced water will show them to be equal within a milligram. Rather, we are claiming that they will be equal within a small margin due to measurement error, the effects of wind and wave, motion of the ship, etc. The theory does not attempt to explain the causes of this margin of error, which in practice might be of the order of 0.1%. We might then say the theory is "accurate" within 0.1%. It could be argued that the theory does indeed claim exact equality, at least under certain ideal and probably unattainable conditions such as zero wind, zero motion, no surface tension, etc., and that it is unfair to regard measurement errors and deviations caused by inevitable disturbances as inaccuracies of the theory. But a theory which asserts a conclusion only under forever unattainable conditions is empty, since it can never apply to real data. It may alternatively be suggested that if careful measurement is made, it will be found that the weight of the ship, plus any downwards pull of surface tension, minus any hydrodynamic wave force, etc., etc., will exactly equal the weight of water displaced within measurement error. But this is not the same theory — it is a more elaborate and perhaps more accurate theory.

In a slightly different vein, the Newtonian equation for the kinetic energy of a moving mass, $E = \frac{1}{2}mv^2$, can be said to have inaccuracies due not only to the kind of error and unobserved effects described above, but also an error of order $\frac{1}{2}v^2/c^2$ because it ignores relativistic effects. If the speed v is less than 1000 km/sec, this inaccuracy is less than 0.01%. Whatever the sources of error, it seems plausible that all theories will fail to match our data exactly, but that some will be more accurate than others. Exactly how we can most usefully quantify the inaccuracy of a theory will be discussed later.

1.5 Explanation

A dictionary definition of the word "explain" is "to make plain or under-standable". We take an explanation of a body of data to be a demonstration that the data are not unexpected given a relatively small set of premises. By "not unexpected" we mean that the premises either imply the data proposi-tions, or, more commonly, imply close approximations to the data. Two cases need to be distinguished. In some explanations, the necessary premises are already known and accepted by the reader of the explanation. In this case, the explanation is purely a deductive demonstration that the data should be expected to be more or less as they are, given what is already known. We will not be interested in such explanations. In other explanations, not all of the necessary premises are known *a priori*. Rather, the explanation proposes one or more new premises, and then goes on to show that the new premises, combined with ones already known and accepted, imply or approximately imply the data. In forming such explanations, the new premises are an inductive inference from the data. Typically, they are general propositions which cannot be deduced from the data and premises already known. However, if they are assumed to be true, the data is found to be unsurprising.

Two imaginary examples may clarify the distinction we wish to draw.

First, suppose there is an amateur carpenter who knows and is familiar with concepts of length, area and angle, is competent at arithmetic and elementary algebra, but who has never studied geometry. The carpenter notices that it is possible to make right-angled triangular frames whose sides are integral numbers of decimetres, but only if the numbers are multiples of a few sets such as $\{3, 4, 5\}$, $\{5, 12, 13\}$ and $\{15, 8, 17\}$. In seeking an explanation of these observations, he might, given time, deduce Pythagoras's theorem from the premises he knows and accepts about lines, areas and angles, then deduce that these sets of integers satisfy the theorem but most others do not. He might even be able to deduce that any such integer set must have the form $\{(a^2 - b^2), 2ab, (a^2 + b^2)\}$, where a and b are any unequal positive integers. This would be an explanation of the first kind: a demonstration that what has been observed is not surprising given what the carpenter already believed. No new premise is required and nothing is inductively inferred from the data.

Now imagine an ancient Egyptian surveyor who was a competent user of geometry and knew many of the simple properties of triangles, but otherwise knew no more than his fellows. In particular, he knew that the sum of the three angles of a triangle equals two right angles (180°). As he rose in the surveyors' hierarchy, he noticed minor inconsistencies appearing in the data and ordered the large-scale resurveying of the kingdom. He was surprised to find that in the largest triangles covered by the survey, the sum of the angles consistently exceeded 180° by a small amount, which seemed to be proportional to the area of the triangle. After much reflection, he finds that the data can be explained if he supposes that the world is not flat, as everyone had thought, but spherical, with a diameter of about 7000 miles. If he adds this premise to what he knows of geometry, he can deduce that the sum of the angles of a triangular piece of land should exceed 180° by about five thousandths of a degree for every thousand square miles of area. This deduction agrees well with the survey data, so he accepts the explanation.

This explanation requires a new premise in addition to what the surveyor knew. The new premise, that the world is a sphere of 7000 miles diameter, is more complex than the old implicit assumption of a flat earth, and involves a number. It could not be deduced from the data. Rather, it was derived by induction from the data, and the diameter estimated from the data. The new premise is falsifiable — new data could conceivably show it to be untrue — and is actually false. The world is not quite spherical, nor is its diameter 7000 miles. However, the explanation is good. At the expense of one inferred premise of no great complexity, the deviations of the data from what is expected are greatly reduced. Henceforth, we will restrict the term *explanation* to this second sort, which involves the inductive inference of a new premise or theory and/or the estimation of unknown quantities.

1.5.1 Explanatory Power

Our view is that an inductive inference from a body of data is a premise which, if assumed, allows the data to be explained. Other propositions already known and accepted may be involved in the explanation, but are by themselves insufficient to allow a satisfactory explanation of a purely deductive kind.

To develop this view into an account of when an inductive inference can be regarded as satisfactory and how competing inferences may be compared, it is necessary to develop a quantitative measure of the merit of an explanation, or at least of the relative merits of competing explanations. We have suggested that the explanatory power of an inductive inference or theory increases with the volume of data explained and the accuracy of the explanation. It decreases with the complexity of the theory, the number of inexplicable parameter values appearing in the theory, and (we will see later) the precision with which these quantities must be specified in order to achieve an accurate explanation. In short, a good inductive inference is one which explains much by assuming little. Other considerations, such as causal structure, have been proposed as contributing to or necessary for explanatory power. At least for now, we will not consider them, and rather discuss only what follows from the criteria above. We now propose a step towards quantifying these considerations.

First, we will simplify the problem by assuming that all the inductive inferences to be assessed apply to the same body of data. The extension to situations where one theory explains more data than another is easy but is best treated later.

For a given fixed body of data, we propose to recast all competing explanations into the same canonical form.

1.5.2 The Explanation Message

An *explanation message* of a body of data comprises two parts. The first is a statement of all the inductively derived premises used in the explanation, including numeric values for quantities assumed in these premises (the diameter of the earth, for example). The second part of the explanation is a statement of all those details of the data which cannot be deduced from the combination of the induced premises and such other premises as are already known, accepted, and not in question. Let us call the already-known premises the *prior premises*. Being already known to the receiver, no statement of the prior premises need appear in the message.

First, note that a person knowing only the prior premises and not the data is able to recover the original body of data exactly from this message. The first part tells him to assume the truth of the new premises — the "theory". From these and the prior premises, he can then deduce much about the data. The second part completes his knowledge of the data by telling him all the details which he could not so deduce. Thus, the explanation message may be regarded as a restatement of the data without loss of any information or detail. The restatement is in a "coded" form which can be "decoded" by anyone with knowledge of the prior premises. Another way of regarding the explanation message is that it states a theory about the data, then states the data itself in a condensed form which assumes the truth of the theory.

We will argue that the best explanation of the data is the one leading to the shortest explanation message, and that the best inductive inference which can be drawn from the data is the inference used in the shortest explanation message. That is, we claim the shortness of the explanation message using an inferred theory is a measure of its explanatory power. Henceforth, we will not distinguish between an explanation message and an explanation expressed in other forms. When we refer to an explanation or its length, we mean the explanation message or its length.

Even from the above informal account, it is clear that the length of an explanation takes into account all the factors affecting explanatory power. The length of the first part, which states the inductively inferred premises or theory, will be longer for a complex theory than for a short one. Its length increases with every quantity assumed, as the first part must state its assumed numeric value. Its length increases with the precision to which these values need be specified.

On the other hand, the length of the second part decreases with the scope and accuracy of the theory. Data falling outside the scope of the theory must be stated in full, since nothing about such data can be deduced from the theory. Hence, the greater the scope of the theory, the less data need be stated fully. Typically, the theory, together with the prior premises, will not allow exact deduction of the data as observed. For quantitative data, the best that we can hope is that values may be deduced close to but not exactly equaling the measured values. The observed value may be corrupted by measurement error, and the deduced value will often be deduced in part from other values in the data, and hence itself be corrupted by error. And of course the theory and its parameters may only be approximate. Thus, for quantitative data within the scope of the theory, the second part of the message must at least record the differences between deduced and measured values. The more accurate the theory, the smaller will be these differences, and hence the shorter will be their representation in the second part.

Similarly, for data which refers to discrete values rather than real-valued quantities, there remains some possibility of errors in observation. If, further, the theory is not entirely accurate, the deduced values will sometimes differ from correctly recorded observations. Thus, for discrete data, the second part will typically require a list of "exceptions" recording which observed values differ from the predictions of the theory, and the actual observations in these exceptional cases. The more accurate the theory, the fewer will be these exceptions, and the shorter will be the second part.

Overall, the "best" inference or theory, as assessed by its explanation length, will be a compromise between complexity on the one hand, and scope and accuracy on the other. An overly complex theory may be slightly more accurate, and hence give a slightly shorter second part, but will require a long description in the first part. An overly simple theory will require only a short first part, but will be relatively less accurate or have narrower scope, and hence leave more errors and exceptions to be stated in the second part.

Note that if a proposed theory is particularly poor or overly-complex, or if the data is very sparse, the length of the explanation may exceed the length of a message which simply records the data as measured, with no attempt at explanation. In such a case, we regard the proposed theory as unacceptable.

This informal discussion suggests that the two-part explanation model conforms qualitatively with what we expect and require of inductive inference. Inferences regarded as good in this model must have content, because the second part will be shorter than a simple transcription of the given data only if the first part implies something of substance about the data. The model has an inherent balance between the complexity of the theory and its having meaningful implications about the data. It provides a criterion for rejecting a theory as useless and for comparing the merits of competing theories. The model accommodates the observation that theories are not necessarily rejected on the grounds of a single or a few contradictory measurements: such data can be flagged as "exceptional" in the second part and recorded as measured. The "flagged" record of inexplicable data is slightly longer than a bare transcription of the data, so if too many observations disagree with the theory, the explanation may become longer than a copy of the data as given, and so be rejected, but small amounts of conflicting data need not lead to rejection. Finally, an acceptable theory in this model necessarily makes testable predictions about new data on the same subject. It predicts that the data will be such that after taking into account what can be deduced from the theory and prior premises, the remaining details of the new data can be specified more briefly than the data as measured.

The reader may have noticed a loophole in the discussion of whether an explanation is acceptable. It may be the case that the prior premises alone, without any inductively-derived theory, imply enough about the data to allow it to be restated more briefly. In this case we could have an explanation message shorter than the original statement of the data as measured, yet containing no first part, i.e., no inference from the data. Such an explanation is not ridiculous and is often met in practice. It is an example of the first kind of "explanation" mentioned above, in which it is shown that the data is not surprising given what is already known and accepted. Since our interest is in explanations involving inductive inference, we wish to exclude such explanations from discussion even though they have a useful place in human discourse. They also provide a base length, shorter than the original data, against which inductive explanations should be compared.

We therefore modify slightly our definition of "acceptable" theories and explanations. We will require the length of the explanation message which states and uses an acceptable theory to be shorter than any message restating the data using only the implications of prior premises. That is, we shift the target for an inductive theory by allowing for the implications of prior premises. This modification affects only the criterion for acceptability of a theory. It does not affect comparisons among competing theories. In a sense, we have simply redefined the "null hypothesis" against which all theories must compete. Rather than taking as the null hypothesis the assumption that the data as given shows no regularities at all which might be exploited to recast it more briefly, we now take as the null hypothesis the assumption that the only regularities in the data are those implied by the prior premises.

Qualitatively, our model has much to recommend it. However, if it is to be anything more than an aphorism, it must be given quantitative substance. That is, we must be able to put numeric values on the lengths of explanations. The necessary tools are described in the next chapter, and show a close relation between this account of inductive inference and Bayesian statistical inference.

1.6 Random Variables

A *random variable* has a value which is not known with certainty. For instance, if a coin is tossed, its attitude when it comes to rest may be assigned one of the two values Head and Tail. If the coin has not yet been tossed, or if it has landed under a table and no-one has yet looked to see how it lies, the value of the toss (i.e., the attitude of the coin) is not as yet known and could be either Head or Tail. The value may then be represented by a random variable, say, "v". (For no good reason, this book departs from a common convention, and usually names random variables by lower-case letters.) Then the equation "$v = $ Head" is the proposition that the coin will be found to land, or to have landed, with its head uppermost.

Note that we are using a subjective interpretation of random variables. If the coin has been tossed and come to rest under a table, and my friend has crawled under the table and had a look at it but not yet told me the outcome, I am still justified (in this interpretation) in representing the value of the toss as a random variable, for I have no certain knowledge of it. Indeed, my knowledge of the value is no greater than if the coin had yet to be tossed. Other definitions and interpretations of the idea of a random variable are possible and widely used. We will later hope to show that, for our purposes, the differences in definition lead to no differences in the conclusions drawn from any statistical enquiry. For the moment, let it suffice that the subjective interpretation adopted here is convenient for our present exposition.

The *range* of a random variable is the set of values which the variable (as far as we know) might equal. Thus, the range of the coin-toss variable v is the set {Head, Tail}. The *value* of a random variable is the actual (but as yet unknown) value denoted by the variable.

Random variables may be either *discrete* or *continuous*. A discrete random variable has a range which is a discrete set of values, e.g., { Head, Tail } or { Married, Single, Divorced }. The range may be countably infinite. That is, it may include an infinite number of values, but if so there must be a rule for establishing a one-to-one correspondence between the values in the range and the positive integers. For instance, the range of a random variable "s" might be the set of all non-empty finite sequences of symbols 0 and 1. There is an infinite number of such sequences, since the length of a sequence is unbounded, but they can be placed in one-to-one correspondence with the integers, as shown in Table 1.1.

Integer	Sequence
1	0
2	1
3	00
4	01
5	10
6	11
7	000
8	001
9	010
10	011
etc	etc

Table 1.1. An enumeration of binary strings

A continuous random variable has a range which is a continuum or part of one. For instance, if the random variable "m" denotes the mass (in grams) of a raindrop, its range might be all the real numbers between 0.01 and 20.0. This range is uncountably infinite. That is, this range cannot be placed in 1-to-1 correspondence with the positive integers.

A *vector-valued* continuous random variable has a range which is (part of) a continuum of more than one dimension. For instance, if "p" denotes the point of impact of a dart on a dart board, the range of p is the two-dimensional area of the face of the board, and two real numbers (say the height of the point and its East-West location across the width of the board) are needed to specify a value of the variable. In general, the value of a vector-valued variable with a D-dimensional continuum range is a vector, or ordered list, of D real numbers. We may also have vectors which are ordered lists of discrete variables. In either case the individual simple variables in the list are termed *components* of the vector.

1.7 Probability

Consistent with our subjective definition of a random variable, we adopt a subjective definition of probability. Let v be a discrete random variable, with a range

$$\{v_1, v_2, \ldots, v_i, \ldots, v_R\}$$

Note that here the symbols v_1, v_i, etc. are not random variables. Rather, they represent the known values forming the range of v, and any one of them might be the actual value of v.

The *probability* of the proposition "$v = v_1$" is a real number between zero and one representing how likely it is that the value of v is v_1. We use the notation $\Pr(\text{proposition})$ or sometimes $\text{Prob}(\text{proposition})$ or just $P(\text{proposition})$ to mean the probability of the proposition. Thus, we write the probability that v has value v_2 as $\Pr(v = v_2)$. In contexts where the identity of the random variable is obvious, we may abbreviate this notation simply to $\Pr(v_2)$. (Note that were v continuous, no non-zero probability could attach to the proposition "$v = v_1$" for arbitrary v_1.)

A probability of one represents certain knowledge that the proposition is true. Zero represents certain knowledge that the proposition is false. A probability of $1/2$ represents complete uncertainty: we consider the proposition equally likely to be true or false. In general, the higher the probability, the more likely we consider the proposition.

We require the numerical values assigned to probabilities to satisfy certain axioms set out below. These axioms are also satisfied by "probabilities" defined in other, non-subjective, ways. In what follows, X, Y, etc. denote propositions. $\bar{X}$ denotes the negation of X. That is, $\bar{X}$ is true if and only if (iff) X is false, and vice versa. $X.Y$ denotes the proposition that both X and Y are true. $X \vee Y$ denotes the proposition that X is true, or Y is true, or both are true.

Axiom 1 $0 \leq \Pr(X) \leq 1$
Axiom 2 $\Pr(X|Y) + \Pr(\bar{X}|Y) = 1$
Axiom 3 $\Pr(X.Y) = \Pr(Y)\Pr(X|Y)$

where $\Pr(X|Y)$ is the probability we assign to proposition X if we know Y is true. $\Pr(X|Y)$ is read "the probability of X given Y", and is called a *conditional probability*.

Notation: We will often write $\Pr(X.Y)$ as $\Pr(X, Y)$. Also, we often write $\Pr(X|Y.Z)$ as $\Pr(X|Y, Z)$. This is the conditional probability of X given that both Y and Z are true.

From the above axioms and the axioms of Aristotelian propositional logic follow the identities:

$$
\begin{aligned}
\Pr(X, Y) &= \Pr(Y, X) = \Pr(Y)\Pr(X|Y) = \Pr(X)\Pr(Y|X) \\
\Pr(X, Y, Z) &= \Pr(X)\Pr(Y|X)\Pr(Z|X, Y) \\
\Pr(X \vee Y) &= \Pr(X, \bar{Y}) + \Pr(\bar{X}, Y) + \Pr(X, Y) \\
&= \Pr(X) + \Pr(Y) - \Pr(X, Y) \\
\Pr(X|Y) &= \Pr(X)\Pr(Y|X)/\Pr(Y) \qquad \text{(Bayes' Theorem)}
\end{aligned}
$$

(The singular case $\Pr(Y) = 0$ will not arise in our use of Bayes' theorem.)

If $\{X_1, X_2, X_3, \ldots, X_i, \ldots\}$ is a set of propositions which are exhaustive and mutually exclusive, so that one and only one of them must be true, then $\sum_i \Pr(X_i) = 1$. Hence, if $\{v_1, v_2, \ldots, v_i, \ldots\}$ is the range of discrete random variable v,

$$
\sum_i \Pr(v = v_i) = 1
$$

or, in abbreviated notation where the identity of v is obvious,

$$
\sum_i \Pr(v_i) = 1
$$

1.8 Independence

If two propositions X and Y are such that $\Pr(X|Y) = \Pr(X)$, then the truth or falsity of Y does not affect the probability of X. If this is so,

$$
\Pr(X, Y) = \Pr(X|Y)\Pr(Y) = \Pr(X)\Pr(Y)
$$

Also,

$$
\begin{aligned}
\Pr(X, Y) &= \Pr(Y|X)\Pr(X) \\
\Pr(X)\Pr(Y) &= \Pr(Y|X)\Pr(X) \\
\Pr(Y) &= \Pr(Y|X)
\end{aligned}
$$

Hence, the truth or falsity of X does not affect the probability of Y. Such propositions are called "statistically independent" of one another, or simply "independent".

1.9 Discrete Distributions

For a discrete random variable v with range $\{v_1, v_2, \ldots, v_i, \ldots\}$, it is convenient to represent the probabilities of the various values in its range by a function of the integer "i" indexing the values. Generally, we can write such a function as a function of either the index or the value, whichever is the more convenient:

$$\Pr(v = v_i) = f(i) \quad \text{or} \quad f(v_i) \qquad \forall i$$

Since v must equal some value in its range,

$$\sum_{\text{range of } v} f(v_i) = 1$$

Such a function is called a probability distribution function, or simply a distribution, and will here usually be denoted by the letter f. If $f(i)$ gives the probability that $v = v_i$ for all values in the range of v, f will be called the "distribution of v".

Often, we may know that the distribution of a variable v is one of a parameterized family or class of functions, but we may not know which one. In such cases, we will write the distribution of v as the function

$$\Pr(v = v_i) = f(i|\theta) \quad \text{or} \quad f(v_i|\theta)$$

meaning that the distribution of v is that function in the family identified by the parameter value θ. Sometimes, the family of functions will have more than one parameter, i.e., it will be a family of two or more dimensions. In such case we may still write the distribution as $f(v_i|\theta)$, but read θ as a vector whose components are the several parameters, or we may write the distribution as $f(v_i|\alpha, \beta, \ldots)$ where $\alpha, \beta, \ldots$ are the parameters.

Note the use of the vertical bar in the same sense as it is used in the notation for probabilities. $\Pr(X|Y)$ means the probability of X *given* that Y is true. Similarly, $f(v|\theta)$ means the distribution of v *given* that the parameter has the value θ. The symbols after the bar show what is assumed to be known or fixed that can affect the probability or distribution. They are said to *condition* the probability or distribution.

Finally, be warned that in discussing distribution, we often treat the random variable v as an ordinary algebraic variable, and use it as the argument of the distribution function. Thus, we will write

$$f(v|\theta)$$

and treat it as an ordinary function of two algebraic variables, v and θ.

The algebraic variable v has the same range of values as the random variable v, and the algebraic variable θ ranges over the values identifying members of the family of distributions.

1.9.1 Example: The Binomial Distribution

A series of N trials is conducted. Each trial either succeeds or fails. It is believed that $\Pr(\text{trial succeeds})$ has the same value θ $(0 \leq \theta \leq 1)$ for all trials, and that the trials are independent. Let n be the number of successes. Then n is a random variable with range $\{0, 1, \ldots, N\}$. The distribution of n is called a "Binomial distribution" and depends on the parameter θ.

$$f(n|\theta) = \binom{N}{n}\theta^n(1-\theta)^{(N-n)}$$

Here, $\binom{N}{n}$ is the mathematical notation for the number of ways of selecting a subset of n things out of a set of N things. The distribution is a member of a family of distributions, each characterized by a different value of the parameter θ. For example, if $\theta = 0.3$ and $N = 100$,

$$\Pr(20 \text{ successes}) = \binom{100}{20}0.3^{20}0.7^{80}$$

and the distribution of n is

$$f(n) = f(n|0.3) = \binom{100}{n}0.3^n0.7^{(100-n)}$$

Note that the number of trials N is not normally regarded as a parameter, even although it enters into the distribution function. Strictly, we should write the distribution of n as $f(n|\theta, N)$, meaning the probability of getting n successes given that there were N trials and that each trial had success probability θ. However, values such as N describing known, fixed conditions under which the random variable will be observed are often not treated as parameters of the distribution. The status of a parameter is often reserved for those quantities affecting the distribution whose values might well not be known.

1.10 Continuous Distributions

Suppose v is a continuous random variable, i.e., one which can take any value in a continuous range, and let a be a value in that range. Then, in general, one cannot usefully define the probability $\Pr(v = a)$. For instance, if it is believed that v could equally well have any value in the real interval $1 < v < 2$, the probability that v will have exactly the value a should equal the probability that v will equal b, for any pair of values a and b in $(1, 2)$. But there are infinitely many values in this interval. If all are to have equal probability and the probabilities are to sum to 1, each probability must be infinitesimally small. Thus, we cannot define the kind of distribution function used for discrete random variables.

Instead, for a scalar continuous random variable v, we define a *probability density function*, or simply *density*, by

$$\Pr(a < v < b) = \int_a^b f(v)dv$$

For sufficiently well-behaved variables,

$$\lim_{\delta \to 0} \frac{\Pr(a < v < a + \delta)}{\delta} = f(a)$$

That is, $\delta f(a)$ gives the probability that the value of v will lie in a small interval of size δ near the value a.

Note that we will often use the same function symbol $f(\cdot)$ for either a discrete distribution or a density. As with discrete variables, the density may be one of a family of densities with parameter θ, where we write the density as $f(v|\theta)$, or perhaps as $f(v|\alpha, \beta, \ldots)$ if there are several parameters.

If v is a vector-valued continuous random variable, the same notation is used, but the element dv must be interpreted as an element of area, volume, etc. rather than as an element of the real line. For instance, if v is a 2-vector with components x and y, we may write the density of v either as $f(v)$ or as $f(x, y)$. The probability that v lies within a region R of the (x, y) plane can be written as either

$$\int_{v \in R} f(v)dv \quad \text{or} \quad \int\int_{(x,y) \in R} f(x, y)dxdy$$

1.11 Expectation

If v is a discrete random variable with range $\{v_1, v_2, \ldots, v_i, \ldots\}$ and distribution $f(v)$, and if $g(v)$ is some function defined for all values in the range of v, then we define the "expectation" of $g(v)$ as

$$\sum_{\text{range of } v} f(v_i)g(v_i)$$

and symbolically denote it as $Eg(v)$ or simply Eg.

In situations where we can obtain or observe an unlimited number of instances or realizations of v, the long-term average of $g(v)$ over many instances of v will approach $Eg(v)$. In N instances, we expect v to take the value v_1 about $Nf(v_1)$ times, and in each of these occasions g will take the value $g(v_1)$. Similarly, we expect g will take the value $g(v_2)$ about $Nf(v_2)$ times, and so on. Thus, the sum of all the g values obtained in the N instances should be approximately

$$\sum_{\text{range of } v} Nf(v_i)g(v_i) = N\sum f(v_i)g(v_i)$$

Dividing by N to obtain the average gives

$$\sum f(v_i)g(v_i) = \text{E}g$$

Thus, $\text{E}g$ represents the average value we expect to get for $g(v)$ over many instances.

Even when there is no possibility of observing many instances of v, i.e., when it is not possible to interpret $f(v_i)$ as a long-term average frequency of getting $v = v_i$, the expected value $\text{E}g$ still usefully summarizes what a rational person might consider to be a "fair average" value for $g(v)$, assuming that the distribution $f(v)$ properly represents his uncertainty about v. For instance, suppose I arranged a wager with you, that I will pay you 3.50 dollars, we toss a 6-sided dice, and then you pay me 1 dollar for each dot showing on the uppermost surface of the die. Let $v_i = i$ be the number shown by the die, with the range $\{1, 2, \ldots, 6\}$. Let $g(v_i) = g(i)$ be my net monetary gain from the wager. Then we have

$g(1) = -2.50$ dollars (I paid 3.50 dollars and got 1 dollar back)
$g(2) = -1.50$ dollars
$\vdots \qquad \vdots$
$g(6) = +2.50$ dollars (I paid 3.50 dollars and got 6 dollars back)

Consider my situation before the die is cast. I do not know what the value of my net gain $g(v)$ will be, and there will be no repetitions of the wager, so there is no consideration of a long-term average gain over many tosses. However, if I believe that all numbers in the range 1 to 6 are equally likely to come up, so $f(v_i) = 1/6$ for all i, then the expectation $\text{E}g(v_i)$ is

$$\sum_{i=1}^{6} f(i)g(i) = \sum_{i=1}^{6} \frac{1}{6}(i - 3.5) = 0 \text{ dollars}$$

The "expected" net gain is zero, and we would normally regard the wager as a "fair bet": neither you nor I have an advantage.

By contrast, if there are grounds for believing the die to be biased so that number 5 is twice as likely to occur as any other, then my "expected" net gain is

$$\frac{1}{7}(-2.5) + \frac{1}{7}(-1.5) + \frac{1}{7}(-0.5) + \frac{1}{7}(0.5) + \frac{2}{7}(1.5) + \frac{1}{7}(2.5) = 0.214 \text{ dollars}$$

The positive "expectation" of g or "expected net gain" is an indication that the wager is biased in my favour. Were it to be repeated 1000 times, I could expect to win about 214 dollars, and even though there will in fact be only the one wager, Eg represents a fair assessment of how beneficial (in purely monetary terms) the wager is likely to be to me.

Note that this use of the term "expectation" (and "expected value") is a technical definition which does some violence to the normal English meaning of the word. In the biased-die case, the definition gives an expected value $Eg = 0.214$ dollars, or about 21 cents. However, I certainly do not expect (in the usual sense) my net gain to be anything of the sort. My net gain can only be one of the values -2.5, -1.5, -0.5, 0.5, 1.5 or 2.5, and cannot possibly be 21 cents. The value $Eg(v)$ might better be termed an "average in probability" rather than "expected value" or "expectation", but the latter terms are entrenched in the statistical literature and will be used in their technical sense in this work.

The definition of expectation extends in the obvious way to functions of continuous random variables. If v is a continuous random variable with probability density $f(v)$, and $g(v)$ is a function of v, then the expected value of $g(v)$ is defined as

$$Eg(v) = \int f(v)g(v)dv$$

Conditional expectations are also useful: the expected value of $g(v)$ given some value or proposition y which affects the distribution of v is defined as

$$E(g(v)|y) = \sum_{\text{range of } v} f(v_i|y)g(v_i) \quad \text{(discrete)}$$

or

$$E(g(v)|y) = \int f(v|y)g(v)dv \quad \text{(continuous)}$$

Finally, we may be interested in the expected value of v itself (called the *mean* of v), in which case $g(\cdot)$ is the identity function. Then

$$Ev = \sum_{\text{range of } v} f(v_i)v_i$$

$$E(v|y) = \sum_{\text{range of } v} f(v_i|y)v_i$$

or for continuous v:

$$Ev = \int f(v)\, v\, dv$$

$$E(v|y) = \int f(v|y)\, v\, dv$$

1.12 Non-Bayesian Inference

Given some data and a set of probabilistic models for the data, we would like to be able to infer some statement about which model or subset of models should be preferred in the light of the data. We will represent the data by a proposition or vector of values x, and the set of models by Θ, with θ denoting a model in the set.[1] Note that we use the term "model" to denote a fully-specified probability distribution over the possible range of the data, with no parameters left free. We use "family of models" or "model family" to refer to a set of models individually identified by different parameter values.

For each model θ, we assume the probability of getting data x given model θ is known, and we write it as $\Pr(x|\theta)$.

Classical non-Bayesian inference attempts to draw some conclusion about the model using only x and the probabilities $\{\Pr(x|\theta) : \theta \in \Theta\}$. The difficulty of doing so is evident if we consider the simplest of examples. Suppose there are only two models considered possible, θ_1 and θ_2, and that data x is obtained such that

$$\Pr(x|\theta_1) = 0.001 \qquad \Pr(x|\theta_2) = 0.01$$

Armed only with those facts, what can be said about the true source of the data, θ? Clearly, we cannot deduce any statement of the form "$\theta = \theta_1$", since the data does not logically exclude either model. Nor do the axioms of probability and logic allow us to deduce a probability for any such statement. That is, we cannot obtain from these facts any value for $\Pr(\theta = \theta_1)$ (which we can abbreviate to $\Pr(\theta_1)$).

The inequality $\Pr(x|\theta_2) > \Pr(x|\theta_1)$ is the only deducible statement which distinguishes between θ_1 and θ_2 in the light of the data. We may well feel that the inequality favours belief in θ_2 over belief in θ_1, but the strength of this preference cannot be quantified as a probability.

When the set of possible models Θ is a family with a single real parameter, the situation is superficially improved but actually no better. Let θ denote the parameter identifying a member of the family, and $\Pr(x|\theta)$ be the probability of obtaining the data given the parameter value θ. If we consider a proposition such as A: "$a < \theta < b$", we cannot even deduce the probability $\Pr(x|A)$, let alone the probability $\Pr(A)$. However, at least in some cases, it appears that we can deduce something useful about such a proposition.

Suppose that Θ is the family of Normal densities with standard deviation 1.0 and mean θ (the parameter). Let the data be a sample of 8 values independently drawn from the density, $x_1, x_2, \ldots, x_8$. Let $\bar{x}$ be the sample mean $\frac{1}{8}\sum_{i=1}^{8} x_i$, and suppose $\bar{x} = 1.272$. It is easily proved for the Normal density

---

[1] For notational simplicity, we let x denote either scalar or vector data values, and do not use $\underline{x}$ to distinguish the vector case.

that the mean of N independent random values drawn from $\text{Normal}(\mu, \sigma^2)$ is a random value drawn from $\text{Normal}(\mu, \sigma^2/N)$. Thus, in this case, $\bar{x}$ is a random variable drawn from $\text{Normal}(\theta, 1/8)$. Equivalently, we may say that $(\bar{x} - \theta)$ is a random variable drawn from $\text{Normal}(0, 1/8)$. Then, since we know $\bar{x}$ from the data, cannot we infer some probabilistic statement about $(\theta - \bar{x})$, and hence θ? For example, tables of the Normal distribution function show that a random value from $\text{Normal}(0, 1/8)$ has probability 0.99 of lying in the range ± 0.911. Can we not then say that the proposition B: "θ lies in the range 1.272 ± 0.911" has probability 0.99, i.e., $\Pr(B) = 0.99$? Unfortunately, we cannot.

If we know nothing except that $\bar{x}$ is the mean of 8 values drawn from $\text{Normal}(\theta, 1)$, then the proposition C: "$|\bar{x} - \theta| < 0.911$" is distinct from B and has probability 0.99, i.e., $\Pr(C) = 0.99$. However, in the present case we know more, namely the 8 data values. We have no right to assume that C is independent of the data, i.e., that $\Pr(C) = \Pr(C|x_1, \ldots, x_8)$. In some special cases, C may be approximately independent of the data, but in general it is not.

For this simple example, we can be sure that C is independent of some aspects of the data. The Normal distribution has the property that if the standard deviation is known, then for any set of data values, $\Pr(x_1, \ldots, x_N|\theta)$ can be factorized into two probabilities:

$$\Pr(x_1, \ldots, x_N|\theta) = \Pr(x_1, \ldots, x_N|\bar{x}) \Pr(\bar{x}|\theta)$$

where the first factor does not depend on θ. Hence the probability of any proposition about θ or $(\bar{x} - \theta)$ may depend on the data only via $\bar{x}$, and all other aspects of the data must be irrelevant. A function of the data, such as the mean $\bar{x}$ in this example, which allows the probability of the data given the parameters to be factorized in this way, is called a "sufficient statistic". The value of a sufficient statistic contains all the information about the parameters which can be recovered from the data.

However, it remains possible that proposition C is dependent on $\bar{x}$, and hence that $\Pr(\bar{x} - 0.911 < \theta < \bar{x} + 0.911) \neq 0.99$. An example may clarify this possibility. Suppose that, in addition to knowing the data to be drawn from $\text{Normal}(\theta, 1.0)$, we also happen to know that θ is positive, although its value is unknown. This additional knowledge does not affect the probability of the proposition that the mean of 8 values drawn from the Normal distribution will lie within ± 0.911 of the true mean θ. However, it does change the probability, given $\bar{x}$, of proposition C, that θ lies within ± 0.911 of $\bar{x}$. Although $\theta > 0$, it is quite possible that the data will give a negative sample mean, say, $\bar{x} = -0.95$. If we observe such data, then we will be sure that proposition C is false, since the range $\bar{x} \pm 0.911$ becomes the range -1.861 to -0.039, and we are sure $\theta > 0$ and hence not in this range. Thus, if $\bar{x} \approx -0.95$, $\Pr(C|\bar{x}) = 0$. Again, suppose the data yielded $\bar{x} = -0.90$. Then proposition C becomes "$|\theta + 0.90| < 0.911$" or "$-1.811 < \theta < 0.011$". Combined with our knowledge that $\theta \geq 0$, proposition C now implies "$0 \leq \theta < 0.011$". While this proposition is not known to be false, we would be hesitant to regard it as having probability 0.99, i.e., very probably true, but we have no means of calculating its probability from what is known. More generally, if we know almost anything about θ in addition to the observed data, we must conclude that the probability of proposition C is dependent on $\bar{x}$, even if we cannot compute it.

The impossibility of deducing probabilities for propositions such as "θ lies within ± 0.911 of the sample mean", using knowledge only of the model probability distribution, is well known. However, a range such as $\bar{x} \pm 0.911$ is often stated as an inference from the data, and is called a "confidence interval". Rather than claiming that the proposition "θ lies within ± 0.911 of $\bar{x}$" has *probability* 0.99, the proposition is said to have *confidence* 0.99. "Confidence" is not the same as probability, no matter how the latter term is defined. It is rather unsatisfactory that starting with assumptions stated in terms of probability, one can only make an inference stated in terms of the even more problematic concept of "confidence".

1.12.1 Non-Bayesian Estimation

Given data x believed to be drawn from a source modelled by some distribution in a known family with unknown parameter θ, one might wish to infer a "best guess" value of θ, accepting that no probabilistic statement will be possible about its accuracy. Such a "best guess" is called an *estimate*. In a non-Bayesian framework, the raw material for forming an estimate comprises the observed data x and the function $f(x|\theta)$ giving the probability of obtaining data x from a source with parameter value θ. With such limited raw material, the most general process available for forming an estimate appears to be the "Maximum Likelihood" (ML) method. This method chooses as the estimate that value of θ which maximizes $f(x|\theta)$.

Viewed as a function of θ with given data x, the function $f(x|\theta)$ is called the *likelihood* of θ. Clearly, for given x, $f(x|\theta)$ is not a probability distribution or density for θ. No such distribution or density can be inferred from x and $f(x|\theta)$. The ML method is just an extension to a parameterized family of models of the simple preference scheme first discussed for two models: prefer the model having the highest probability of giving the data.

If no information other than the data and the function $f(x|\theta)$ is to be used in the estimation, it is difficult to see any alternative to ML. How else can one obtain a value within the range Θ of the unknown parameter from such raw material? Indeed, so long as $f(x|\theta)$ has only a single local maximum as a function of θ, the ML estimate can be shown to have several desirable properties. It depends on the data only via functions of the data which are sufficient statistics, and so uses no information in the data which is not relevant to θ. For many model families, ML is *consistent*. That is, if more and more data is obtained, the ML estimate can be expected to approach the true value of θ. Also, for many model families, ML is *efficient*, meaning that it uses all the information in the data which is relevant to the value of θ. However, ML is not unique among estimation methods in possessing these features. In some cases, ML can be improved upon by allowing a little more information into the non-Bayesian framework.

Up to this point, we have tacitly treated the value of θ identifying a member of a family of models simply as an identifying label, with no quantitative interpretation. However, in many cases, the parameter(s) of a family of models are quantities whose numeric values are meaningful. For instance, the mean diameter of a population of sand grains is itself a diameter, to be stated in some physical unit such as millimetres, and its value may be important in, say, calculations of wind erosion. Similarly, the mean kinetic energy of a collection of gas molecules is itself an energy, and has physical meaning related to the temperature of the gas. In such cases, we may well import into the statistical inference process arguments based on the quantitative difference between the true parameter value and the estimate. Let θ_0 denote the true value and $\hat{\theta}$ an estimated value based on data x. We write the estimate as a function of the data: $\hat{\theta} = m(x)$. The function $m()$ is called an *estimator*.

The *bias* of an estimator is the expected difference between the true value θ and the estimate:

$$
\begin{aligned}
B(m, \theta_0) &= \mathrm{E}(\hat{\theta} - \theta_0) \\
&= \int f(x|\theta_0)\,(m(x) - \theta_0)\,dx
\end{aligned}
$$

Clearly, the bias in general depends on the estimator and on the true parameter value θ_0, and where possible it seems rational to choose estimators with small bias.

Similarly, the *variance* of an estimator is the expected squared difference between the true parameter value and the estimate:

$$
\begin{aligned}
V(m, \theta_0) &= \mathrm{E}(\hat{\theta} - \theta_0)^2 \\
&= \int f(x|\theta_0)\,(m(x) - \theta_0)^2\,dx
\end{aligned}
$$

Again, it seems rational to prefer estimators with small variance.

In general, it is not possible to base a choice of estimator on a preference for small bias and/or variance, because both B and V depend on θ_0, which is unknown. However, for a few particularly well-behaved model classes, it is possible to choose an estimator which has zero bias and minimal variance whatever the value of θ_0. Such estimators are called "Minimum Variance Unbiased". More generally, even when the model class is such that no minimum variance unbiased estimator exists, it may be possible to choose an estimator which is, say, less biased than the ML estimator, yet which retains all the favourable properties of the ML estimator. A familiar example is the estimation of the variance (squared standard deviation) of a Normal distribution with unknown mean μ and unknown variance σ^2. Given data comprising N values $(x_1, x_2, \ldots, x_N)$ drawn from Normal(μ, σ^2), the ML estimate of σ^2 is

$$\hat{\sigma}_{ML}^2 = \sum_i \frac{1}{N}(x_i - \bar{x})^2$$

which has bias $(-\sigma^2/N)$. The estimate

$$\hat{\sigma}_{UB}^2 = \sum_i \frac{1}{N-1}(x_i - \bar{x})^2$$

however, has zero bias for any true μ, σ, and is usually preferred.

Note that considerations of bias and variance apply only to a particular parameterization of the model family. For instance in the above example, while $\hat{\sigma}_{UB}^2$ is an unbiased estimate of σ^2, $\hat{\sigma}_{UB}$ is not an unbiased estimate of σ. Also, the few model families which admit of Minimum Variance Unbiased estimators have such estimators for only one parameterization. Thus, the usefulness of these considerations is quite limited.

1.12.2 Non-Bayesian Model Selection

One of the most difficult targets for statistical inference is to make a choice among two or more model families, where each family has a different mathematical form and/or a different number of unknown parameters. For example, given data comprising N values from a univariate distribution, we might like to suppose whether the data comes from a Normal density with unknown μ and σ^2, or from a Cauchy density

$$f(v|c, s) = \frac{s}{\pi(s^2 + (v - c)^2)}$$

with unknown location c and spread s. This problem is analogous to the choice between two simple models for the data, but more difficult because the parameter values for either family are unknown. The only choice criterion available without using additional premises is to compare the maximum likelihood obtained in one family with the maximum likelihood obtained in the other, and to choose the family giving the greater. This process is equivalent to choosing a single model from each family by ML estimation of the unknown parameters, and then comparing the two resulting models.

An important subclass of this type of inference problem arises when one family of models is a subset of the other. Typically, the restricted family is defined by fixing one or more of the unknown parameters in the more general family, so the smaller family has fewer unknown parameters, and these are a subset of the parameters of the larger family. The smaller family is then called the *null hypothesis*, and one seeks to infer whether the data give reason to suppose that they come from some model in the full family rather than some model in the null hypothesis. A simple comparison of the maximum likelihoods obtained within the two families will no longer suffice. Since the full family includes all the models in the null hypothesis, the maximum likelihood in the full family must be at least as great as that in the subset family, and will almost always be greater.

The typical non-Bayesian approach to this problem is to devise some statistic, i.e., some function of the data, whose distribution can be calculated on the assumption that the data comes from a model in the null hypothesis, but whose distribution does not depend on the parameters of that model. The statistic is also chosen so that its distribution for models not in the null hypothesis is different, typically in such a way that the statistic is likely to take values which would be improbable under the null hypothesis. The value of this *test statistic* is then computed from the data, and the null hypothesis rejected if the value obtained is deemed sufficiently improbable.

A classic example of this inference technique was devised for choosing between the full family of Normal models Normal (μ, σ^2) with μ and σ unknown, and the null hypothesis Normal(μ_1, σ^2) where μ_1 is a known value but σ is unknown. Thus, the desired inference is to decide whether or not the true mean equals μ_1. Given data comprising N values $(x_1, \ldots, x_N)$ drawn from the distribution, the test statistic

$$ t = \frac{\bar{x} - \mu_1}{\sqrt{\dfrac{1}{N-1} \sum_i (x_i - \bar{x})^2}} $$

is computed. If $\mu = \mu_1$, the numerator of this expression has a distribution of form Normal$(0, \sigma^2/N)$. The denominator has a more complex distribution with mean σ, but both distributions have widths proportional to σ. Thus, the distribution of t does not depend on σ. Its mean is zero. Further, if the true mean differs from μ_1, the distribution of t no longer has zero mean. Thus, if the observed value of t is far from zero, this event is improbable if $\mu = \mu_1$, but not improbable for some other values of μ, and the null hypothesis is rejected.

There is an extensive literature on the construction and use of test statistics for many inference problems of this general class. Many of the test statistics are related to a rather general test statistic called the *log likelihood ratio*, and defined as

$$ \lambda = \log \left(\frac{\max_{\text{full family}} f(x|\theta_f)}{\max_{\text{null}} f(x|\theta_n)} \right) $$

It is the natural logarithm of the ratio between the maximum likelihood obtained in the full family and the maximum likelihood obtained in the null family. Here, θ_f denotes the full set of parameters and θ_n the restricted set. Under certain regularity and smoothness conditions on the form of the function $f(x|\theta_f)$, it can be shown that for large data samples and assuming the data comes from a model in the null hypothesis, the statistic 2λ has a distribution close to the ChiSquared form with d degrees of freedom (written χ_d^2), where d is the number of parameters having fixed values in the null hypothesis. Thus, if the null hypothesis is true, 2λ is expected to have a value around d (the mean of χ_d^2) and rarely to have values very much greater.

The log likelihood ratio can be interpreted in another way. Suppose data x is obtained from an unknown member of a family with general distribution form $f(x|\theta)$, and that the family has n parameters, i.e., θ is a vector with n components. Let θ_0 be the true parameter values. Then, under regularity and smoothness conditions, the random variable

$$v = 2\log\left(\frac{\max_{\Theta} f(x|\theta)}{f(x|\theta_0)}\right)$$

has asymptotically a distribution close to χ_n^2 form, and $E(v) = n$. Let $\hat{\theta}$ be the maximum likelihood estimate of θ, so $v = 2\log f(x|\hat{\theta}) - 2\log f(x|\theta_0)$.

The value of $\log f(x|\theta_0)$ is of course unknown, but may be guessed as being roughly given by

$$\log f(x|\theta_0) \approx \log f(x|\hat{\theta}) - \frac{1}{2}E(v)$$
$$\approx \log f(x|\hat{\theta}) - \frac{n}{2}$$

where the value of $\hat{\theta}$, and hence of $f(x|\hat{\theta})$, is calculated from the given data x. Now suppose that it is believed that the data comes from one of two different families, with forms $f(x|\theta)$ and $g(x|\phi)$, and that they have respectively n_1 and n_2 parameters. We do not assume that one family is a subset of the other.

In comparing the two families f and g, it can be argued that if the data comes from an unknown model in family f, then the log of its probability given that model is roughly $\log f(x|\hat{\theta}) - n_1/2$. Similarly, if the data comes from an unknown model in family g, its log-probability under that model is roughly $\log g(x|\hat{\phi}) - n_2/2$. Hence, to choose between the two families, one might prefer the larger of these two quantities. Equivalently, we might prefer family f if

$$\log f(x|\hat{\theta}) - \log g(x|\hat{\phi}) > \frac{1}{2}(n_1 - n_2)$$

i.e., if

$$\log \frac{f(x|\hat{\theta})}{g(x|\hat{\phi})} > \frac{1}{2}(n_1 - n_2)$$

Thus, we are again led to the log likelihood ratio test, as $(n_1 - n_2)$ is the number of parameters for family f over and above the number for family g, but it is no longer required that g be a subset of f.

In the above rule, the value of the test statistic (the log likelihood ratio) is compared with $\frac{1}{2}(n_1 - n_2)$, which would be its expected value were g a subset of f and the data came from a model in g. We do not require the first of these conditions, and certainly do not know whether the second is true. Hence, it is not obvious that $\frac{1}{2}(n_1 - n_2)$ is necessarily the best value against which to compare the test statistic. It has been argued by Akaike [1] that it is better to compare the log likelihood ratio against $(n_1 - n_2)$ rather than $\frac{1}{2}(n_1 - n_2)$. That is, the log likelihood of each family should be "penalized" by 1 for each free parameter, rather than 1/2. Later we will see arguments for a "penalty" which is not a constant for each parameter, but rather a value of order $\frac{1}{2}\log N$, where N is the sample size of the data. It is known [5, 40, 60] that for some pairs of families f and g, the Akaike criterion is inconsistent. That is, no matter how much data is acquired, it can be expected to show a preference for the more complex model family even when the data comes from the simpler family.

1.13 Bayesian Inference

We have seen that non-Bayesian statistical inference cannot lead to probabilistic statements about the source of the data. In choosing among competing models, or model families, we can at best compare the likelihoods of the competing models, and more generally only estimates or bounds on these likelihoods. The form of statement which can be deduced from these comparisons is usually equivalent to the statement below.

"Something has happened which would be surprising if I knew the data came from model/family A, but less surprising if I knew it came from model/family B." Since the substance of such a statement is conditional upon conditions which are not true, the statement is counterfactual and a leap of faith is required to translate it into what we want to hear, which is something like the statement that the source of the data is probably in family B, or at least that family B is a better guess than family A.

The introduction of new premises into the argument can allow the deduction of rather more meaningful conclusions. We now outline Bayesian statistical inference, in which we assume that probabilistic knowledge about the source of the data is available independent of, or prior to, the observed data.

We will begin by discussing inference problems in which the set of possible models is discrete. Model families with unknown real valued parameters will be treated later. Let Θ be the set of models with identifying labels $\{\theta_1, \theta_2, \ldots\}$. The Bayesian approach assumes that, even before the data x is known, we have reason to assign a probability $\Pr(\theta_i)(\theta_i \in \Theta)$ to each competing model. This is called the *prior probability* of the model, since it is the probability we assign to the proposition that the data come from the model before, or prior to, our seeing the data. The probability distribution $\{P(\theta_i), \theta_i \in \Theta\}$ is called the *prior distribution* or simply the *prior*. As in the classical non-

Bayesian approach, we also assume the data probability distribution function or likelihood $\Pr(x|\theta_i)$ is known for every model.

Then, using Bayes' theorem we write

$$\Pr(\theta_i|x) = \frac{\Pr(x|\theta_i)\Pr(\theta_i)}{\Pr(x)} \qquad \forall \theta_i \in \Theta$$

Here, $\{\Pr(\theta_i|x)\}$ is a new probability distribution over the possible models, called the *posterior distribution* or simply the *posterior*. $\Pr(x)$ is the marginal data probability given by

$$\Pr(x) = \sum_{\theta_i} \Pr(x|\theta_i)\Pr(\theta_i)$$

$\Pr(x)$ acts as a normalizing constant for the posterior distribution. The value of $\Pr(\theta_i|x)$ for model θ_i is called its *posterior probability*, interpreted as the probability that model θ_i is indeed the source (or an accurate model of the source) of the observed data x. Loosely, it is the probability, given the data, that model θ_i is "true". (Since $\Pr(x)$ does not depend on θ, it need not be calculated if we are only interested in comparing the posterior probabilities of different models for fixed data.)

The Bayesian approach thus makes possible the inference of probabilistic statements about the source of the data. The posterior probabilities, unlike measures of "confidence", obey the usual axioms of probability. For instance, the posterior probability that the source was either θ_i or θ_j is

$$\Pr((\theta_i \vee \theta_j)|x) = \Pr(\theta_i|x) + \Pr(\theta_j|x)$$

Also, if two sets of data x and y are obtained from the same unknown source, then

$$\Pr(\theta_i|y, x) = \frac{\Pr(x|\theta_i, y)\Pr(\theta_i|y)}{\sum_j \Pr(x|\theta_j, y)\Pr(\theta_j|y)}$$

If the data sets are independent, that is, if for a *known* model θ the probability of its yielding data x is not affected by knowing that it has also yielded data y, then $\Pr(x|\theta_i, y) = \Pr(x|\theta_i)$ and we have

$$\Pr(\theta_i|y, x) = \frac{\Pr(x|\theta_i)\Pr(\theta_i|y)}{\sum_j \Pr(x|\theta_j)\Pr(\theta_j|y)} = \frac{\Pr(x|\theta_i)\Pr(y|\theta_i)\Pr(\theta_i)}{\sum_j \Pr(x|\theta_j)\Pr(y|\theta_j)\Pr(\theta_j)}$$

$\Pr(\theta_i|y)$ is the posterior probability assigned to θ_i after seeing data y. The above equation shows that, in considering further data x, $\Pr(\theta_i|y)$ plays the role of the prior probability of θ_i, and leads to the posterior probability $\Pr(\theta_i|y, x)$ assigned to θ_i after seeing the additional data x.

Note that the final posterior distribution $\Pr(\theta|y, x)$ is independent of the order in which the two data sets are considered. Whether or not the data sets are independent for a given model, we have

$$\Pr(\theta_i|x, y) = \frac{\Pr(x, y|\theta_i)\Pr(\theta_i)}{\sum_j \Pr(x, y|\theta_j)\Pr(\theta_j)}$$

which is a symmetric function of x and y.

The above results show that, given a prior probability distribution over a discrete set of possible models, the Bayesian method makes it possible to choose that model which is most probably the source of the data, and to make statements about the probability that this choice, or any other, is correct.

Henceforth, we will usually denote the prior on a set of possible models by the generalized function "$h(\theta)$". Here, $h(\theta)$ may represent a probability, if θ is discrete, or, as discussed below, a probability density.

We now consider problems in which Θ, the set of possible models, is a continuum rather than a discrete set. Let θ denote the real-valued parameter or vector of parameters identifying a particular model. Since Θ is a continuum, prior knowledge cannot assign a non-zero prior probability to every member of Θ. Instead, we may have a prior probability *density* over Θ. (It is of course possible to have a prior distribution which assigns a non-zero probability to every member of some countable subset of Θ, and a probability density over the remainder of Θ, but this possibility introduces only mathematical complexity to the problem, and nothing new in principle. We shall ignore it for the time being.)

We write the prior density over Θ as $h(\theta)$, with the meaning that the prior probability that the true model lies in some region R of Θ is given by

$$\Pr(R) = \int_R h(\theta)d\theta$$

where $d\theta$ is an element of line, area, volume, etc., depending on the number of scalar parameters, i.e., depending on the dimension of the continuum Θ. Given data x and the data probability function $\Pr(x|\theta)$, the Bayes identity now allows the calculation of a *posterior density* on Θ:

$$p(\theta|x) = \frac{h(\theta)\Pr(x|\theta)}{\Pr(x)}$$

where the marginal probability of data x is now

$$\Pr(x) = \int_\Theta h(\theta)\Pr(x|\theta)d\theta$$

(Henceforth, all integrals or summations with respect to θ will be over the set Θ of possible models unless otherwise shown.)

Note that the prior density $h(\theta)$ and posterior density $p(\theta|x)$ are often called simply the prior and posterior, as is also done with distributions over a set of discrete models.

The posterior density can be used to calculate the probability, given the data x, that the source is in some region R of Θ:

$$\Pr(R|x) = \int_R p(\theta|x)d\theta$$

and so to attach posterior probabilities to such propositions as "$a < \theta < b$", "$\theta < 0$", "$\log \theta > 1$", etc. for a scalar parameter θ. Similarly, if θ has two or more components, $\theta = (u, v)$ say, then by integrating the posterior density $p(\theta|x) = p(u, v|x)$ with respect to some components, one can compute posterior probability densities involving only the other components. Thus, the posterior probability density of component u is given by

$$\int_{v \in \Theta} p(u, v|x)dv$$

However, the posterior density of θ does not allow us to attach a non-zero probability to individual values of θ. Being a density, it cannot give a probability to a proposition of the form "$\theta = a$".

Thus, the Bayesian argument, when applied to a continuum of possible models, does not lead directly to a simple rule for choosing a "best guess" estimate of the parameter. When Θ is discrete, one may choose the discrete model of highest posterior probability, but when Θ is a continuum, no one model has any posterior probability. It might be thought that an obvious estimation rule analogous to the simple rule for discrete Θ would be to estimate θ by that value having the largest posterior density given the data, that is, to choose the mode of the posterior density:

$$\hat{\theta} = \text{Mode}\, p(\theta|x)$$

This rule is in general unacceptable, as it depends on the particular parameterization or coordinate system used for identifying members of Θ. Instead of identifying models by the parameter θ, we could equally well use any other quantity which is a one-to-one function of θ, say

$$\phi = g(\theta) \qquad \theta = g^{-1}(\phi)$$

Assuming for simplicity that the function g is differentiable, we can then use the standard rules for transforming probability densities to obtain the prior and posterior densities of ϕ:

$$\text{Prior density } h_\phi(\phi) = h(\theta)\frac{d\theta}{d\phi} = \frac{h(\theta)}{\dot{g}(\theta)}$$

where

$$\dot{g}(\theta) = \frac{d}{d\theta}g(\theta)$$

$$\text{Posterior density } p_\phi(\phi|x) = \frac{h_\phi(\phi)\Pr(x|g^{-1}(\phi))}{\Pr(x)}$$

$$= \frac{h(\theta)}{\dot{g}(\theta)} \frac{\Pr(x|\theta)}{\Pr(x)}$$

$$= \frac{p(\theta|x)}{\dot{g}(\theta)}$$

In general, the mode of $p_\phi(\phi|x)$ will not correspond to the mode of $p(\theta|x)$

$$\text{Mode } p_\phi(\phi|x) \neq g(\text{Mode } p(\theta|x))$$

The modes can only correspond if $\dot{g}()$ happens to have zero slope at the mode.

A similar objection can be raised to using the mean of the posterior density as an estimate: it also is not invariant under a change in the choice of parameter. However, if θ is a single scalar parameter, there is some logic in choosing as the estimate the median of the posterior density. One can say that the true parameter is equally likely, given the data, to be above or below this value, and the median is invariant under a (monotonic) change in the choice of parameter. However, the median is not defined when there are two or more parameters, and the posterior medians of single components of a vector parameter are not invariant.

The difficulty outlined above in obtaining a parameter estimate from the Bayesian argument alone is one of the problems we believe has been overcome in the new approach to be developed here.

We now consider Bayesian choice among two or more model families, perhaps with different numbers of parameters. One family may or may not be a subset of another. Let the families be indexed by j, and let the probability of data x for a model in family j be $f_j(x|\theta_j)$, where parameter variable θ_j is the parameter for family j. It is convenient to describe the prior distribution by the notation

$$\text{Prior } = h(j, \theta_j) = h(j)h(\theta_j|j)$$

Here $h()$ is a generalized function symbol denoting prior probability or density, $h(j)$ is the prior probability that the data source is a model in family j, and $h(\theta_j|j)$ is the prior density of parameter θ_j, given that the model is in family j. Then

$$\sum_j h(j) = 1; \quad \int h(\theta_j|j) \, d\theta_j = 1 \text{ (for all } j)$$

Given data x, the posterior can be similarly described by a generalized function symbol $p()$.

$$p(j, \theta_j|x) = \frac{f_j(x|\theta_j)h(j, \theta_j)}{\Pr(x)}$$

$$= p(j|x) \, p(\theta_j|j, x)$$

Here, $\Pr(x)$ is the marginal probability of obtaining data x from any model in any family

$$\Pr(x) \;=\; \sum_j \int f_j(x|\theta_j)\, h(j,\theta_j)\, d\theta_j$$

$$=\; \sum_j h(j) \int f_j(x|\theta_j)\, h(\theta_j|j)\, d\theta_j$$

The latter form can be recognized as summing (prior probability that the family is j) times (marginal probability of getting x from some model in family j). Thus, we can write

$$\Pr(x) = \sum_j h(j)\, \Pr(x|j)$$

where

$$\Pr(x|j) = \int f_j(x|\theta_j)\, h(\theta_j|j)\, d\theta_j$$

The posterior probability that the data comes from family j is

$$\Pr(j|x) = \frac{h(j)\, \Pr(x|j)}{\Pr(x)}$$

and the posterior density of parameter θ_j, given or assuming that the data comes from family j, is

$$\Pr(\theta_j|j,x) = \frac{f_j(x|\theta_j)\, h(\theta_j|j)}{\Pr(x|j)}$$

The posterior distribution $\{p(j|x)\}$ over the families behaves exactly as the posterior distribution over a discrete set of unparameterized models. It allows us to make probability statements about families or groups of families in the light of the data, and suggests a simple choice rule for inferring the family when a choice must be made, namely, choose the family of highest posterior probability. However, the difficulty of estimating parameter values, i.e., of choosing the "best" model within a family, remains. Nor is it entirely obvious that the "best" model, however that is defined, will necessarily be a member of the family with highest posterior probability.

1.14 Bayesian Decision Theory

In certain circumstances, the inability of the Bayesian argument by itself to pick a "best" estimate of parameter values is of no consequence. It may be that the objective of the statistical investigation is to choose among a set of possible actions whose favourable or unfavourable consequences depend on the true state of the source of the data. A much-studied example is the problem of deciding on the treatment of a patient on the basis of the patient's symptoms. Here, the set of "models" is the set of possible causes of the symptoms, i.e., Θ is a set of possible diseases or conditions. Let θ be some disease in Θ, x be the observed symptoms and $\Pr(x|\theta)$ be the probability that someone suffering θ would show x. The prior $h(\theta)$ may reflect the known frequency of disease θ in the population to which the patient belongs. Then the Bayesian argument allows us to calculate the posterior

$$\Pr(\theta|x) = \frac{h(\theta)\,\Pr(x|\theta)}{\Pr(x)}$$

for any θ.

Suppose now that there is a set A of actions $\{a_1, a_2, \ldots, a_k, \ldots\}$ which may be taken, i.e., a set of treatments which could be given, and that previous experience has shown that the cost, in terms of money, time, suffering and final outcome, of treating a sufferer from θ with treatment a_k is $C(\theta, a_k)$. (In general, C would usually represent an expected or average cost, as θ and a_k might not fully determine the consequences.) Then the expected cost of taking action a_k given the data x is

$$
\begin{aligned}
\mathrm{E}(C(a_k)|x) &= \sum_\theta \frac{h(\theta)\ \Pr(x|\theta)\ C(\theta, a_k)}{\Pr(x)} \\
&= \sum_\theta \Pr(\theta|x)\, C(\theta, a_k)
\end{aligned}
$$

It is then rational to choose that action $\hat{a}$ for which $\mathrm{E}(C(a_k)|x)$ is least.

This technique of using the posterior to decide on the action of least expected cost, given the data, is known as *Bayesian decision analysis*, and has been much studied. Although presented above in terms of a discrete set of possible models and a discrete set of possible actions, the extension to parameterized models and/or a continuum of possible actions is immediate and raises no problems. If Θ is a continuum with parameter θ, the expected cost of action a given data x is

$$\mathrm{E}(C(a)|x) = \int p(\theta|x) C(\theta, a)\, d\theta$$

where $p(\theta|x)$ is the posterior density of θ. This expression is valid whether the set A of possible actions is continuous or discrete. When A is continuous, $C(\theta, a)$ remains a "cost", e.g., a sum of money, and is not a density over A. Similarly, $\mathrm{E}(C(a)|x)$ is not a density, and its minimum is invariant under changes of parameterization of Θ and A.

We see that, given the addition of a cost function, Bayesian inference leads to a rational basis for decision which is valid whether or not the set of possible models is or contains a continuum family with unknown parameters. Where a cost function is available, we can see no conceptual objection to Bayesian decision analysis, and the approach developed in this work adds nothing to

Bayesian decision theory save (we hope) some useful comments about prior probabilities.

Bayesian decision theory does not require or involve making any statement about the "best" or most probable model of the data, nor does it involve any estimation of parameters. In fact, it really makes no inference about the source of the data other than a statement of the form: "Our knowledge of the source of the data is such that the action of least expected cost is $\hat{a}$", which is not about the source, but about our knowledge of it.

In some circumstances, Bayesian decision analysis can appear to lead to estimates of parameters. Consider a problem where the set of possible models is a single family with parameter θ, and the set of possible actions is also a continuum, having the same dimension as the family of models. Then it may be that the choice of action can be expressed in a form similar to a statement about the model. For example, imagine a traveller faced with the problem of what clothes to pack for a visit to Melbourne. Her extensive wardrobe offers an almost unlimited range of garments, each suitable for one particular narrow range of temperatures. Prior knowledge and meteorological data allow the traveller to infer a posterior density over θ, tomorrow's temperature in Melbourne. In this case, the action chosen, i.e., the selection of clothes, can also be expressed as a temperature: the temperature θ_a for which the chosen clothes are ideal. The cost $C(\theta, \theta_a)$ is related to the difference between the temperature θ_a and the actual temperature θ to be encountered in Melbourne.

In cases such as the above, where the parameters of the action or decision are commensurable with the unknown parameters of the model, the action parameters θ_a of the least-cost action are sometimes considered to be an estimate of the model parameter θ. This possibility arises most commonly when the cost function can be expressed as a function $C(\theta - \theta_a)$ of the difference between "estimate" and true value. In our view, this use of the term "estimate" is misleading. It is not difficult to imagine cases where the least-cost action is described by an action parameter θ_a such that $\Pr(x|\theta = \theta_a) = 0$. For instance to use a fanciful example, the meteorological data might show quite conclusively that the temperature in Melbourne will not be in the range 22–23°C, but has high posterior probability of being either about 20° or about 25°. The least-cost action might then be to pack clothes ideal for 22.5°C, even although the traveller is certain that this will not be the temperature. Of course, it is more usually the case that the least-cost θ_a is a reasonable estimate of θ, even if it is not so intended.

The above remarks are no criticism of the Bayesian decision-making process. They aim only to argue that the process need not and does not make any estimate about the source of the data. No assertion, however qualified, is made about the "true state of the world".

Similar but stronger remarks can be made about the outcome of Bayesian decision when the set of possible models includes different parameterized families. The process makes no assertion about which family contains the source of the data. Even if the chosen action is described in the same terms as are used to name a family, we cannot properly regard the terms describing the least-expected-cost action as an assertion or guess about the "true" family.

If Bayesian decision analysis is accepted as a sound basis for choosing actions, as we argue, there may seem no need ever to go beyond it in any analysis of data. Whenever data is collected, observations made, model families devised and inferences drawn, it is for some purpose or purposes. These purposes may be vital or frivolous, but almost always some action will be taken in the light of the data to further these aims. If Bayesian decision can reliably guide these actions, need we ever worry about weighing one theory against another, or trying to discover precise values for unknown parameters? The decision process automatically takes into account all the theories, models and possible parameter values we are prepared to consider, weights each in due proportion to the support it gets from the data, and leads to an appropriate choice of action.

If the Bayesian decision process were universally feasible, we would have indeed no reason other than idle curiosity to pursue any other kind of statistical inference. Unfortunately, the limitations of human reason and the division of responsibilities in human societies make the process infeasible in all but a few arenas. An obvious obstacle is that the investigator who collects and analyses the data is rarely in a position to know what decisions will be made using the conclusions of the investigation. Some of the immediate objectives of the study may be known, but many inferences, especially those widely published, will be used by persons and for purposes quite unknown to the investigator. The investigator will often have even less knowledge of the cost functions involved in the decisions based on his work. It is unlikely that Kepler could have had any idea of the consequences of accepting his inference that planetary orbits are elliptical, or that Millikan could have put a cost to any error in his estimation of the electronic charge.

Because an investigator is not usually in a position to know the uses which will be made of his conclusions and the cost functions relevant to those uses, one can argue that a statistical analysis of the data should proceed no further than the calculation of the posterior distribution. Any agent who, faced with a range of decisions and a cost function, wishes to use the results of the investigation will find all the relevant results expressed in the posterior distribution. The posterior encapsulates everything about the source of the data that can be inferred within the assumptions of the investigation (i.e., within the constraints imposed by the set of models considered). On the other hand, if the investigator goes beyond the posterior, and states as his inference a "best guess" of model family and parameter values, he is in general censoring and/or distorting the data. Except in those simple cases where a small set of sufficient statistics exists, no one model can accurately summarize all the information in the data which may be relevant to future decisions.

There are thus good reasons to consider the task of a statistical investigation to be complete when the posterior has been calculated. Further inference, e.g., the choice among actions or the prediction of future events, can be based on the posterior but require additional information not normally known to the original investigator.

In the real world, however, the above policy is a counsel of perfection, rarely able to be carried out and rarely meeting practical needs. There are two difficulties. First, in even modestly complicated problems, the set of models considered possible may be quite large and have a fairly complex structure. For instance, it may commonly include several structurally different families of models with different numbers of parameters. The result can be that the posterior is a complicated function, difficult both to express and to evaluate. Indeed, in some cases an accurate specification of the posterior is just as lengthy as the body of data on which it is based. "Conclusions" of such a sort are hard to communicate and give little or no insight into the nature of the data source.

The second difficulty is that posterior distributions of any complexity are often too difficult to use in further reasoning. Consider a study on the effect of various inputs on the yield of a crop. The data might include observed yields under a range of fertilizer and rainfall inputs, soil types, temperature profiles over the growing season, etc. The models considered might range from simple linear families through ones allowing for interaction of different inputs and saturation effects to model families which incorporate detailed modelling of photosynthesis and other biochemical reactions. The posterior distribution over these families and their numerous parameters would be a very complicated function. Even within a single family, such as the family of non-linear regression models, the posterior densities of the unknown parameters would have complex shapes and correlations. It would be extremely difficult to use such a function in answering practical questions such as "If November is unusually cool and dry, will fertilizer applied in December have any useful effect?" Here, the most useful result that could be obtained from the study is a single model, preferably of no great mathematical complexity, which captured most of the effects visible in the data with reasonable accuracy.

A further striking example where a posterior would be almost useless is the "computer enhancement" of noisy images such as satellite photographs. Here, the "models" are all those 2-dimensional brightness patterns which the subject of the photo might have had, the data is the brightness pattern of the available image(s), the data probability function describes the probability distributions of the various noise and distortion processes which corrupt the image, and the prior should reflect what is already known about the subject. The processing of the given image to provide a clearer picture of the subject is essentially a statistical inference. However, to give as the result a posterior distribution over all possible pictures would be ludicrous: what is needed is a single picture showing the "best guess" which can be made about the subject. Merely to enumerate all the "models" of the subject would be infeasible for an image comprising a million pixels. It is likely that the set of models with high posterior probability would allow at least two brightness values for each pixel independently, giving at least $2^{1000000}$ models to be enumerated.

1.15 The Origins of Priors

The Bayesian argument takes a prior probability distribution over a set of models, adds information gleaned from data whose source is believed or assumed to be in the model set, and results in a new, posterior, probability distribution over the set. We have seen that the new distribution can properly be taken as the prior in analyzing more data from the same source, giving a new posterior which in turn serves as the prior when yet more data is obtained. Well-known convergence results assure us that as more and more data is obtained, the final posterior will almost certainly converge towards a distribution placing all the probability on the model of the data source, provided that this model is in the set considered possible by the original prior. Thus, to some extent we can regard the choice of prior as unimportant. Provided the prior is not extreme, i.e., does not give zero prior probability to any model (or interval of models in the case of a continuum), the evidence gained from sufficient data can always overwhelm the prior distribution. In real life, however, the option of collecting more and more data is not always feasible and economic, and the data which is available may not suffice to overwhelm the effects of a misleading prior. In practice, the prior should be treated seriously, and care should be taken to ensure that it represents with some fidelity what we really know and expect about the source of the data.

Here, we should remark that the Bayesian approach seems to demand that prior probabilities have a meaning distinct from a naive frequentist interpretation of probabilities. First, the very concept of a "probability" attaching to a proposition such as "giant pandas evolved from the bear family" is inadmissible if probability is to be interpreted in naive frequentist terms. The proposition either is or is not true. There is no useful sense in which it can be said to be true 83% of the time, or in 83 out of 100 cases on average in this universe. The frequentist interpretation has been extended to accommodate probabilities for such statements by such devices as an ensemble of possible universes, but we do not follow this line. Rather, we believe this objection to prior (and posterior) probabilities is empty if probabilities are regarded as measures of subjective certainty. Later, we will address the question of how such a subjective interpretation can be reconciled with the apparently objective probabilities involved in statements like: "Every plutonium atom has probability 0.5 of decaying radioactively in the next 12,000 years".

A more fundamental problem with priors is that the Bayesian argument cannot explain how a prior distribution is arrived at. More generally, it fails to explain how we come to have varying degrees of certainty about any propositions. Given some initial distribution of belief or probability, the Bayesian approach shows us how observed data should lead us to modify that distribution. But in the absence of a prior distribution, the standard Bayesian argument cannot lead us to infer any probability distribution over alternative models, no matter how much data we obtain. Whence, then, come our priors?

There are several possible answers to this question.

1.15.1 Previous Likelihood

The prior could be based on previous data relevant to the current problem. Without a pre-existing distribution, it could not be obtained from the data by a Bayesian inference, but by a leap of faith the prior over a discrete set of models could be obtained from the likelihood function $\Pr(z|\theta)$ for previous data z. The likelihood as it stands is not a probability distribution over the possible models, i.e., the possible (discrete) values of θ, but a normalized form

$$h(\theta_j) = \frac{\Pr(z|\theta_i)}{\sum_j \Pr(z|\theta_j)}$$

could usually be constructed and used as a prior. (For some model families, $\Pr(z|\theta)$, the sum in the denominator, may be infinite.) In effect, the prior is obtained by stretching non-Bayesian inference and regarding the likelihood of a model on data z as being an indication of its probability.

Formally, the above device is equivalent to assuming an original, primitive prior in which all models have equal prior probability, then using the previous data z to obtain a posterior by a conventional Bayesian argument. Hence, deriving a "prior" by appeal to previous data in this way is no more or less defensible than supposing all models to have equal prior probability.

When this kind of process is used to construct a prior over a continuum of models with parameter θ, a further leap of faith is required. For some previous data z, the likelihood function $\Pr(z|\theta)$, when regarded as a function of θ with given z, is not a density. To infer from it a prior density over the continuum of the form

$$h(\theta) = \frac{\Pr(z|\theta)}{\int \Pr(z|\theta)d\theta}$$

could be justified only if one is persuaded that the variable θ is the "correct", or "most natural", parameter for the continuum. Use of the same device to obtain a "prior" for some other parameter ϕ, non-linearly related to θ, would give a prior for ϕ implying quite different prior expectations about the model.

1.15.2 Conjugate Priors

The prior could be based on the mathematical properties of the data distribution function (or density) $\Pr(x|\theta)$. For instance, it is common to choose a "conjugate prior" $h(\theta)$ which has the property that, when used with data from the distribution family $\Pr(x|\theta)$, it gives a posterior of the same mathematical form as the prior. That is, $h(\theta)$ and $\Pr(\theta|x) = h(\theta)\Pr(x|\theta)/\Pr(x)$ belong to the same mathematical family, whatever the value of x. An example of a conjugate prior is the Beta prior density for the parameter θ of a Binomial distribution. If the data x is the number of successes in N trials, where each trial has independently an unknown but constant probability θ of succeeding, then

$$\Pr(x|\theta) = \binom{N}{x}\theta^x(1-\theta)^{N-x}$$

If the prior density $h(\theta)$ has the Beta form

$$h(\theta) = \frac{\theta^{\alpha-1}(1-\theta)^{\beta-1}}{B(\alpha,\beta)} \qquad (0 \le \theta \le 1)$$

where $B(\alpha,\beta)$ is the Beta function $\Gamma(\alpha)\Gamma(\beta)/\Gamma(\alpha+\beta)$ and α, β are constants greater than zero, then the posterior density $p(\theta|x)$ is given by

$$p(\theta|x) = \frac{\theta^{x+\alpha-1}(1-\theta)^{N-x+\beta-1}}{B(x+\alpha, N-x+\beta)}$$

which is also of Beta form.

Priors which are normalized forms of the likelihood function derived from some real or imagined pre-existing data z are always conjugate. The prior $h(\theta)$ then has the form

$$h(\theta) = K\Pr(z|\theta)$$

where K is a normalization constant. The posterior given data x becomes

$$\begin{aligned} p(\theta|x) &= \frac{K\ \Pr(z|\theta)\ \Pr(x|\theta)}{\Pr(x)} \\ &= K_2\ \Pr(x,z|\theta) \end{aligned}$$

where K_2 is a new normalization constant. Thus, the posterior $p(\theta|x)$ is a normalized form of the likelihood function arising from the combined data x and z, and so is of the same mathematical form as the prior.

A useful listing of the general forms of conjugate priors for many common probability models is given in Bernardo and Smith's book on Bayesian Theory [4]. They also give the resulting forms for the posterior distributions.

Conjugate priors are mathematically convenient but any argument for basing the prior on the mathematical form of the data distribution family $\Pr(x|\theta)$ must be suspect.

Under our subjective view of probabilities, the prior should represent what we expect about the model *before* the data to be analysed is known. Only if this is so are we justified in using the Bayes identity. We also consider that an honest choice of prior should be independent of the data probability distribution given the model. We may properly think of the collection of observable data as an (imperfect) procedure designed to measure the unobservable parameter θ. The kind of data collected and its probabilistic dependence on θ determine the function $\Pr(x|\theta)$. That is, this function describes the procedure. Our prior beliefs about θ are not, or should not be, modified by knowledge of the procedure available to measure θ.

1.15.3 The Jeffreys Prior

There is a weak argument which might be used by a statistician who is given data to analyse and who is unable to extract from his client any coherent statement about prior expectations. The statistician could argue that, if the client has chosen a procedure characterized by $\Pr(x|\theta)$ to investigate θ, the choice might reflect what the client originally believed about θ. Thus, if the form of $\Pr(x|\theta)$ is such that the distribution of x is a rapidly varying functional of θ only when θ is in some range $a < \theta < b$, the statistician might argue that the client must have expected θ to lie in this range, or he would have chosen a better procedure. The most general expression of this line of argument is the "Jeffreys" prior, which is a density proportional to $\sqrt{F(\theta)}$, where $F(\theta)$ is the "Fisher Information", a generalized measure of how sensitive the probability of x is to variation of θ (described more fully later in Section 5.1).

This argument is not very convincing. It assumes that a range of "procedures" was available to the investigator, each procedure being well-matched to some set of prior beliefs about θ. The real world is rarely so obliging. Jeffreys, while noting the interesting mathematical properties of the Jeffreys prior, did not advocate its use as a genuine expression of prior knowledge (or ignorance).

Our prior beliefs about the model will usually be fairly vague. Even when we have reason to expect one model, or one range of parameter values, to be more likely than others, the strength of this prior certainty would rarely be quantifiable more precisely than within a factor of two, and more commonly only within an order of magnitude. So there is usually no value in attempting to specify our prior distribution with great precision. If we can find a conjugate or other mathematically convenient prior distribution which roughly accords with our prior beliefs, and excludes no model or parameter value which we consider at all possible, then there is no reason not to adopt it. For instance, the Beta prior is convenient for the parameter of a Binomial distribution, and by choosing appropriate values for the constants α and β we can find within the Beta family distributions expressing a wide range of prior beliefs. Thus, if we set $\alpha = \beta = 0.1$, the Beta density diverges at $\theta = 0$ and $\theta = 1$, expressing a strong expectation that θ is extreme, i.e., we expect that either success is very common or failure is very common. Conversely, setting $\alpha = 4$ and $\beta = 3$, the density has a broad peak at $\theta = 0.6$. If we can find values of α and β which give a density doing no violence to our prior knowledge, we might well use it in preference to some other mathematical form marginally more in line with our expectations.

1.15.4 Uninformative Priors

A prior may be formulated to express our ignorance of the source of the data rather than substantial prior belief. That is, when almost nothing useful is known or suspected about the source of the data, we can try to form a prior which says nothing about θ. Such *colourless* or *uninformative* priors have been well described in the literature. To take a simple example, suppose that the data comprises the locations of serious cracks in a long straight road. If we believe the cracks are the result of minor earth tremors due to some deep geological fault, we might take as the model family a Normal distribution of locations with unknown mean μ (presumably over the fault) and unknown standard deviation σ. For such a distribution, μ is known as a parameter of *location* and σ as a parameter of *scale*. These terms are used whenever, as in this case, the probability density of a value y can be expressed as

$$\text{Density}(y) = \frac{1}{\sigma} G\left(\frac{y - \mu}{\sigma}\right)$$

where the function G does not otherwise depend on y, μ or σ. In this case, if the location y of a crack has the density Normal(μ, σ^2) then the linearly transformed value $(y - \mu)/\sigma$ has the density Normal$(0,1)$.

A location parameter may have no "natural" origin of measurement. The location y of a crack can equally well be expressed in kilometres North or South of town X as in km N or S of town Y. We may feel that whether X or Y is taken as the origin of measurement for y (and hence for μ) has no bearing on our prior expectations about the numeric value of y. Equivalently, we may feel that our prior knowledge is so irrelevant that we would be no more and no less surprised to learn that μ was 10 km N of X than we would be to learn that μ was 10 km N of Y, or anywhere else along the road for that matter. In such a case, we may argue that the prior density for μ, $h(\mu)$, should be a Uniform density, i.e., that $h(\mu)$ should be a constant independent of μ. This prior density then expresses no preference for any value of μ over any other.

Similarly, if we have no useful information about the depth and activity of the suspected fault, we may feel totally ignorant about the expected spread of the crack locations, i.e., about σ. We may feel that we would be no more or less surprised to learn that σ was 1 m, 10 m, 100 m or 10 km. If so, knowing the unit of measurement for σ (m, km, mile, etc.) would not lead us to modify our prior expectations as to the numerical value of σ. Being wholly ignorant of the location parameter μ means our $h_\mu(\mu)$ should be unaffected by a shift of origin, hence $h_\mu(\mu) = h_\mu(\mu - a)$ for any a, and hence $h_\mu(\mu)$ must be uniform. Being wholly ignorant of the scale parameter σ means that our prior $h_\sigma(\sigma)$ should be unaffected by a change of units, hence $h_\sigma(\sigma) = bh_\sigma(b\sigma)$ for any positive b. The factor b arises from the rule for transforming densities. If x has density $f(x)$ and variable $y = g(x)$, then the density of y is

$$\frac{f(g^{-1}(y))}{\left(\frac{dy}{dx}\right)}$$

The only prior satisfying $h_\sigma(\sigma) = bh_\sigma(b\sigma)$ for all $b > 0$ is $h_\sigma(\sigma)$ proportional to $1/\sigma$. Equivalently, we may suppose the prior density of $\log \sigma$ to be uniform, since a change in units for σ is equivalent to a change in origin for $\log \sigma$.

The above arguments, or variants, support the common practice that prior ignorance of a parameter of location is expressed by a uniform prior density, and prior ignorance of a parameter of scale is expressed by a $(1/\theta)$ prior density or, equivalently, by a uniform prior density for $\log \theta$. There is an objection to these uninformative priors: they cannot be normalized. If we are wholly ignorant of μ, its possible range is $\pm\infty$, and the integral of any non-zero uniform density over this range cannot be finite. Similarly, if we are wholly ignorant of σ, its possible range is $[0, \infty)$, and the integral of any non-zero density proportional to $1/\sigma$ over this range cannot be finite. The fact that these priors are improper is a reminder that we are never wholly ignorant *a priori* about any quantity which we hope to measure or estimate. This point will be elaborated in the next chapter.

In practice, the use of these improper priors is often admissible. Although the priors are improper, when combined with some real data they usually lead to posterior densities which are proper, i.e., normalizable. Conceptually, we may argue that when we use, say, a uniform prior for a location parameter μ, we really mean that the prior density of μ is almost constant over some large but finite interval of width W, i.e., we really mean something like

$$h(\mu) = \frac{1}{W}; \quad (a - W/2 < \mu < a + W/2)$$

Provided that the data x is such that $\Pr(x|\mu)$ falls to negligible values for μ outside the range $a \pm W/2$, the values of the constants a and W will have negligible effect on the posterior density of μ, so we need not specify them.

The arguments for these uninformative priors are special cases of a more general "argument from ignorance". If we feel that our expectations about some parameter θ are unchanged by any operation on θ in some group of operations, then the prior we use for θ should have at least the same symmetry as the group. Here, the terms *group* and *symmetry* are used in the sense of group theory. Thus, if we feel our expectations about μ are unchanged by the group of translation operators $\mu \to \mu + a$, $h(\mu)$ should have the symmetry of this group, and hence must be uniform.

If some trial can produce outcomes A, B or C, and we feel that our prior knowledge about A is the same as about B or C, then our expectations should be unchanged by any permutation of the outcomes. Our prior for the probabilities of the outcomes should then have the same symmetry, so we should choose a prior with the property that

Prior density (Prob (A)) = Prior density (Prob (B)), etc.

1.15.5 Maximum-Entropy Priors

The prior may be chosen to express maximum ignorance subject to some constraints imposed by genuine prior knowledge. The most well-known form is called a *maximum entropy* prior. Suppose we have an unknown parameter θ about which we know little, but we do have reason to believe that $h(\theta)$ should satisfy one or more equations of the form

$$G(h(\theta)) = C$$

where G is a known operator taking a distribution or density as argument and giving a numeric value, and C is a known constant. The maximum entropy approach chooses $h(\theta)$ to maximize

$$H = \begin{cases} -\int h(\theta)\, \log(h(\theta))\, d\theta & h() \text{ a density} \\ -\sum_{\Theta} h(\theta)\, \log h(\theta) & h() \text{ a discrete distribution} \end{cases}$$

subject to the constraints $\{G_k(h(\theta)) = C_k, k = 1, 2, \ldots\}$

The quantity H is called the *entropy* of the distribution $h()$. At least when θ ranges over a fixed discrete set of values, i.e., when $h()$ is a discrete distribution, it can plausibly be argued that, of all distributions satisfying the constraints, the maximum entropy distribution implies the least amount of information about θ. The theory is given in the next chapter. Note that the constraints

$$G_0(h()) = \sum_{\theta} h(\theta) = C_0 = 1; \qquad h(\theta) \geq 0 \quad (\text{all } \theta)$$

are always imposed.

If no other constraint is imposed, H is maximized by choosing $h(\theta)$ to be constant.

The term "entropy" derives from thermodynamics, and we now give an example with a thermodynamic flavour. Suppose that our data x relates to the vertical velocity θ of a gas molecule, and we have a known data probability function $\Pr(x|\theta)$. Our prior knowledge is that the molecule has been selected from a fixed body of gas by some process which should not be affected by θ, the body of gas as a whole has no vertical momentum, and the mean squared vertical velocity (which can be inferred from the temperature) is known. All molecules have the same mass. Assuming the subject molecule to be randomly selected, we can then argue that $h(\theta)$ should satisfy the constraints

$\int h(\theta)\theta \, d\theta = 0$ (No average momentum)
$\int h(\theta)\theta^2 \, d\theta = T$ (Known average squared vertical velocity)
$\int h(\theta) = 1$ (Normalization)
$h(\theta) \geq 0 \quad \forall \theta$ (No negative density)

Using the method of indeterminate (Lagrange) multipliers, it is then easily shown that H is maximized by the Normal prior

$$h(\theta) = \frac{1}{\sqrt{2\pi T}} \exp\left(-\frac{\theta^2}{2T}\right) = \text{Normal}(0, T)$$

When, as in this example, $h()$ is a density rather than a discrete distribution, the maximum entropy construction unfortunately depends on the chosen parameterization. If we ask for the prior which is least informative about, say, θ^3 rather than θ, then under exactly the same constraints we get a different prior for θ. Like many of the arguments outlined above for original priors, the result depends on the choice of parameterization of the continuum of possible models. These arguments can command our support only to the extent that we are persuaded that the chosen parameters give the most "natural" way of specifying a particular model in the continuum. The only argument so far presented which is exempt from this criticism is the argument based on symmetry. This argument requires as its premise a prior belief that our expectations should be unaffected by a certain group of operations on the continuum. It does not depend on how we parameterize the continuum (provided we appropriately modify the parametric description of the group), but leaves unexplained how we might ever come to believe in the premise. In general, it also fails fully to determine the prior.

1.15.6 Invariant Conjugate Priors

The conjugate prior $h(\theta)$, which is the normalized form of the likelihood of θ given some prior (possibly imaginary) data z

$$h(\theta) = \text{constant} \times \Pr(z|\theta)$$

has the unfortunate property that for fixed z, the prior distribution over models depends on how the models are parameterized. However, this objection does not apply to a prior constructed as the posterior density of θ given some initial prior $h_0(\theta)$ and some real or imagined prior data z. If we define the normalized prior density as

$$h(\theta) = \text{constant} \times h_0(\theta) \Pr(z|\theta)$$

then $h(\theta)$ is invariant under transformations of the parameter, providing $h_0(\theta)$ is appropriately transformed. That is, if $\phi = g(\theta)$ where the transformation $g()$ is invertible, and the "initial" priors for θ and ϕ obey

$$h_{0\theta}(\theta)d\theta \;=\; h_{0\phi}(\phi)d\phi$$
$$=\; h_{0\phi}(g(\theta))\left(\frac{d}{d\theta}g(\theta)\right)d\theta$$

then the priors for θ and ϕ resulting from the "prior data" z are also equivalent:

$$h_\theta(\theta)d\theta \;=\; \mathrm{Const}\,P_\theta(z|\theta)h_{0\theta}(\theta)d\theta$$
$$=\; \mathrm{Const}\,P_\phi(z|\phi)h_{0\phi}(\phi)d\phi$$
$$=\; h_\phi(\phi)d\phi$$

for $\phi = g(\theta)$.

This construction may seem only to defer the problem, leaving us still to choose the "initial" prior $h_0(\theta)$. However, for model families possessing simple conjugate prior forms (essentially the exponential family of distributions), the "uninformative" priors of Section 1.15.4 usually turn out to be degenerate or extreme members of the conjugate family, and by virtue of their genesis in the symmetry or transformation properties of the model family, inherently transform properly. Thus, prior densities defined as the "posteriors"

$$h(\theta|z) = \text{constant} \times h_0(\theta)\Pr(z|\theta)$$

where $h_0(\theta)$ is the "uninformative" prior, are also of conjugate form, but unlike those based purely on the likelihood, transform correctly. The uninformative prior $h_0(\theta)$ is usually improper, but provided the "prior data" z are sufficiently numerous, the resulting posterior, which we then take as the prior for analysing new data, is normalizable. We will call such priors "invariant conjugate". Note that although convenient, they have no other special virtue or claim to credence.

1.15.7 Summary

To summarize this introduction to priors, we have so far found no explanation in either Bayesian or non-Bayesian inference of how our prior experience and observations could lead us to formulate a prior over a continuum of models. Even over a discrete set of models, the symmetry and maximum-entropy arguments so far presented can suggest how we might come to formulate a prior which assigns every model the same probability, but cannot suggest how we might rationally come to any other prior belief except by starting with an equiprobable prior and using some pre-existing data. This last exception is more rarely available than might appear, since it applies only when the pre-existing data is believed to come from the same source as the data to be analysed.

Despite this gloomy picture, it is often, perhaps usually, found that rational persons with similar background knowledge of the subject of an investigation can agree as to their priors, at least within an order of magnitude, and agree that some compromise prior does little violence to anyone's expectations. We will later attempt to give an account of how such agreement is not only possible but to be expected. However, for the time being let us simply assume that prior expectations are not wholly irrational and/or idiosyncratic, and that a prior distribution in rough conformity with the expectations of most rational investigators will be available as a premise for a Bayesian analysis.

1.16 Summary of Statistical Critique

The foregoing sections do not do justice to the statistical inference methods described. However, they do suggest that classical non-Bayesian methods are incapable of obtaining conclusions which (a) have a well-defined interpretation in terms of probability, and (b) are not conditional on propositions not known to be true. Classical Bayesian inference, because it assumes more, can infer more. Probabilities, at least of the subjective kind treated in this work, can be calculated for propositions about the source of the data, and in particular, it is possible to choose that model or model family among a discrete set of competitors which has the highest probability given the data.

Neither classical approach can offer a convincingly general solution to the estimation of real-valued unknown parameters. The non-Bayesian approach can at best derive assertions about intervals of possible parameter values, framed in terms of the rather vague concept of "confidence". In a few particularly simple cases, estimators of no bias and/or minimal variance can be deduced, but these properties apply only to one special parameterization of the model family. The Bayesian approach can deduce a posterior density over parameters, but offers no general method of selecting a "best" estimate which is not tied to a particular parameterization of the model family.

If a "cost function" is added to the Bayesian premises, it becomes possible to deduce actions or decisions of least expected cost. It may be possible to argue that making an estimate $\hat{\theta}$ of a parameter θ is an "action", and that, lacking more specific knowledge of the uses to which the estimate will be put, it is reasonable to assume that the "cost" of an estimate will be some function such as $(\hat{\theta} - \theta)^2$ or $|\hat{\theta} - \theta|$ reflecting the error of the estimate. However, such "least cost" estimates again depend on the parameterization. Where genuine cost functions are known, Bayesian decision analysis is justified. However, it does not lead to genuine estimates of parameter values.

In the absence of a cost function, it can be argued that Bayesian inference is complete when it has obtained a posterior distribution over the possible models and their parameters. In effect, it is argued that this is as far as statistical inference can or should go; the rest is up to the user. This program is in principle defensible but impractical except for very simple sets of models. Generally, the user of the results can absorb and work with only a single model and wants it to be the best available in the light of the data. The classical Bayesian approach gives no convincing way of choosing this model.

Finally, we note that any Bayesian analysis assumes the availability of a prior probability distribution. While in practice there may be no great difficulty in obtaining agreement among reasonable people as to what distribution to use in a specific case, the classical Bayesian argument gives no basis for this agreement. A prior distribution over models may be refined in the light of data to give a posterior distribution, which can serve as the prior in the analysis of further data, but neither the non-Bayesian nor the Bayesian approach gives any grounds for choosing an original prior.

2. Information

This chapter has three sections. The first gives a short introduction to Shannon's theory of information, or at least those aspects which relate to coding theory. The second introduces the theory of algorithmic complexity arising from the work of Kolmogorov, Chaitin and others. These sections could be skipped by readers familiar with the material, as they contain nothing novel. The third section connects these two approaches to the measurement of information with Bayesian statistics, and introduces some slight but useful restrictions on the measure of algorithmic complexity which assist in the connection. It should be of interest to most readers.

2.1 Shannon Information

For our purposes, we define *information* as something the receipt of which decreases our uncertainty about the state of the world. If we are planning a visit to Melbourne, we may not know whether it is raining there, and be wondering whether to take an umbrella. A phone call to a weather service can inform us of the present and forecast weather in Melbourne. Even if we do not wholly believe the report, its receipt has reduced our uncertainty: it has given us information. Information is not directly related to physical quantities. It is not material, nor is it a form of energy, although it may be stored and communicated using material or energy means (e.g., printed paper, radio-waves etc.). Hence, it cannot be directly measured with instruments, or in units, appropriate for physical quantities. The measurement of information is most conveniently introduced in the context of the communication of information. That is, we will look at what happens when information is passed from a *sender* to a *receiver*.

The communication of information normally involves the transfer from sender to receiver of some material object (a magnetic disc, a handwritten letter, etc.) or some form of energy (sound waves, light or radio waves, electrical currents on a phone line etc.) but the choice of vehicle is not important for our purposes. In fact, many communications in everyday life involve several transformations of the information as it is transferred from vehicle to vehicle. Consider just a part of the communication taking place when we hear the race results summarized on the radio. The summary may be given to the news reader as a typed sheet of paper: patterns of black ink on a white substrate. These patterns modify the light waves falling on them from the ceiling lights. Some of the light waves enter the reader's eyes and trigger electro-chemical pulses in her retina, which trigger many other pulses in her brain, and eventually pulses in nerves controlling her lungs, voice box, mouth and tongue. The resulting muscular motions cause pressure waves in the surrounding air. These waves are detected by a microphone which generates an electrical voltage varying in sympathy with the air pressure. The voltage eventually has the effect of controlling the strength or frequency of a rapidly oscillating current in the radio station's antenna, causing modulated electromagnetic waves to be emitted. These waves are detected in a radio receiver and decoded by a rather complex process to produce an electric voltage varying in roughly the same way as that produced by the studio microphone. This voltage drives a cone of stiff paper into rapid vibration, causing sound waves which our ears detect. More electro-chemical pulses travel to and around our brains, and suddenly we are fairly certain that Pink Drink won the 3.20 race at Caulfield. Through all these transformations in representation and vehicle (which we have grossly over-simplified) somehow this information has been preserved and transmitted.

The full theory of information stemming from the work of Shannon [41] has much to say about all the processes involved in the above scenario, although those taking place in human brains remain largely mysterious. However, we need only consider an abstract and simple view of what takes place in such communications.

We will view all representations of the information, be they printed page, waves, nerve impulses, muscle movements, pictures or whatever, as sequences of *symbols*. A symbol is an abstract entity selected from a discrete set (usually finite) called an *alphabet*. The use of the words in this technical sense is based on their familiar use in relation to written or typed text in a natural language. The race summary was first seen as a sheet of paper marked with letters (symbols) selected from the alphabet of written English. The arrangement of the letters on the paper implied by convention a sequence of presentation: left to right, then top to bottom. The colour and chemistry of the ink, the size and font style of the printing etc. are all unimportant for our purposes. What matters is the sequence of symbols. (Note that the alphabet must be considered to comprise not only the 26 ordinary letters, but also (at least) the symbols "space" and "full stop". These punctuation symbols must appear in the sequence to delimit words and sentences. They carry an essential part of the information.)

It is less obvious that a pressure wave in the air or a microphone voltage can be regarded as a sequence of symbols. In both cases, the sequence is clear enough — it is the temporal sequence of the changing pressure or voltage. However, a pressure or voltage can take infinitely many values within a finite range, so it appears that the "alphabet" is infinite. Further, the pressure or voltage is constantly changing, so it takes infinitely many different values in the course of a single second, so it appears that for such media, the communication can only be represented by an infinite sequence of "symbols", each selected from an infinite alphabet. Fortunately these difficulties are only apparent. One of the early results of Information Theory shows that when information is conveyed by a time-varying physical quantity such as pressure or voltage, all the information conveyed can be recovered from measurements of the quantity at regular, discrete intervals, and that these measurements need be made only to a limited precision. When information is conveyed by speech, the words spoken and in fact the whole sound of the utterance can be recovered from pressure or microphone-voltage measurements made about 40,000 times per second, and each measurement need only be accurate enough to distinguish about 65,000 different pressure or voltage values. Speech (or any other audible sound) can thus be represented without loss of information by a sequence of symbols where the sequence contains about 40,000 symbols for every second of sound, and each symbol is selected from an alphabet of size 65,000.

Such a representation of sound as a symbol sequence is routinely used on Compact Discs and increasingly in telephone networks. The measured values of pressure can themselves be represented as decimal numbers, e.g., one of the numbers 0-64,999. Thus, the whole sequence representing a passage of sound can be recorded as a sequence of decimal digits in which groups of 5 consecutive digits represent pressure measurements. In practice, binary rather than decimal numbers are used, so each pressure would be represented as a group of 16 binary digits indicating a number in the range 0, ..., 65,535. Using binary digits allows the entire sequence to be a binary sequence, using only the symbols 0 and 1. The length of the sequence required for one second of sound is then 40,000 x 16 = 640,000 binary digits.

Using similar techniques, pictures, measurements and texts can all be translated (or rather their information can be translated) into binary sequences. As far as is now known, information in any physical medium can be so translated in principle, although no adequate translation mechanisms yet exist for some media such as smells. Since the kinds of data presented for systematic scientific analysis are almost always available in a symbolic form, we will assume henceforth that all information of concern will be representable as sequences of binary digits. We now consider how these sequences may be constructed.

2.1.1 Binary Codes

A binary sequence conveying some information is called a *message*. Clearly, there must be some agreement between sender and receiver as to how to represent information in binary form. That is, they must agree on the meaning of the binary sequences. Such an agreement defines a *code*, which we may also think of as a kind of language. Often, a complete message will convey several pieces of information of the same kind, one after another. If these pieces are independent of one another, in the sense that knowing one piece gives us no hint about what the next might be, each piece may well be independently encoded. That is, the message will comprise several binary subsequences concatenated together. Each subsequence conveys one piece of information, and the same code may be used repeatedly for each piece of information in turn. For example, suppose the information as presented is an apparently random sequence of 2000 Roman letters. To encode this message in binary form, we might agree that, rather than deciding on a binary sequence for each of the 26^{2000} possible messages, we will encode each letter in turn as a 5-binary-digit subsequence and then form the whole message by concatenating the subsequences. Five binary digits per letter will suffice, because there are $2^5 = 32$ possible subsequences. A simple code might be:

$$A \rightarrow 00000 \quad B \rightarrow 00001 \quad C \rightarrow 00010 \quad D \rightarrow 00011 \dots$$
$$M \rightarrow 01100 \dots P \rightarrow 01111 \dots Z \rightarrow 11011$$

If the message began

$$CBPM \dots$$

the binary sequence would begin

$$00010000010111101100 \dots$$

When messages are encoded piece-by-piece in this way, each subsequence is often known as a *word* and the code definition requires only that the binary subsequence for each word be defined. There is no hard-and-fast distinction between messages and words. We introduce the notion of words merely to emphasize that messages *may* be encoded piecemeal, and that a meaningful binary sequence may be followed by more binary symbols representing more information. It will be convenient in the following discussion to consider the construction of a code for words rather than for complete messages, although there is no real difference in the two problems. We will use the term "word" to indicate both the fragment of message being encoded and the sequence or string of binary digits by which it is represented.

Assume we wish to construct a code for a known finite set of words. Let the number of different words be N. (In the example above, N=26.) The code is a mapping from the set of words to a set of finite non-empty binary strings. We assume the mapping to be one-to-one, as for our purposes there is nothing to be gained by allowing several binary representations of the same word. The strings in the code must obviously be all distinct. If two words were represented by the same string, the receiver would have no way of knowing which of the two was meant by the string. Further, we will require the code to have the *prefix property*: no string is a prefix of any other. The reason for this requirement is easily seen. Suppose in the random-letters example above, the code for A was 0000, the code for B was 00001 and the

code for Z was 11001. Then if the message began AZ ...the binary form
would begin 00001 ... The receiver of the binary sequence would have no
way of knowing whether the message began with A followed by some letter
whose binary string began with 1, or began with B. A code having the prefix
property is called a *prefix code*. (It is possible to define codes which lack the
prefix property yet allow unambiguous decoding. However, such codes have
no advantages for our purposes.)

A binary prefix code with N distinct words can easily be constructed. We
can find the lowest integer k such that $2^k \geq N$ and let all words have k binary
digits. Since all words have the same length, the prefix property is obvious.
Since there are 2^k distinct sequences of k binary digits, and $2^k \geq N$, each
of the N words can be assigned its own binary sequence or word according
to some convenient convention. Such equal-length codes are commonly used
because of their simplicity. A well-known example is the ISO-7 code whose
"words" all have seven digits, and are used to represent 128 different letters,
digits, punctuation marks and other symbols. However, equal-length codes
are not the only possible codes, and we shall be interested in codes whose
words are of unequal length.

Any binary prefix code for a set of N words can be represented by a *code
tree*. A code tree is usually drawn as an inverted tree with the root at the top.
The *nodes* of the tree are all those points where branches meet or end, so the
root is a node. Every node save the root is at the lower end of some branch,
and every node may have 0, 1 or 2 branches depending from it. Nodes having
no dependent branches are called *leaves*. Each leaf corresponds to, and can
be labelled with, one of the N words. Every word labels exactly one leaf, so
there are N leaves. We will be mainly interested in trees in which all non-
leaf nodes have exactly two dependent branches. Example: a possible binary
prefix code for the set of 5 words { A B C D E } is

$$A \to 0 \quad B \to 100 \quad C \to 101 \quad D \to 110 \quad E \to 111$$

This code can be shown as the tree in Figure 2.1. The rule for reading the
binary strings from the tree is as follows. To find the binary string for a
word, follow a path from the root to the node labelled by the desired word.
The path must follow branches of the tree. Whenever a branch to the left
is taken, write down a "0". Whenever a branch to the right is taken, write
down a "1". When the desired labelled node is reached, the digits written
down will form the binary string for the desired word. For instance, to get
from the root to the leaf labelled "C" we branch right, left, and right, so the
string for C is 101. Note that words of the given set label only leaf nodes.
This fact guarantees the prefix property, since the path from the root to a
leaf cannot continue on (downwards) to another leaf.

The code tree is convenient for decoding a message comprising several words. Suppose the message were BABD. Then the binary form is
1000100110.

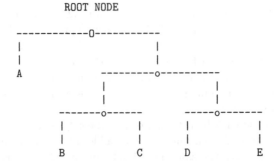

Fig. 2.1. Tree for prefix code.

To decode it, begin at the root of the code tree and follow left or right branches downward, guided by successive digits of the message string. Follow a left branch if the digit is "0", right if it is "1". Thus, the first three digits lead us to the leaf labelled "B", so the first letter of the message is B. Having reached a leaf and decoded a word, start again at the root with the next digit of the string (in this case the fourth digit). As it is "0" follow the left branch from the root to leaf "A". Starting again from the root, the next three digits (100) lead again to "B" and so on.

We need not be concerned with codes whose code tree contains a node having just one dependent branch. Any such node and its dependent branch may be deleted from the tree, and the branch leading to it joined directly to the node at the lower end of the deleted branch. The resulting tree preserves all the leaves of the original, it still has the prefix property (i.e., only leaves are labelled as words) and the code strings for some words are now shorter than in the original code. Such a deletion of a useless node is shown in Figure 2.2.

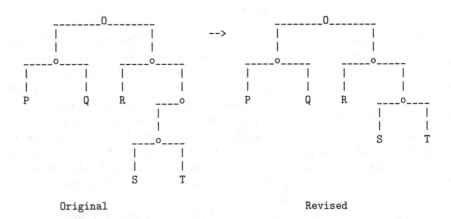

Fig. 2.2. Deletion of a useless node.

The revision shortens the strings for words S and T from 4 digits to 3, leaving the strings for other words unchanged. Since there is no advantage in making the binary strings of a code longer than they need be, we assume all codes and trees to have no nodes with just one dependent branch. All interior (non-leaf) nodes have two dependent branches, and leaves of course have none.

2.1.2 Optimal Codes

Codes using strings of different lengths for different words are useful because they offer the possibility of making binary message strings short. When a body of information is to be stored in or communicated via some medium, there are obvious economic advantages in minimizing the use of the medium, which may be expensive. Other things being equal, a code which allows the information to be represented in a short string is preferable to one requiring a longer string. We are thus led to ask whether the code for a given set of words can be chosen to minimize the length of the binary string. Suppose that the set of possible words has N members. It is obvious that, whatever the word actually communicated, there is a code which encodes this word with a single binary digit. For instance, if the set of words is {A B C D E}, there is a code encoding A with one digit and all other words with 3 digits. The same is true for every other word in the set. This fact is of no use in practice, because sender and receiver must agree on the code to be used *before* the information to be communicated is available. There is no point in agreeing to use a code which minimizes the length of the string for "A" if we have no grounds for supposing that A will be sent rather than B C D or E. (It is of course possible for sender and receiver to agree that one of a set or family of codes will be employed, to be determined by the sender after the information is available. But if it is so agreed, the binary string must begin with some additional digits identifying the chosen code. Although this technique will later be seen to be very important, it really amounts to no more than an agreement to use a certain single rather complicated code. It would certainly not permit a word from the 5-member set to be encoded with one digit, no matter what that word might be.)

Since it is meaningless to ask for the code which gives the least string length, we may choose to seek the code which gives the least *maximum* string length. That is, we can choose the code to minimize the length of the longest string for any word. It is easily shown that this code assigns strings of the same length k to some words of the set and length $(k-1)$ to the remainder, where k is the smallest integer satisfying $2^k \geq N$. If $2^k = N$, all strings have length k. Essentially, the code of minimal maximum length is an equal-length code with some nodes deleted if $N < 2^k$.

When the sender and receiver negotiate the choice of code, they do not yet know the information to be sent. However, they may have grounds to believe that some words are more probable than others. That is, they may agree on

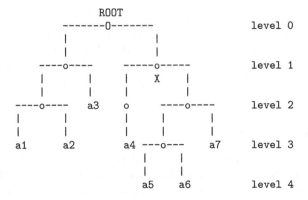

Fig. 2.3. Sample binary tree for N words.

a non-uniform prior probability distribution over the set of words. Let the set of words be $\{a_i : i = 1, \dots, N\}$ and the probability of word a_i be p_i for all i. This probability is the subjective probability held by the communicating parties that the word to be sent will be a_i. Then we may seek a code which minimizes the *expected* length of the code string. If the chosen code encodes word a_i with a string of length l_i, the expected string length is

$$E(l) = \sum_i p_i l_i$$

Codes which minimize $E(l)$ are called *optimal* codes.

Before considering how $E(l)$ can be minimized, we must establish a bound on the choice of the string lengths. The chosen code can be represented by a code tree, i.e., a rooted binary tree with N leaves corresponding to the N words. Consider any such tree, for example the tree in Figure 2.3.

Each node may be placed on a *level* corresponding to the number of branches on the path from the root to the node. In particular, the level of a leaf node equals the length of the string encoding the word at that leaf. For instance, the leaf for word a_2 is on level 3, and the tree shows that the string for a_3 is 001, of length 3.

Now give each node of the tree a "weight" according to the following rule: the weight of a node on level l is 2^{-l}. Then

(a) The weight of a node is greater than or equal to the sum of the weights of its children, where the *children* of a node are the nodes (if any) at the lower ends of the branches depending from the node.

(b) It follows by mathematical induction from (a) that the weight of a non-leaf node is greater than or equal to the sum of the weights of all the leaves of the subtree rooted at the node. For example, the weight of the node X in the diagram must be not less than the sum of the weights of the leaves a_4, a_5, a_6 and a_7.

(c) The weight of the root is 1.

Hence, from (b) and (c):

$$\sum_i 2^{-l_i} \leq 1$$

where the summation is over all leaves, i.e., over all words of the code. This inequality is a form of Kraft's inequality. Equality is reached unless the tree contains one or more nodes having just one dependent branch. The example tree above has such a node, but as we have mentioned before, there is no reason to use codes having such nodes. For codes of interest, we will usually have the equality $\sum_i 2^{-l_i} = 1$

We can now proceed to choose the string lengths l_i to minimize the expected string length $E(l) = \sum_i p_i l_i$ subject to the above constraint. Using the method of indeterminate multipliers:

$$\frac{\partial}{\partial l_i} \left(\sum_i p_i l_i - \lambda (\sum_i 2^{-l_i} - 1) \right) = p_i + \lambda (\log 2) 2^{-l_i}$$

Equating to zero gives

$$2^{-l_i} = -\frac{p_i}{\lambda \ln 2}$$

Using $\sum_i 2^{-l_i} = 1$ and $\sum_i p_i = 1$ gives $2^{-l_i} = p_i$, so $l_i = -\log_2 p_i$.

That is, the expected string length is minimized when a word of probability p_i is encoded using a string of length $-\log_2 p_i$ binary digits. Probable words are encoded with short strings, improbable ones with long strings. Except in the unlikely event that all the word probabilities are negative powers of 2, it will not be possible to choose string lengths (which are of course integers) exactly satisfying $\sum_i 2^{-l_i} = 1$. For the moment, we will ignore this complication and assume that either the probabilities are negative powers of 2 or that we can somehow have non-integer string lengths.

Assuming that for every word, $p_i = 2^{-l_i}$, it is clear that the "weight" of the leaf node for a word equals its probability. The probability that some interior node of the code tree will be visited in decoding a string equals the sum of the probabilities of all the leaf words in the subtree rooted at that node. Thus, since we exclude trees with nodes having only one child, the probability of visiting any interior or leaf node equals its weight. Further, for any interior node Y at level l, the probability of visiting Y is 2^{-l} and the probability of visiting its left child is $2^{-(l+1)}$. Thus, the probability of visiting the left child given that the decoding process has reached Y is $2^{-(l+1)}/2^{-l} = 1/2$. But, if we have reached Y, its left child will be visited if and only if the next digit in the string being decoded is "0". Hence, given that we have reached Y, the next digit is equally likely to be "0" or "1". But this is true for any interior node Y. Reaching Y while decoding a string is determined by the digits already decoded: these determine the choice of left or right branch at each step, and hence determine what node is reached. Hence, regardless of what digits have already been decoded, i.e., regardless of what node the decoding has reached, the next digit has probability $1/2$ of being "0". Thus, in the strings produced by an optimal binary code, each binary digit has equal probability of being 0 or 1 independently of the other digits. Knowledge of the early digits of a string gives no clue about the values of later digits. It follows that each binary digit in the string optimally encoding a word of probability p is distinguishing between two equally probable possibilities.

2.1.3 Measurement of Information

The above considerations have led to the idea that the amount of information conveyed by a word (or message) can usefully be equated to the number of binary digits in a string optimally encoding the word or message. The unit of this measure is the amount of information conveyed by naming one of two equally probable alternatives. As we have seen, every binary digit of the string conveys just this amount of information. This unit of information is conventionally named the *bit*. The word is a contraction of "binary digit", but its meaning in Information Theory is distinct from the meaning of "binary digit". The latter is a symbol having value either 0 or 1. The bit is an amount of information which, in an optimal code, is encoded as a single binary digit.

In other codes, a binary digit can convey other amounts of information. For instance, we could arrange to communicate every day a single binary digit showing whether rain had been recorded in Alice Springs during the preceding 24 hours. Let 0 mean no rain, 1 mean rain has fallen. Alice Springs has a very dry climate, so the great majority of transmitted digits would be zero, and the subjective probability that tomorrow's digit will be 1 would be perhaps 0.1. Then the information conveyed by each digit would not be one bit, since the two possibilities distinguished by the digit are not equally probable.

If we take the information conveyed by a word as being the number of binary digits in a string optimally encoding the word, then the information conveyed by a word of probability p_i is $-\log_2 p_i$. This definition is unambiguous for a set of words admitting an ideal optimum code, i.e., a set in which all words have probabilities which are negative powers of 2. We will assume for the time being that it can be adopted whatever the probabilities of the words. For instance, we will define the information conveyed by a word of probability 0.01 as $-\log_2 0.01 = 6.6438\ldots$ bits. Since this number is not integral, it cannot equal the number of binary digits in a string optimally encoding the word. However, the extension of the definition to arbitrary probabilities can be justified on two grounds. First, the number of digits in an optimal encoding of the word is usually close to $-\log_2 p_i$. Second, and more persuasively, we will later show that when any long message of many words is optimally encoded, the length of the resulting string is expected to be within one digit of the sum of the information contents of its words as given by this definition.

This definition of information content satisfies several intuitive expectations.

(a) A message announcing something we strongly expected has little information content.

(b) A message announcing a surprising (low probability) event has high information content.

(c) If two messages convey independent pieces of information, then the information content of a message announcing both pieces is the sum of the information contents of the two separate messages. If message M_1 announces event E_1 of probability $P(E_1)$, and message M_2 announces event E_2 of probability $P(E_2)$, then their information contents are respectively $-\log_2 P(E_1)$ and $-\log_2 P(E_2)$. A message announcing both events has content $-\log_2 P(E_1, E_2)$ and since E_1, E_2 are independent, $P(E_1, E_2) = P(E_1)P(E_2)$, so $-\log_2 P(E_1, E_2) = -\log_2 P(E_1) - \log_2 P(E_2)$.

(d) If two propositions E_1, E_2 are not independent, the information in a message asserting both does not equal the sum of the information in two messages asserting each proposition singly, but does not depend on the order of assertion. Suppose a message optimally encodes E_1 and E_2 in that order. The first part, asserting E_1, has length $-\log_2 P(E_1)$ binary digits (ignoring any rounding-off to integer values). The second part encodes E_2. But, by the time the receiver comes to decode the second part of the message, the receiver knows proposition E_1 to be true. Hence, the probability he gives to E_2 is not in general $P(E_2)$ but $P(E_2|E_1)$, the probability of E_2 given E_1. An optimal encoding of the second part will therefore have length $-\log_2 P(E_2|E_1)$. The length of the whole optimally coded string is therefore

$$-\log_2 P(E_1) - \log_2 P(E_2|E_1)$$
$$= -\log_2 P(E_1)P(E_2|E_1)$$
$$= -\log_2 P(E_1, E_2) \equiv -\log_2 P(E_2, E_1)$$

which is independent of the order of E_1 and E_2.

By definition, the information in the message asserting E_1 and E_2 is the length of the coded message in binary digits. Hence, the above equations apply to the information contents as well as to the string lengths. This result is in accord with our intuition. If we are told E_1, and E_2 is to be expected if E_1 is true, then the additional information we get when told E_2 is small. In particular, if E_1 logically implies E_2, $P(E_2|E_1) = 1$ so once we have been told E_1, a further assertion of E_2 adds nothing to our information.

(e) The information in a message relates in an obvious way to the ability of the message to distinguish among equiprobable possibilities. Intuitively, we would regard a message which names one of 1000 possibilities as being more informative than a message which names one of only five possibilities. The relationship between information and number of possibilities is easily seen to be logarithmic. Imagine a message which names one of 12 possibilities. We can think of the twelve possibilities as being arranged in a table having 3 rows and 4 columns. Any possibility can be named by naming its row and column. Thus, the message could be made of two words: the first naming one of 3 possible rows, the second naming one of 4 possible columns. The information conveyed in the two words together names one of 12 possibilities. Generalizing to an N-by-M table of (NM) possibilities, we see that

Information to name one of $N \times M$ possibilities

= Information to name one of N possibilities (the row)

+ Information to name one of M possibilities (the column)

If $I(N)$ denotes the information needed to name one of N equiprobable possibilities, we have

$$I(N \times M) = I(N) + I(M)$$

The only function $I()$ satisfying this relation for all positive integers is the logarithmic function $I(N) = \log(N)$. This argument does not fix the base of the logs, but if we note that one of 2^k possibilities can be named by a k-digit binary number, we must have

$$I(2^k) = k \quad \text{bits}$$
$$I(N) = \log_2 N \quad \text{bits}$$

Since the N possibilities are assumed equally likely, the probability of each is $1/N$. Hence, we can write

$$
\begin{aligned}
I(N) &= \log_2 N \\
&= -\log_2(1/N) \\
&= -\log_2 (\text{probability of thing named})
\end{aligned}
$$

(f)   The information conveyed by a word or message announcing some event or proposition depends on only the probability p of the event announced. What other events might have been announced by the word, and the distribution of the remaining probability $(1 - p)$ over them, are of no consequence. The amount of information in the message depends on what did happen, not on what did not. This observation has the practical consequence that, in discussing the information content of a message, we need only establish the probability of the event or proposition conveyed by the message. We need not enumerate the entire set of messages which might have been sent but were not, nor need we calculate the probabilities of the events that might have been announced.

2.1.4 The Construction of Optimal Codes

An optimal binary code for a set of words minimizes the expected string length $\sum_i p_i l_i$ where p_i is the probability of the ith word of the set and l_i is the length, in binary digits, of the string encoding that word. Given a set of words and their probabilities $\{p_i : i = 1, 2, \ldots\}$, we now consider how an optimal code can be constructed. We have seen that in the ideal case where all probabilities are negative powers of two, the optimal code obeys the relation

$$\forall i \quad l_i = -\log_2 p_i$$

giving expected length

$$\sum_i p_i l_i = -\sum_i p_i \log_2 p_i$$

When the probabilities are not all negative powers of two, this relation cannot be observed because all lengths l must be integral.

First, observe that whatever the probabilities, we can easily design a code such that

$$\sum_i p_i l_i < -\sum_i p_i \log_2 p_i + 1$$

That is, we can devise a code with expected string length exceeding the ideal value by less than one digit. For every word in the set (assumed finite), let q_i be the largest negative power of two not exceeding p_i. Then $p_i \geq q_i > p_i/2$ for all i, and $-\log_2 q_i$ is an integer satisfying

$$-\log_2 q_i < -\log_2 p_i + 1$$

In general, $\sum_i q_i < 1$. Add to the set of words some additional dummy words with "probabilities" which are negative powers of 2, such that for the augmented set of words $\sum_i q_i = 1$. Then design an optimal binary code for the now-ideal set of probabilities $\{ q_i : i \text{ covers real and dummy words } \}$. This code can then be used as a code for the given set of words. The expected string length is

$$-\sum_i p_i \log_2 q_i \qquad \text{(summation over real words only)}$$

$$< \quad \sum_i p_i(-\log_2 p_i + 1)$$

$$< \quad \sum_i p_i log_2 p_i + \sum_i p_i$$

$$< \quad -\sum_i p_i \log_2 p_i + 1$$

The code has the capacity to encode the dummy words. As this capacity will never be used, the code is obviously not the most efficient possible, but

A:	0	0	0	
B:	0	1	0	0
C:	0	1	0	1
D:	0	1	1	
E:	1	0	0	
F:	1	0	1	
G:	1	1		
H:	0	0	1	

Fig. 2.4. Sample code table.

none the less has an expected string length exceeding the ideal value by less than one. We now describe two constructions which do better than this crude approach.

A *Shannon-Fano code* [41, 14] is based on the principle that in an optimal code, each digit of a code string should have equal probability of being zero or one. The construction builds the code tree from the root downwards. The first digit of a string shows whether the word encoded lies in the left subtree depending from the root, or the right subtree. In order to make these two possibilities as nearly as possible equiprobable, the set of words is divided into two subsets with nearly equal total probabilities. Words in the first subset are assigned to the left subtree and have code strings beginning with zero. Those in the other subset are assigned to the right subtree and have code strings beginning with one. The construction then proceeds recursively. Words in the left subtree are divided into nearly equiprobable subsets, and the subsets assigned to different subtrees, and so on until all words are fully coded. For example, consider the set of words and probabilities below:

$$A\ (1/36)\quad B\ (2/36)\quad C\ (3/36)\quad D\ (4/36)$$
$$E\ (5/36)\quad F\ (6/36)\quad G\ (7/36)\quad H\ (8/36)$$

One way of dividing the set into equiprobable subsets is {A, B, C, D, H} , {E, F, G}.

The left subset can be exactly equally divided into {A, H}, {B, C, D}, each subset having probability 9/36. The right subset cannot. The best that can be done is the split {E,F}, {G}. Proceeding in this way we obtain the code in Figure 2.4. It can be seen that the choice of an equiprobable split for the early digits can lead to unfortunate consequences later in the construction. For instance, the subtree of words beginning "00" contains only A and H with probabilities 1/36 and 8/36. Although these are far from equal, there is no choice but to split them, with the result that the most-probable and least-probable words are coded with strings of the same length. The expected string length is 2.944 digits. This may be compared with the ideal expected string length, or *entropy* defined as $-\sum_i p_i \log_2 p_i = 2.7942$ bits.

A *Huffman code* [19] is also based on an attempt to have each binary digit decide between equiprobable subsets, but the construction is bottom-up, moving from the leaves of the code tree towards the root. It is both easier

to construct and more efficient than the Shannon-Fano scheme. We begin
with a set of nodes comprising leaves of the tree labelled with the words and
their probabilities. Then this set of nodes is searched to find the two nodes of
smallest probability. These two nodes are joined in a subtree by introducing
a new node which has the two nodes as children. The child nodes so joined
are removed from the set of nodes and replaced by the new node. The new
node is labelled with a probability equal to the sum of the probabilities of its
children. Then again the set of nodes is searched for the two nodes of smallest
probability, these are joined and replaced by their parent, and so on until the
set of nodes is reduced to a single node, which is the root of the tree. Using
the same example as above:

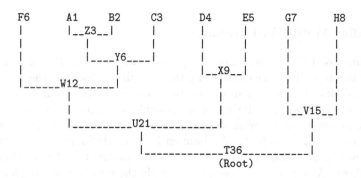

Fig. 2.5. Construction of a code tree.

The first step joins nodes A and B, replacing them with a new node (say Z)
of probability $(1 + 2)/36$. Then Z and C are joined and replaced by a parent
Y of probability $(3 + 3)/36$, and so on. Schematically, the construction can
be shown as in Figure 2.5. Probabilities are shown as times 36.

The leaves were slightly rearranged to avoid branches of the tree crossing
one another. Using the convention that left branches are labelled "0", the
resulting code is as shown in Figure 2.6.

A:	0	0	1	0	0
B:	0	0	1	0	1
C:	0	0	1	1	
D:	0	1	0		
E:	0	1	1		
F:	0	0	0		
G:	1	0			
H:	0	1			

Fig. 2.6. Rearranged code table.

The improvement over the Shannon-Fano code is evident. The string lengths are now monotonically decreasing with increasing word probability. The expected string length is 2.833 digits, about 0.11 less than for the Shannon-Fano code but still, of course, greater than the entropy. The Huffman construction can be proved to give optimal codes.

For both constructions, and indeed for any binary code tree, the labelling of the left and right branches depending from a node with the digit values 0 and 1 is arbitrary. A different choice will give different code strings for the words, but the lengths of the strings and the expected string length are unchanged. Thus, there are many "Huffman" codes for the same set of words. All are optimal, and the choice of one of them can be made by some convention.

2.1.5 Coding Multi-Word Messages

We have discussed the construction of optimal codes for "words" or "messages" indifferently. If we are considering the encoding of long messages conveying many bits of information, it may be inconvenient to attempt an optimal code construction for the set of all possible messages. There may be many millions of possible messages any one of which could be sent, each having a very small probability. A Shannon-Fano or Huffman code for the whole set would be efficient, but very tedious to construct by the methods outlined above. A common approach is to divide the message into sequence of "words" each of which can take only a manageable number of values. The simplest situation, which we will consider first, arises when the message can be expressed as a sequence of words in such a way that all words are selected from the same set of values, and the probabilities of these values are known constants which do not depend on the position of the word in the message or on the values of preceding words. For instance, it may be possible to express the message as a sequence of letters where each letter is one of A, B or C, and the probabilities of these letter values are in the ratio 1:4:5 independently of what preceding letters may be. In this situation, we can construct an optimum code just for the *alphabet*, or set of possible letters. To encode a message of many letters, one simply concatenates the binary strings representing each letter of the message in turn. For the 3-letter alphabet { A, B, C } above, with the probabilities as given, an optimal (Huffman) code is

$$A : 00 \qquad B : 01 \qquad C : 1$$

with expected string length 1.5 digits. In a long message of, say, 1000 letters, one could expect each letter to occur with a frequency nearly proportional to its probability, so we would expect the length of the coded message to be about 1500 binary digits. However, the Huffman code for the alphabet, while as good as can be achieved, is not very efficient. For the given letter probabilities, the entropy, or ideal expected string length $-\sum_i p_i \log_2 p_i$, is

AA	0.01	0 0 0 1 0 0
AB	0.04	0 0 0 1 0 1
AC	0.05	0 0 0 0 0
BA	0.04	0 0 0 0 1
BB	0.16	0 0 1
BC	0.20	1 0
CA	0.05	0 0 0 1 1
CB	0.20	1 1
CC	0.25	0 1

Fig. 2.7. Code table for pairs.

only 1.361, so the expected information content of a 1000-letter message is only 1361 bits. The use of the Huffman coding of each letter in turn wastes about 139 binary digits.

This inefficiency can be reduced if we make a small step towards coding entire messages rather than individual letters. The simplest modification is to devise a Huffman code for pairs of letters rather than individual letters. The set of possible pairs, their probabilities and a Huffman code for the set of pairs is shown in Figure 2.7.

This code has expected string length 2.78 binary digits per letter pair, or 1.39 digits per letter. Hence, if we code the letters of a 1000-letter message in pairs, the expected length is reduced from 1500 digits to 1390 digits, only 29 digits longer than the ideal minimum.

A further increase in efficiency might be obtained by devising a Huffman code for the 27 possible triplets of letters, or the 81 possible quadruplets. However, in this case most of the achievable improvement has been gained simply by coding pairs.

2.1.6 Arithmetic Coding

An ingenious coding technique has been devised by Langdon *et al.* [37] called *arithmetic coding*. This technique achieves a coding efficiency very close to the theoretical optimum which would be reached by a Huffman code for the entire set of possible messages. In practice, arithmetic coding easily achieves a string length for a long message within 0.01% of the information content of the message. We give only an outline of the method. Its full implementation is well described in [59]. While not unduly complicated, it requires careful attention to details of computer arithmetic which are outside our present interest.

Arithmetic coding is applicable to messages presented as a sequence of words. The set of possible words is not necessarily fixed: it may depend on the position of a word in the message and on the preceding words of the message. Also, the probability distribution over the set of possible words may depend on position in the message and on the preceding words. These dependencies must be known *a priori* to the receiver. It is also necessary that the set of possible next words, given the preceding words, be ordered in some way known *a priori* to the receiver. Thus, at any point along the message, we have an ordered set of possible next words with known probabilities. Whatever the set might be, let us label its members A, B, C, etc. Then any message can be represented as a sequence of label letters. The same letter, D say, may stand for different words at different places in the message, and may have different probabilities, but the rules for determining the word corresponding to D and its probability are implied by the rules already known by the receiver for determining the ordered set of possible words and their probabilities, given the preceding words. This labelling scheme for representing a message as a sequence of letters allows us to define a lexical ordering on the set of all possible messages. A message beginning Z A P Q B V . . . is lexically earlier than a message beginning Z A P Q B W . . . . That is, the relative order of two messages is determined by the relative alphabetic order of the first letter in which they differ. Note that we need no rule to determine the relative order of A X B and A X B C. The set of messages must have the prefix property, as otherwise the receiver could not know when the message was finished. Hence, the above two sequences could not both be possible messages.

As well as establishing a lexical order in the set of possible messages, we can in principle calculate the probability of each message. The probability of a message C A B A is

$$P(C)P(A|C)P(B|CA)P(A|CAB)$$

where each of these probabilities is derivable from the known rules of the message set.

With these preliminaries, we can now define the following representation of the set of possible messages. Consider the interval $[0, 1]$ of the real line. We will call this interval the *probability line*. It can be divided into a number of small intervals, each corresponding to a possible message. Let the interval for a message have a length equal to the probability of the message. Since the sum of the probabilities of all messages is one, the message intervals exactly fill the probability line. Further, let the message intervals be arranged in lexical order, so that the lexically earliest message ("AAA . . .") is represented by an interval beginning at zero, and the lexically last by an interval ending at one. The diagram in Figure 2.8 shows the resulting probability line for a set of two-word messages in which the set of possible words is { A, B, C } with constant probabilities { 0.4, 0.3, 0.3 }.

It is now obvious that any message in the possible set can be specified by specifying a number between 0 and 1 which lies in its interval. Thus, message AB could be identified by any number x in the range $0.16 < x < 0.28$. Arithmetic coding chooses a binary fraction to identify a message. That is, it encodes a message as a binary string which, when interpreted as a binary fraction, lies in the interval corresponding to the message. The binary fractions must be somewhat constrained to ensure that the code strings have

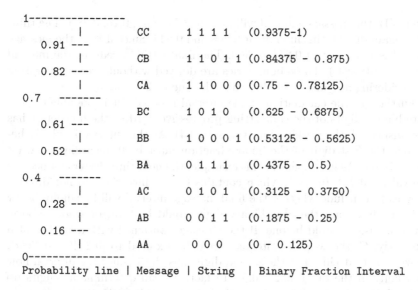

```
1----------------
      |           CC    1 1 1 1   (0.9375-1)
   0.91 ---
      |           CB    1 1 0 1 1 (0.84375 - 0.875)
   0.82 ---
      |           CA    1 1 0 0 0 (0.75 - 0.78125)
  0.7    -------
      |           BC    1 0 1 0   (0.625 - 0.6875)
   0.61 ---
      |           BB    1 0 0 0 1 (0.53125 - 0.5625)
   0.52 ---
      |           BA    0 1 1 1   (0.4375 - 0.5)
  0.4    -------
      |           AC    0 1 0 1   (0.3125 - 0.3750)
   0.28 ---
      |           AB    0 0 1 1   (0.1875 - 0.25)
   0.16 ---
      |           AA    0 0 0     (0 - 0.125)
  0----------------
Probability line | Message | String  | Binary Fraction Interval
```

Fig. 2.8. Probability line for a set of two-word messages.

the prefix property. This is ensured by regarding a binary fraction with n binary digits as denoting, not a point on the probability line, but an interval of size 2^{-n}. For instance, the code string 010 is interpreted as the interval $1/4 \leq x < 3/8$, and 0111 is interpreted as the interval $7/16 \leq x < 1/2$. Then the arithmetic coding of a message chooses the shortest binary string representing an interval lying wholly within the message interval. The diagram above shows the resulting codes for this simple example. In the worst case, the size of the interval represented by the code string can be almost as small as a quarter of the size of the message interval. Thus, the length of the code string for a message can exceed $-\log_2$(Probability of message) by almost 2. This excess is negligible for a long message, but it might seem that arithmetic coding is no more than a rather roundabout implementation of Shannon-Fano coding, as each digit of the code string effectively divides an interval of the probability line into equiprobable subsets.

The advantage of arithmetic coding is that it is not necessary ever to calculate the probabilities of all possible messages. One need only compute the boundaries on the probability line of the interval representing the given message. This calculation can be done progressively, dealing with each word of the message in turn. For example, consider the message A B C A, where the alphabet {A, B, C} and their probabilities {0.4, 0.3, 0.3} remain fixed as in the example above.

The first letter divides the probability line into three intervals with boundaries at 0.4 and 0.7. Since the first letter is A, the message interval will be somewhere in [0, 0.4). The second letter divides this interval into three, in the ratio 0.4:0.3:0.3 with boundaries at 0.16 and 0.28. Since the second letter is B, the message interval will be somewhere in [0.16, 0.28). The third letter again divides this interval at 0.208 and 0.244. Since it is C, the message interval is reduced to [0.244, 0.28). The final letter A reduces the interval to [0.244, 0.2584). These boundaries are derived without ever enumerating or considering any message other than the one to be sent.

Further, as the calculation of the interval proceeds, it is possible to generate binary digits of the code string progressively. After the first letter has been inspected, the working interval is [0, 0.4). As this interval is wholly below 0.5, the first digit of the binary fraction must be 0, regardless of what letters follow. As it happens in this example, the next three letters reduce the interval to [0.244, 0.2584), which contains 0.25. Thus, they do not directly allow us to conclude whether the final message interval will be wholly below 0.25, in which case the next binary digit would be zero, or wholly above, in which case it would be one. If the message continued with a further fifth letter, say, C, the working interval would be reduced to [0.25408, 0.2584), allowing us to decide that the second digit must be 1, since the interval now lies wholly in the range [0.25, 0.5). In fact, we can determine all digits up to the sixth, since the interval now lies wholly within [0.25, 0.265625), or in binary, [0.010000, 0.010001). Thus, the binary string must begin 010000 . . . . Proceeding in this way, arithmetic coding deals with the words of the message in turn, and produces binary digits of the code string whenever the working interval is sufficiently reduced. At the end of the message, some small number of additional binary digits may be needed to specify a binary interval wholly within the final message interval.

It may appear from the above that arithmetic coding requires arithmetic of arbitrarily high precision in order to calculate the interval boundaries. However, it turns out to be possible to do the calculations with a fixed and quite modest precision by periodically rescaling the numbers involved. There is a slight loss of efficiency involved, but use of 32-digit binary arithmetic typically gives code strings whose length exceeds $- \log_2$(Probability of message) by less than 0.1%.

The importance of arithmetic coding for the present study is that it gives a constructive demonstration that it is not only possible in principle but also feasible to encode messages as binary strings whose length is essentially equal to the information content of the message.

2.1.7 Some Properties of Optimal Codes

If an optimal code is used to encode a message, the length of the code string is (within one digit) given by $- \log_2$ (Prob. of message), and so is almost exactly equal to the information (in bits) conveyed by the message. Thus, each binary digit conveys one bit on average. For this to be possible, each binary digit must, independently of all others, be equally likely to be one or zero. This property, as we have seen, forms the basis for optimal code constructions. When asserting that each digit is equally likely to be 0 or 1, there are two senses in which these equal probabilities may be understood. First, we may mean that, if a message is in some way randomly selected from the set of possible messages in accordance with the assumed probability distribution over that set, then each digit of the resulting message is randomly and equiprobably selected from the values of 0 and 1. That is, the digits are the result of a random process with probability $1/2$ of giving a one.

Even if the observational protocol which gave rise to the message to be sent cannot properly be regarded as such a random selection, there is another sense in which the assertion may be read. We assume that the probability distribution over messages used in constructing the code accurately models the receiver's expectations about what message might be received. In this case, however the message was in fact acquired by the transmitter, it will still be true that, after receiving some digits of the coded message, the receiver will still have no reason to expect the next digit to be more likely one than zero, or vice versa. That is, from the receiver's point of view, the digits are no more predictable than those generated by an independent equiprobable random process.

Indeed, if a receiver who is expecting a series of messages in some optimal code instead receives a random stream of digits, the receiver will still decode the stream as a series of messages, and find nothing in the messages to show that they are garbage. When an optimal code matched to the receiver's expectations is used, any long string of digits decodes to a perfectly meaningful series of messages.

Suppose the set of possible messages has a probability distribution $\{p_i : i = 1, 2, \ldots\}$ and for all i, message i is encoded by a string of length $l_i = -\log_2 p_i$ digits. Then the probability that a stream of random, equiprobable binary digits will begin with the string for message i is just the probability that its first l_i digits will match those of the message string. Each random digit has probability $1/2$ of matching the corresponding message digit, so the probability of a match to message i is $(1/2)^{l_i} = p_i$. That is, message i will appear at the beginning of the random stream with precisely the probability that the receiver expects.

A further consequence is that if a random stream of digits is decoded, the expected length of the first message decoded is $\sum_i p_i l_i$, which, for an optimal code, is nearly equal to $-\sum_i p_i \log_2 p_i$.

2.1.8 Non-Binary Codes: The Nit

All the above discussion of codes generalizes easily to codes using more than two digit values. Most obviously, if the coded string uses the four digits 0, 1, 2 and 3 rather than just the binary digits 0 and 1, the length of the code string can be halved. Each base-4 digit can represent two binary digits using the mapping

```
00 -> 0
01 -> 1
10 -> 2
11 -> 3
```

Similarly, if base-8 digits are used, allowing the digit values $0, 1, \ldots, 7$, the code string will be only one third the length of the equivalent binary string, since each base-8 digit can represent a group of three binary digits. Generally, if N digit values are allowed, where N is a power of 2, each can represent $\log_2 N$ binary digits. Hence,

$$\frac{\text{length of binary string}}{\text{length of base N string}} = \log_2 N$$

This result can also be seen to follow from our conclusion that naming one of N equiprobable possibilities involves an information of $\log_2 N$ bits. That is, one base N digit can convey $\log_2 N$ bits.

For long messages, the above ratio between the lengths of binary and base N representations holds approximately for any N, even if N is not a power of 2. The Huffman and Arithmetic constructions generalize directly to any $N > 2$, with the result that the length of a base N string optimally encoding a message of probability p satisfies

$$\begin{aligned} \text{length} \ &\leq \ -(\log_2 p)/(\log_2 N) + 1 \\ &\leq \ -\log_N p + 1 \end{aligned}$$

We will have little or no need to consider the use of non-binary message strings. However, it will later be convenient to use a measure of information which can be thought of as the "length" of a message string employing $e = 2.718\ldots$ possible digit values, where e is the base of natural logarithms. The information conveyed by one such "digit" is

$$\log_2 e = 1.44\ldots \quad \text{bits}$$

This unit is termed the *nit* (for "natural bit") or, by some writers, the *nat*. It is often a more convenient unit than the bit. The information conveyed by a message of probability p is $-\ln(p)$ nits. Often, approximations used in computing p and its logarithm yield expressions directly involving natural logarithms rather than logs to base 2. There is no distinction in principle between information measured in nits and information measured in bits. It is merely a matter of choice of unit.

2.1.9 The Subjective Nature of Information

The information content of a message, as we have defined it, is the negative log of its probability, i.e., the probability of the event or data or proposition it conveys. As our view of probability is subjective, it follows that the information content is also a subjective measure. A message which tells us something we already know conveys no information, a message which tells us something we regarded as improbable gives us a great deal of information. This view may appear to be at odds with objective concepts such as the information-carrying capacity of a telegraph channel, or the information-storage capacity of a magnetic disc. However, such measures relate to the physical capability of a medium to store or transmit information. Whether the medium actually conveys or stores that much information depends on the code used and the prior expectations of the receiver. What the physical limit means is that it is the maximum amount of information which can be conveyed using a code which maximizes the expected amount of information given the prior expectations of the receiver. If some other code is employed, the amount of information which one can expect to store or convey will be less than the physical capacity. The actual information conveyed in a particular instance may in fact exceed the physical capacity, but this can occur only if the code gives high probability, and hence a short code string, to a proposition which the receiver regarded as improbable. If the receiver's prior expectations are not unfounded, such an accident is expected to be rare.

Note that a receiver may well be able to decode information coded using a code which is far from optimal, given his prior expectations. The receiver must normally have prior knowledge of the code in order to be able to decode the strings he receives. Whether or not he regards the code as efficient is immaterial. A receiver may even be able to decode strings in a code of which he has imperfect prior knowledge. Humans often use codes, such as natural languages, which are not optimal for *any* set of prior expectations. In such codes, many strings may have no meaning, i.e., be nonsense conveying no message, and the meaningful strings may exhibit regularities of structure such as rules of grammar and spelling, which do not serve to convey information. Given messages in such a *redundant* code, a receiver may be able to discover much of the coding rules, especially if he has strong and well-founded expectations about the content of the messages, and the code is not optimized for these expectations. For instance, a coded message may announce that gold has been found in Nowheresville. The message may be intended for an audience with little reason to expect such an announcement, and hence be coded as a rather long string. A receiver who happens already to know of the gold discovery, but does not know the code, may be able work out the code using this knowledge. Cryptoanalysis, the art of "breaking" codes designed to be obscure, relies heavily on the redundancies of natural languages and on prior expectations about the content of the messages.

There are good reasons for natural languages to be less than "optimal" when regarded as codes. One reason is that spoken language is transmitted from speaker to listener via a *noisy* channel. The listener may not be able to hear correctly all the words spoken. Codes with substantial redundancy can tolerate some degree of corruption without becoming unintelligible, whereas optimal codes, in which every digit matters, are very sensitive to corruption. Most possible symbol sequences (possible vocal utterances in the case of speech) are ruled out by the grammar and vocabulary of natural languages. Thus, if a listener hears something like "Today the weather is very kelled", the listener knows the sentence must have been mis-uttered or misheard, because it does not conform to the rules of English. The listener may guess (probably correctly) that he should have heard "Today the weather is very cold", because this sentence, out of all legal sentences, sounds most like what he heard. Artificial codes for information transmission and storage are often designed so that legal strings conform to a strict pattern while most strings do not. A corrupted received string can then be corrected by replacing it by the nearest legal string, provided the degree of corruption is not too great. Although of great practical importance, such error-correcting codes are outside our concerns.

Another reason for the inefficiency of natural languages is that they must serve for communication among the members of a large and diverse population. It is not normally possible for a speaker to have detailed knowledge of the prior expectations of her listener(s), and quite impossible for a broadcaster or journalist even to know who are the listeners and readers. Thus, the use of an optimal code is impossible, since a code can be optimized only with knowledge of the receiver's subjective probabilities for possible messages. A code intended for receipt by a wide and imperfectly known audience cannot be based on strong assumptions about the probabilities of different messages. Rather, it must allow every message a string length comparable to the negative log of the lowest probability accorded the message by any receiver. A receiver of a message whose length is less than the negative log of the probability which that receiver gave to the message will tend to find the message unintelligible or unbelievable. Thus, we find that users of natural languages will typically tend to frame the meaning they wish to convey as a long utterance or text when addressing an unknown or unfamiliar audience, but in a much shorter form when addressing a person whose prior expectations are well-known. Also, allowing some redundancy can make coding and decoding much simpler.

The implications for our present concerns are that, when discussing or calculating an amount of information, we must be careful to specify what prior knowledge we are assuming on the part of the receiver. We must be prepared to enter arguments as to whether it is or is not reasonable to assume certain prior knowledge or expectations and to modify our calculations of information content accordingly. However, for our purposes we need not consider errors in the transmission of messages, and hence need not be concerned with error-correcting codes.

To summarize this section, the information content of a message is the negative log of the (subjective) probability of the propositions conveyed. With negligible error, the content in bits equals the length of the binary string encoding the message in an optimal code. We are suspicious of measures advanced as measures of "information" unless it can be demonstrated that the measure equals the length of a message conveying a specified proposition to a receiver with specified prior expectations. To illustrate the grounds for this suspicion, and the importance of specifying the receiver, we give an example discussed by Boulton [6].

2.1.10 The Information Content of a Multinomial Distribution

Suppose we have a bag containing a very large number of balls of K different colours. Balls of the same colour are identical for our purposes. Let N balls be drawn in sequence from the bag, where N is a very small fraction of the number of balls in the bag. The result of the draw can be represented as a sequence of N colour symbols, each naming one of the K colours. Suppose the draw yielded M_k balls of colour k ($k = 1, \ldots, K$, $\sum_k M_k = N$) so the sequence has M_k symbols for colour k. At least two expressions have been proposed and used as measures of the "amount of information" in such a multinomial sequence. At least two others have some claim to the title. By examining these measures and attempting to relate them to the lengths of optimally coded messages, we can show the importance of the message length concept in measuring information.

Measure A is given by

$$A = \log \frac{N!}{\prod_k M_k!}$$

It is the logarithm of the number of ways the N balls could be arranged in a sequence. A is the log of the number of colour sequences which could be drawn to yield M_k balls of colour k for all k. If the balls are picked in such a way that each ball remaining in the bag is equally likely to be picked next, each of those colour sequences is equally likely, so A is the log of a number of equally likely events. Thus, A matches one of the ideas described before as a measure of information.

Suppose that the sequence of colours drawn from the bag is to be encoded as a message. Then it is certainly possible for a message of length A to inform a receiver of the sequence, *provided* that the receiver already knows N and the colour counts $\{M_k : k = 1, \ldots, N\}$. The message could be encoded in several ways. One way would be simply to number all the possible colour sequences with a binary number, using some agreed enumeration, and then to transmit the binary digits of the number specifying the actual sequence obtained. Another way of some interest is to construct the message from N segments, each segment giving the colour of the next ball in the sequence. Suppose that n balls have already been encoded, of which m_{nk} had colour k ($k = 1, \ldots, K$). Then the number of balls remaining is $N - n$, of which $M_k - m_{nk}$ are of colour k. The probability that the next ball will be of colour k is then $(M - m_{nk})/(N - n)$, so a segment announcing that the next colour is k should optimally have length $-\log\{(M_k - m_{nk})/(N - n)\}$. It is easily shown that, using this ball-by-ball encoding, the total message length is exactly A, whatever the order of colours in the sequence. The receiver of the message can decode it segment by segment. After decoding the first n segments, he will know m_{nk} for all k. Knowing $\{M_k\}$ a priori, he can then calculate the probabilities $(M - m_{nk})/(N - n)$ for all k, and hence decode segment $n + 1$, which is encoded using these probabilities.

Thus, A deserves to be regarded as a measure of information: the information needed to convey the sequence to a receiver who already knows the exact number of each colour.

Measure B has the form

$$B = -\sum_k M_k \log r_k$$

where $r_k = M_k/N$. (Here and elsewhere, we take $x \log x$ to be zero when $x = 0$.) It is N times the "entropy" of the frequency distribution of colours in the sequence if the fractions $\{r_k\}$ are regarded as probabilities of the various colours. This form again is familiar as an expression for a message length. It is the length of an optimally coded message conveying a sequence of events (the colours in this case) where each "event" is one of a set of K possible events, and the probability of event type k is r_k independently of whatever events have preceded it. That is, B is the length of a message conveying the colour sequence to a receiver who does not know the exact number of each colour, but who believes the next ball has probability r_k of being colour k regardless of the preceding colours. For instance, the receiver might believe that a fraction r_k of the balls in the bag had colour k. Thus, B is again a message length, and hence a measure of information, but one which assumes different, and weaker, prior knowledge on the part of the receiver. Hence, $B > A$, as we will see below.

Form B is questionable as a measure of the information in the colour sequence, in that it assumes a receiver who happens to believe the population colour probabilities $\{p_k : k = 1, \ldots, K\}$ have values which exactly equal the relative frequencies $\{r_k : k = 1, \ldots, K\}$ of the colours in the actual sequence. More realistically, we might suppose that the sender and receiver know the true bag population colour probabilities $\{p_k\}$, and that the actual colour counts $\{M_k : k = 1, \ldots, K\}$ result from the random sampling of N balls from the bag. In that case, the frequencies $\{r_k\}$ would be expected to differ somewhat from the population probabilities $\{p_k\}$. The message length, using a code optimal for the known probabilities, is then $-\sum_k M_k \log p_k$, which is greater than $-\sum_k M_k \log r_k = -\sum_k M_k \log(M_k/N)$ unless $p_k = r_k$ for all k. For any given set of probabilities $\{p_k\}$, it is easily shown that for a random sample of N balls, the expected difference

$$\mathrm{E}[-\sum_k M_k \log p_k + \sum_k M_k \log r_k]$$

is approximately $(K-1)/2$. It might therefore be argued that a better indication of the information needed to inform a receiver who already knows the probabilities is

$$B_1 = -\sum_k M_k \log r_k + \frac{(K-1)}{2}$$

Forms C and D again differ from A and B in the assumptions made about the receiver's prior knowledge. For both C and D, we will assume that the receiver knows N but neither the exact numbers $\{M_k\}$ nor the probabilities $\{p_k\}$. Rather, we assume the receiver to believe initially that all colour mixtures are equally likely. That is, the receiver has a uniform prior probability density over the continuum of possible probability tuples $(\{p_k\} : \sum_k p_k = 1, p_k > 0$ for all k). It is easily shown that with this prior density, all colour-number tuples $(\{M_k\} : \sum_k M_k = N, M_k \geq 0$ for all k), are equally likely a $priori$. One way of optimally encoding a message conveying the colour sequence to such a receiver is as follows.

A first part of the message can convey the tuple of colour counts $\{M_k : k = 1, \ldots, K\}$. For N balls, the number of possible tuples is $\binom{N+K-1}{K-1}$, and under the prior expectations of the receiver, all are equally likely. Hence, the length of the first part is the log of the number of possible tuples:

$$\log \frac{(N+K-1)!}{(K-1)!N!}$$

Once the receiver has received this part, he knows the colour counts and so can understand a second part framed in form A. The total message length is thus

$$\begin{aligned} C &= \log \frac{(N+K-1)!}{(K-1)!N!} + \log \frac{N!}{\prod_k M_k!} \\ &= \log \frac{(N+K-1)!}{(K-1)! \prod_k M_k!} \end{aligned}$$

An Incremental Code. Another way of encoding the message for such a receiver who knows neither the probabilities $\{p_k\}$ nor the counts $\{M_k\}$ is also of interest. The message can be encoded in segments each giving the colour of the next ball. For the first ball, the receiver must expect all colours with equal probability, so the length of the first segment is $\log K$. However, after receiving the first n segments, the receiver will no longer have a uniform prior density over the possible colour probability tuples. Rather, he will have a posterior density

$$\text{Dens}(\{p_k : k = 1, \ldots, K\}) = G \prod_{k=1}^{K} p_k^{m_{nk}} \qquad (p_k > 0; \sum_k p_k = 1)$$

where G is a normalization constant. He will therefore consider the probability that the $(n+1)th$ ball will have colour k to be

$$P_{n+1}(k) = \underset{\sum_{i=1}^{K-1} p_i < 1, p_i \geq 0}{\int \int \int} \cdots \int \left[\left(G \prod_{i=1}^{K} p_i^{m_{ni}} \right) p_k \right] dp_1 dp_2 \cdots dp_{K-1}$$

$$= \frac{m_{nk} + 1}{n + K}$$

The length of segment $(n+1)$ encoding colour k using an optimal code is thus $-\log((m_{nk} + 1)/(n + K))$. The total message length is the negative log of the product of all the N progressive probabilities used in the N segments. The denominators of the probability fractions range from K to $(N + K - 1)$ as n ranges from 0 to $(N - 1)$. The M_k segments announcing colour k give numerators ranging from 1 to M_k. Hence, the product of all the fractions is

$$\frac{\prod_k M_k!}{K(K+1)(K+2)\ldots(K+N-1)} = \frac{(\prod_k M_k!)(K-1)!}{(N+K-1)!}$$

giving the same measure C as before.

It is important to note that the probabilities $\{P_{n+1}(k) = \dfrac{m_{nk} + 1}{n + K}\}$ used in encoding the colour of the $(n+1)$th ball need not be interpreted as estimates of the population probabilities of the colours. The receiver may, if interested, use some such expression to obtain progressive estimates of the population probabilities but such estimates would be immaterial to the decoding of the message. The probability $P_{n+1}(k)$ is the probability that the next ball will have colour k, given the initial uniform density over colour distributions and the numbers of the different colours $\{m_{nk}\}$ so far known. It is not an estimate.

As form C assumes less prior knowledge on the part of the receiver, we must expect the message to be longer than for forms A or B, as is seen later.

An Explanation Code. Form D is the length of a message which assumes the same prior knowledge as C, i.e., a uniform prior over population probabilities. However, the message is encoded differently. For form D, we assume that the message comprises two parts, as in form C, but their content is different. The first part asserts a set of colour probabilities $\{\hat{p}_k : k = 1, \ldots, K\}$. The second encodes the colour sequence using a code similar to that of form B. That is, the code is optimal if the probability that the next ball is colour

k equals $\hat{p}_k$ and is independent of preceding balls. The length of the second part is thus

$$-\sum_k M_k \log \hat{p}_k$$

The calculation of the length of the first part is not trivial and is discussed in later chapters.

Form D assumes the same prior knowledge as does form C. It is slightly longer than form C because it is somewhat redundant. Once the details of how the coding is to be done have been agreed by sender and receiver, form C provides just one message string for each possible colour sequence (as do forms A and B). Form D, however, permits a colour sequence to be represented by any of several message strings, all intelligible to the receiver. To use form D, sender and receiver must have agreed on a way of encoding tuples of colour probabilities, for use in the first part of the message. When the sender wishes to transmit a colour sequence, there is nothing to stop her choosing *any* codeable probability tuple to be encoded in the first part and used in the second part for encoding the colours, provided only that the chosen tuple does not state a zero probability for some colour which actually appears in the sequence. The length of the second part, $-\sum_k M_k \log \hat{p}_k$, will be minimized if the sender chooses a probability tuple closely matching the actual frequencies of colours in the sequence to be sent, i.e., if she chooses a tuple with $\hat{p}_k \approx r_k$ for all k. However, a message encoding the sequence (albeit a longer message) can still be constructed if she makes some other choice of tuple.

The redundancy inherent in form D can be viewed another way. The receiver of a form-D message can decode it to discover the colour sequence. But he also discovers something else, not deducible from the colour sequence. He discovers that the sender has *estimated* the population colour probabilities to be those she stated in the first part of the message. Thus, form D is a very simple example of what we have defined as an explanation. Its first part asserts a simple "theory" about the origins of the colour sequence, viz., that the colours were drawn from a source which produces colour k with probability $\hat{p}_k$, for all k. The second part of the message then encodes the colour sequence using a code which would be optimal were the theory true: it encodes an occurrence of colour k with a segment of length $-\log \hat{p}_k$.

Strictly speaking, the value of the form-D message length should not be regarded as a measure of the information in the colour sequence, since a form-D message informs the receiver not only about the sequence, but also about a theory which is not implied by the sequence. However, the extra length of form D over form C is small when the best possible "theory" is used. Thus, in situations where it is easier to compute the length of form D than the length of form C, there is little error in using the former as an approximation to the latter.

Comparison of Measures of Information. The values of the message lengths for forms A, B, C and D are shown below. For forms A and C, we have used Stirling's approximation

$$\log(n!) = \left(n + \frac{1}{2}\right)\log n - n + \frac{1}{2}\log(2\pi) + O\left(\frac{1}{n}\right)$$

Here, the logs are natural logs, to base e. The resulting lengths below are given in nits rather than bits, and all logs are natural. We have also used the approximation

$$\log(x + d) \approx \log x + \frac{d}{x} + O\left(\frac{d^2}{x^2}\right)$$

to manipulate the expressions into forms showing their differences most clearly. All the expressions below are accurate to order $1/M_k$, so if every colour appears at least 10 times in the sequence, the error is less than one nit.

Form A: Exact colour counts already known.

$$A \approx -\sum_k M_k \log r_k - \frac{1}{2}\sum_k \log r_k - \frac{K-1}{2}\log N - \frac{K-1}{2}\log(2\pi)$$

Form B: Colour probabilities already known.

$$B_1 \approx -\sum_k M_k \log r_k + \frac{K-1}{2}$$

Form C: Uniform prior on colour probabilities.

$$C \approx -\sum_k M_k \log r_k - \frac{1}{2}\sum_k \log r_k + \frac{K-1}{2}\log N - \\ \frac{K-1}{2}\log(2\pi) - \log(K-1)!$$

Form D: Uniform prior on colour probabilities. Message is explanation using best estimates of probabilities.

$$D \approx C + \frac{1}{2}\log((K-1)\pi) - 1$$

In all of the above, $r_k = M_k/N$ for all k.

All of the expressions A to C are measures of information about the same set of data. They differ only in what is assumed to be already known about the data. They all agree in the dominant term, which is proportional to the size of the data set, i.e., the length N of the sequence. However, the differences are not trivial, being of order $\frac{1}{2}\log N$.

We will consider one more case to emphasize the importance of the receiver's prior knowledge. Here, we suppose that the receiver is already sure that the colours are independently selected from a population with probabilities $\{p_k : k = 1, \ldots, K\}$ which he already knows. In this case the optimal message length for a colour sequence is $-\sum_k M_k \log p_k$. The message can take the form of N segments each coding a ball's colour, where each segment uses a fixed code encoding colour k with a string of length $-\log p_k$. This result obtains regardless of what the sequence of colours might be, or how many times each colour occurs. It might be thought that, if the actual colour frequencies $\{r_k = M_k/N, \text{all } k\}$ differ greatly from the probabilities $\{p_k\}$, a shorter message could be obtained using a version of form C. That is, we can send the message in two parts, the first giving the actual colour counts $\{M_k\}$ and the second encoding the sequence using this information. As shown above, the length of the second part is given by form A: $\log \dfrac{N!}{\prod_k M_k!}$. The length of the first part, however, now depends on the receiver's probabilities $\{p_k\}$. The first part announces that, in a sample of known size N, colour k occurred M_k times (for all k). Given the receiver's beliefs, he expects such an event to occur with probability

$$\frac{N!}{\prod_k M_k!} \prod_k p_k^{M_k}$$

and hence the optimal coding of the first part has length

$$-\log \frac{N!}{\prod_k M_k!} - \sum_k M_k \log p_k$$

The total length of this form of message is thus $-\sum_k M_k \log p_k$ exactly as before. This case is an example of a general result. If a receiver has a fixed belief that the data to be sent in a message will conform to a fixed, probabilistic pattern (or no pattern at all), there is no advantage in terms of message length in using a code which states and then exploits any other probabilistic pattern. Even if the data to be sent appear to exhibit a different pattern, adoption of which could give a short encoding of the data, the message must begin with a statement of this pattern. As the receiver believes the true pattern to be otherwise, he will regard the apparent different pattern as a statistical fluke of low probability. The first part, stating the apparent pattern, will therefore require a long code string, of length the negative log of this low probability. The resulting length of the complete two-part message will thus not be shorter than a message which ignores the apparent pattern and codes the data in accord with the pattern believed by the receiver.

2.1.11 Entropy

Entropy is a term borrowed from classical thermodynamics. As originally used, the term could be defined as follows.

(a) When a body at an absolute temperature T is given a small amount of heat Δ_E, its entropy H increases by the amount $\Delta_H = \Delta_E/T$.

(b) The entropy of a body at absolute zero temperature is zero.

Among the many important properties of entropy is the famous Second Law of thermodynamics: the entropy of a closed system cannot decrease. Later, it was shown that the entropy of a body in thermal equilibrium could be expressed as

$$H = k \log Z$$

where k is Boltzmann's constant and Z is the number of distinct micro-states in which the body might exist consistent with its macroscopic thermodynamic state (macro-state), as defined by its total energy and momentum and any other macroscopically observable features. Here, a *micro-state* of the body means a full specification of the position and velocity of each of its constituent particles.

Shannon noted that $-\log Z$ equals the amount of information which would be needed to specify the micro-state of the body to a receiver who knew only its macro-state (assuming all micro-states consistent with the macro-state to be equally probable). He was thus led to suggest that information could be regarded as negative entropy, if physical units are chosen such that Boltzmann's constant k has the value one. The inevitable increase in the entropy of a closed system undergoing irreversible thermodynamic change is mirrored by the fact that whatever information we may have had about its micro-state is destroyed by the random or quasi-random interactions among the particles in the system. As the system approaches thermal equilibrium, its classical entropy increases towards a maximum, the number of micro-states in which it might be increases, and so an observer measuring only the macro-state becomes more and more ignorant of the micro-state.

The term has been adopted into Information Theory in accord with Shannon's suggestion. However, it should be treated with caution, as it has been used rather loosely, and with a variety of related but not exactly equivalent meanings. Roughly, these meanings have paralleled the various measures of information content of a multinomial sequence discussed above.

(a) The entropy of a multinomial collection. Suppose a collection of N balls is known to contain M_k balls of colour k ($k = 1, \ldots, K$; $\sum_k M_k = N$). The entropy of the collection can be defined as $H_a = \log \dfrac{N!}{\sum_k M_k!}$

In this definition, there is an implicit equation of the set of colour counts $\{M_k\}$ with the macro-state of an isolated thermodynamic body, and of a particular arrangement or sequence of the balls with a particular micro-state. H_a measures the amount of information needed to specify a micro-state given prior knowledge of the macro-state. H is given in bits if the base of logs is 2, in nits if natural logs are used. Pro tem, we assume base 2.

(b) The entropy of a discrete distribution. Consider a random variable x drawn from a discrete probability distribution over the values $1, \ldots, K$, where $\Pr(x = k) = p_k$ $(k = 1, \dots, K; \sum_k p_k = 1)$. The *entropy of the distribution* is often defined as $H_b = -\sum_k p_k \log p_k$.

It equals the expected amount of information needed to specify the value x, given prior knowledge that it is drawn from the distribution. Equivalently, if a message is a sequence of symbols drawn from the set $\{1, \dots, K\}$, and symbol k occurs with probability p_k for all k, independently of previous symbols, then the message can on average convey H_b bits (or nits) of information per symbol.

Again, if a discrete distribution has entropy H_b, the expected length of a message naming a value drawn from the distribution equals H_b when an optimal code is used.

Clearly, H_b is not directly comparable with H_a, since it is a per-instance measure. However, even if applied to a sequence of N instances, giving a total entropy $NH_b = -N\sum_k p_k \log p_k$, this measure still differs from H_a by an amount of order $\log N$, or $(\log N)/N$ per instance.

The quantity NH_b is analogous to a thermodynamic quantity, namely the entropy of a body which is not isolated, but in equilibrium with a "heat bath" of known temperature. In this case, the exact macro-state of the body is unknown, but the expected value of the macro-state variable M_k is $p_k N$, where N is the number of particles in the body and p_k is related to the temperature. In thermodynamic applications, the fractional difference of order $(1/N)\log N$ between H_a and NH_b is negligible, as N is typically very large. However, in dealing with the statistics and information of typical data sets, the difference may be significant.

(c)   The entropy of a density distribution. If x is a real-valued random variable taking values drawn from a probability density $f(x)$, the term "entropy" is sometimes applied to the quantity

$$H_c = -\int f(x)\log(f(x))\,dx$$

H_c may be regarded as the limiting value of the entropy H_b of the discrete distribution

$$\Pr(x_k \leq x < x_k + \delta) = \int_{x_k}^{x_k+\delta} f(x)\,dx \quad (x_k = k\delta,\ k\ \text{integral})$$

as $\delta \to 0$. Its use in this way is unexceptionable, but it must be noted that the interpretation relies on the discretization interval δ being uniform, i.e., not dependent on x. A non-uniform discretization of the density $f(x)$ will yield a different limit for H_b, no matter how fine the discretization. Equivalently, if y is a random variable defined as a monotonic invertible function of x, $y = g^{-1}(x)$, $x = g(y)$ say, the H_c entropy of the density of y will differ from the H_c entropy of the density of x. The density of y is

$$f_y(y) = f(g(y))\dot{g}(y)$$

where $\dot{g}(y) = dx/dy$, and

$$\int f_y(y) \log f_y(y) \, dy \neq \int f(x) \log f(x) \, dx$$

Thus, the entropy H_c of a probability density depends on the choice of random variable. In any use of H_c, one must be careful that the variable x is such that a uniform discretization is appropriate, and that it is to be preferred to any functionally equivalent variable $y = g^{-1}(x)$.

It is this dependence on choice among functionally equivalent variables which vitiates "maximum entropy" arguments for deriving "uninformative" priors for real-valued model parameters. However, the use of a maximum entropy argument for approximating a probability distribution over a data space when the true data distribution is unknown, except for some known constraints, is valid and useful.

Example. It is sometimes stated that the Normal or Gaussian density is the "maximum entropy" density distribution for given mean μ and given second moment $\mu^2 + \sigma^2$. Maximizing $H_c = -\int f(x) \log f(x) \, dx$ subject to the constraints

$$
\begin{array}{ll}
\int f(x) \, dx = 1 & \text{(Normalization)} \\
\int x f(x) \, dx = \mu & \text{(First moment)} \\
\int x^2 f(x) \, dx = \mu^2 + \sigma^2 & \text{(Second moment)}
\end{array}
$$

indeed leads to the solution

$$f(x) = \frac{1}{\sqrt{2\pi}\sigma} e^{-\frac{1}{2}(x-\mu)^2/\sigma^2} = \text{the Normal density } N(x|\mu, \sigma^2)$$

However, were we to define a variable y by $y^3 = x$ (sign y = sign x), and obtain the "maximum entropy" density for y subject to the corresponding constraints

$$
\begin{array}{ll}
\int f_y(y) \, dy & = 1 \\
\int y^3 f_y(y) \, dy & = \mu \\
\int y^6 f_y(y) \, dy & = \mu^2 + \sigma^2
\end{array}
$$

we would obtain a density of the form

$$f_y(y) = \exp(-1 - \lambda_0 - \lambda_1 y^3 - \lambda_2 y^6)$$

implying a density for x of the form

$$\frac{1}{3x^{2/3}} \exp(-a_0 - a_1 x - a_2 x^2)$$

which is clearly not Normal.

In thermodynamics, the fact that the maximum entropy density for x subject to the constraints is Normal has been used to conclude (correctly) that the velocity component distribution in the molecules of a gas in thermal equilibrium has Normal form, since the constraints used correspond to the given macro-state momentum and energy. However, the argument succeeds only because the true distribution (for an enclosed gas) is in fact discrete and the discrete possible velocity components allowed by quantum mechanics are evenly spaced. Under conditions where the discrete values are unevenly spaced, e.g., at temperatures high enough to involve relativistic velocities, the maximum (thermodynamic) entropy distribution is not Normal.

To summarize this section, the importation of the term "entropy" into discussions of information brings some benefits, but can lead to confusion arising from subtly different uses of the term. In particular, the use of H_c in connection with probability densities is dangerous.

In cases where all that is known of a discrete probability distribution can be expressed as one or more constraints on the probabilities of the possible variable values, a distribution which maximizes the H_b entropy of the distribution subject to the known constraints can reasonably be accepted as representing our knowledge of the variable. In a useful sense, it is the weakest assumption that can be made about the distribution of the variable, in that it maximizes the expected amount of additional information needed to specify a value of the variable.

2.1.12 Codes for Infinite Sets

Sometimes the set of possible messages which might be sent in some defined communication is potentially infinite. We say "potentially" because in any real-world communication, the length of the message string must be bounded if it is to be sent and read in bounded space and time. As the number of binary strings of lengths not exceeding some bound is finite, so is the number of different messages which might be sent. However, there are interesting situations in which no bound on message length can easily be set, or where at least the analysis would be complicated by setting a bound. In such cases we may prefer to ignore the limitations of human life and consider a code for an infinite set of messages.

Note that the set must be countable. The set of finite (but unbounded) binary strings can be enumerated, for instance by enumerating them in order of increasing length, as shown in Figure 2.9. The string represented by integer n is just the binary form of n with the leading one deleted.

An uncountably infinite set cannot be mapped 1-to-1 onto the set of finite binary strings.

Given that the set is countable, there is no difficulty in principle in mapping onto a set of finite strings and so producing a code (not necessarily efficient) for the set of messages. We need only ensure that the set of strings has the prefix property, and thereby that the string for each message is unique. Without loss of generality, we can simplify the discussion by first establishing an enumeration of the set of possible messages, i.e., we can label each message

Index	String
1	Λ (the empty string)
2	0
3	1
4	00
5	01
6	10
7	11
8	000
$\vdots$	$\vdots$
15	111
16	0000
$\vdots$	$\vdots$

Fig. 2.9. Set of binary strings in order of increasing length.

with a unique, finite, positive integer. Then the task of constructing a code for the set of messages can be reduced to the construction of a code for the positive integers. Many prefix codes for the integers can be constructed. Two examples are shown below.

2.1.13 Unary and Punctuated Binary Codes

The unary code for integer n is a string of $(n - 1)$ 0s followed by a 1. The string length equals n.

Integer	"Unary" code	"Punctuated Binary" code
1	1	01
2	01	11
3	001	0001
4	0001	0011
5	00001	1001
6	000001	1011
7	0000001	000001
8	00000001	000011
9	000000001	001001
10	0000000001	001011

Fig. 2.10. Unary code and Punctuated-Binary code.

The Punctuated Binary code string (shown in Figure 2.10) for integer n is constructed as follows: First, note that any integer $n > 0$ can be written as a binary integer with digits $d_k, d_{k-1}, \ldots, d_2, d_1, d_0$ such that $n = \sum_{i=0}^{k} 2^i d_i$. The usual binary form uses digit values 0 and 1 for the digits. However, a binary representation of $n > 0$ can also be written using the digit values 1 and 2 as shown in Figure 2.11. (In general for any base B, $n > 0$ can be written using

Decimal Number	Binary Form
1	1
2	2
3	11
4	12
$\vdots$	$\vdots$
47	21111

Fig. 2.11. Binary representation of $n > 0$ using digit values 1 and 2.

digits $1 \ldots B$ rather than the more usual form using digits $0 \ldots (B-1)$.)

To form the Punctuated Binary string for n, express n in the above "inflated binary" form as a string of 1s and 2s. Then change all the 1s to 0s, all the 2s to 1s. Finally, insert a zero after each digit save the last, and add a one after the last digit. These inserted symbols act as punctuation: an inserted zero means the string continues, the final inserted one acts as a full stop. The resulting string length is $2\lfloor \log_2(n+1) \rfloor$, where $\lfloor x \rfloor$ means the greatest integer $\leq x$.

Both the above codes clearly are prefix codes, and include all the positive integers. Many other constructions are possible.

2.1.14 Optimal Codes for Integers

Just as for codes over finite sets, we will be interested in optimal codes for infinite sets, and will discuss optimal codes for the positive integers as examples. As before, we define optimal codes as those of least expected length, given some specified probability distribution over the set of messages (integers). Let p_n be the probability of integer $n > 0$, and l_n the length of the binary code string for n. Then an optimal code minimizes

$$\sum_{n=1}^{\infty} p_n l_n$$

We of course require the probability distribution $\{p_n : n > 0\}$ to be proper, i.e., $\sum_{n=1}^{\infty} p_n = 1$ and $p_n \geq 0$ for all n. (To accommodate the possibility that $p_n = 0$ for some n, we define $p_n l_n$ as zero whenever $p_n = 0$, no matter what the value of l_n.)

For finite sets, an optimal code exists for every proper probability distribution, namely the Huffman code having $l_i \approx -\log_2 p_i$. Such a code gives an expected string length close to the entropy of the distribution, $-\sum_i p_i \log p_i$ However, there exist proper distributions over the integers for which $(-\sum_n p_n \log p_n)$ is infinite. Such *infinite entropy* distributions do not admit of optimal codes, at least in our sense. An infinite entropy distribution may be thought implausible in the real world, since, using any code at all, the expected length of the message announcing a value drawn from the distribution must be infinite. However, for some infinite entropy distributions, there may be a very high probability that, using a suitable code, the announcement of a value will require a string of less than, say, 100 digits. Hence, infinite entropy distributions cannot be totally dismissed as unrealistic.

For codes over finite sets, we have shown that an optimal code represents a value or event i of probability p_i by a string of length close to $-\log_2 p_i$. Equivalently, a code which encodes event i by a string of length l_i is optimal for an implied probability distribution $\{p_i = 2^{-l_i}; \ \forall i\}$. These relations may be extended to cover codes for infinite sets, whether or not the given or implied probability distribution has finite entropy. That is, we may regard use of a code encoding integer n by a string of length l_n as "optimal" if and only if $p_n = 2^{-l_n}$, whether or not $\sum p_n l_n$ is finite. This extension of the notion of optimality to cases where the expected string length is infinite can be rationalized as follows.

Consider some distribution p_n over the integers. For any integer $N > 0$, we may derive a distribution over the finite set $\{n : 1 \le n \le N\}$ defined by $p_n^1 = p_n$ for $n < N$, $p_N^1 = 1 - \sum_{n=1}^{N-1} p_n^1$. This distribution is in effect the distribution obtained by lumping all integers $\ge N$ into the one value "N", which now just means "big". The distribution $\{p_n^1 : n = 1, \ldots, N\}$ is over a finite set and has finite entropy. Hence, it admits of an optimal code in the sense of least expected length, and in such an optimal code,

$$l_n = -\log p_n^1 = -\log p_n \quad \text{for} \quad 0 < n < N$$

That is, if all integers N or greater are lumped together, a code encoding $n < N$ with length $-\log p_n$ is optimal in the usual sense. Since N may be chosen as large as we please, we may always choose N to be greater than any integer n actually encountered, and so conclude that a code is optimal if it encodes n with length $l_n = -\log p_n$ for all finite n. Conversely, we may have reason to adopt a code characterised by the string lengths $\{l_n : n > 0\}$ and be entitled to regard it as good if we are satisfied that the probability of finding integer n is about 2^{-l_n}, even if this implied distribution has infinite entropy.

Note that a distribution of infinite entropy may, but need not, imply a non-zero probability of drawing an infinite integer. The "Unary" code above has $l_n = n$. Hence, it is optimal for, or implies, the probability distribution $p_n = 2^{-n} (n > 0)$. The entropy of this distribution is 2 bits. The "Punctuated Binary" code has $l_n = 2\lfloor \log_2(n + 1) \rfloor$. Hence, all the 2^k integers between $2^k - 1$ and $(2^{k+1} - 2)$ have $l_n = 2k$, $p_n = 2^{-2k}$ for all $k > 0$. They contribute a value

$$2^k (2^{-2k})(2k) = 2k/2^k$$

to the entropy sum. The total entropy is thus

$$2\sum_{k=1}^{\infty} k/2^k = 4 \text{ bits}$$

A Binary Tree Code. We now consider a code for the set of binary trees rather than the set of integers. Here, a binary tree is either a single (leaf) node, or is a node with two dependents, a left subtree and a right subtree, each subtree being itself a binary tree. For this example, the trees

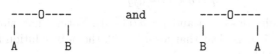

are regarded as different if the subtrees A and B are different. This set of trees corresponds to the set of code trees defining non-redundant binary prefix codes, so a code for this set is also a binary prefix code for the set of binary prefix codes. The particular code for trees which we consider is defined as follows:

The string for a leaf is 0.
The string for a non-leaf tree is 1 followed by the string for its left
    subtree followed by the string for its right subtree.

Some example trees and their code strings are given in Figure 2.12.

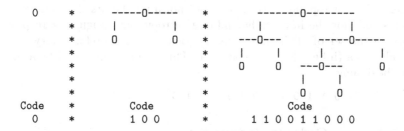

Fig. 2.12. Sample trees and their code strings.

As before, the code implies a probability distribution over binary trees, i.e., that distribution for which the code is optimal. As the code uses one binary digit for each node of the tree, the probability of a tree is $2^{-(\text{number of nodes})}$. Note that the number of nodes, and hence the string length, is always odd.

To calculate the entropy of this distribution, we may note that, as for any optimal code, the entropy is the expected string length. In this case, the string length equals the size of the tree measured in nodes, so we should calculate the expected tree size. Assume for the moment that any node independently has probability a of being a leaf, and $(1 - a)$ of having dependent subtrees. Then the expected size L of a tree is given by

$$L = a \times 1 \text{ (if the node is a leaf)}$$
$$+(1 - a)(1 + 2L) \text{ (if the node has dependents)}$$

Hence, $L = 1/(2a - 1)$ We may also calculate the probability q that the tree is finite, noting that it is finite only if it is a leaf or if both its subtrees are finite. Thus,

$$q = a + (1 - a)q^2$$

This equation has solutions $q = 1$ and $q = a/(1 - a)$. Noting that, since q is a probability, $0 \le q \le 1$, it is clear that for $a > 1/2$, the only solution is $q = 1$. For $a < 1/2$, $q = a/(1 - a)$.

The implied distribution of a code is the distribution for which the code is optimal. But in any string of an optimal binary code, each digit independently of all others has probability $1/2$ of being 0. Since, in the tree code, a node is encoded by the digit 0 if it is a leaf, the probability that a node is a leaf in the implied distribution is also $1/2$. That is, $a = 1/2$. Hence, $L = \infty$, $q = 1$. The expected tree size, and hence the entropy of the distribution, are infinite, but the probability of drawing an infinite tree from the distribution is zero.

This code for binary trees could also be used as a code for the positive integers by labelling every tree with a unique integer. One way of doing this is to label the code strings for the trees in lexographic order, within order of increasing string length. This leads us to the code shown in Figure 2.13 for the integers. The corresponding trees are also shown for the first few numbers.

This code for the integers has infinite entropy, but assigns a zero probability to the set of all infinite integers. Being itself a prefix binary code, the code has a (infinite) binary code tree. This tree is represented by a code string beginning

$$1\ 0\ 1\ 1\ 0\ 1\ 1\ 0\ 1\ 1\ 0\ 1\ 1\ 0 \ .\ .\ .\ .$$

2.1.15 Feasible Codes for Infinite Sets

For some infinite sets, it may be difficult in practice to construct an optimal code. Let the set be the set of integers, with a known probability distribution $\{p_n : n > 0\}$. To construct an optimal code, we would have to choose a unique string for each integer, having length $l_n \approx -\log_2 p_n$, with no string being a prefix of another. If the mathematical form of the distribution is at all complicated, this could be a formidable task, and we may well be content to settle for something less than optimality. In particular, if the distribution $\{p_n : n > 0\}$ has finite entropy, we may be content to use some binary prefix code encoding n with length l_n $(n > 0)$ provided it leads to a finite expected message length. For a given distribution $\{p_n\}$ of finite entropy $H = \sum_n p_n \log_2 p_n$, we will define a code as *feasible* if its expected string

Number	String	Tree

```
Number        String               Tree

  1             0                    0    (The null tree)

  2           1 0 0             ----0----
                                |       |
                                0       0

  3          1 0 1 0 0          ----0----
                                |       |
                                0    ---0---
                                     |     |
                                     0     0

  4          1 1 0 0 0          ----0----
                                |       |
                             ---0---    0
                             |     |
                             0     0

  5         1 0 1 0 1 0 0      ----0----
                                |       |
                                0    ---0---
                                     |     |
                                     0  ---0---
                                        |     |
                                        0     0

  6         1 0 1 1 0 0 0

  7         1 1 0 0 1 0 0

  8         1 1 0 1 0 0 0

  9         1 1 1 0 0 0 0  etc.
```

Fig. 2.13. Code for positive integers and their corresponding trees.

length $E(l) = \sum_n p_n l_n$ is finite. Of course, $E(l) > H$ unless the code is optimal, and feasible codes do not exist for distributions of infinite entropy. It might be thought that if H is finite, any code optimal for some finite-entropy distribution would give finite $E(l)$, but this is not the case. For instance, the Unary code described in Section 2.1.13 is not feasible for the distribution $\{p_n$ proportional to $1/n^2\}$, since the Unary code has $l_n = n$, and $\sum_{n=1}^{\infty} n/n^2$ is infinite. But the distribution $\{p_n$ proportional to $1/n^2\}$ has finite entropy. The punctuated binary code is approximately optimal for it, and gives an expected length of only a few digits. On the other hand, the punctuated binary code is feasible for the distribution implied by the unary code, that is, for the distribution $p_n = 2^{-n}$. It gives an expected string length for this distribution of

$$2 \sum_{n=1}^{\infty} 2^{-n} \lfloor \log_2(n+1) \rfloor \approx 2.52 \text{ digits}$$

which is little more than the 2 digits expected if the optimal unary code is used.

Any finite-entropy distribution over the integers must give integer n a probability p_n which decreases with n for n greater than some integer K. That is, any finite-entropy distribution must eventually "tail-off" for large integers. (Of course, the "tailing off" need not be monotonic: we could have a finite-entropy distribution such that $p_{2n+1} > p_{2n}$ for all integer n.) More precisely, if $\{p_n : n \geq 0\}$ is a finite-entropy distribution, there must exist some finite K and some monotonically decreasing function $f(n)$ such that $p_n \leq f(n)$ for $n > K$. Generally speaking, a code optimal for some distribution $\{p_n\}$ will be feasible for some finite-entropy distribution $\{q_n\}$ only if $\{p_n\}$ "tails off" no more rapidly than $\{q_n\}$.

2.1.16 Universal Codes

An interesting code for the positive integers has been discussed by Leung-Yan-Cheong and Cover (1978). Any integer $n > 0$ can be written in ordinary binary form as a string of k digits, where $k = \lfloor \log_2 n \rfloor + 1$. The first digit of this string is always 1. Of course, these strings themselves do not form a code for the integers, as they lack the prefix property. For instance, the string for $n = 5$ is 101, but this string is a prefix of the string for 10, (1010) among others. To devise a prefix code, first define a function $head(n)$ from integers to strings as:

$$head(1) \quad = \quad \Lambda \text{ (the empty string)}$$

For $n > 1$ a k-digit binary number

$$head(n) \quad = \quad head(k-1) \text{ followed by the } k \text{ digits of } n$$

Thus, for $n = 3$ $(k = 2)$, $head(3) = head(1).11 = \Lambda.11 = 11$.
For $n = 9$ $(k = 4)$, $head(9) = head(3).1001 = 111001$, etc.

The prefix "log *" codeword for n is

Code word for $n = head(n)$ followed by 0

A few examples are shown below

n	:	code string
1	:	0
9	:	1 1 1 0 0 1 0
501	:	1 1 1 0 0 0 1 1 1 1 1 0 1 0 1 0

Decoding a string in this code is easy. The steps are shown below.

1. Set $n = 1$, and begin at the start of the string.
2. If the next digit is 0, exit with n. Otherwise, read this digit and the following n digits as a binary number m.
3. Set $n = m$ and return to Step 2.

Recalling that $k = \lfloor \log_2 n \rfloor + 1$, the length $h(n)$ of the string $head(n)$ is given by

$$h(n) = h(k-1) + k = h(\lfloor \log_2 n \rfloor) + \lfloor \log_2 n \rfloor + 1$$

where we take $h(0) = 0$. For large n,

$$h(n) = h(\lfloor \log_2(\lfloor \log_2 n \rfloor) \rfloor) + \lfloor \log_2(\lfloor \log_2 n \rfloor) \rfloor + 1 + \lfloor \log_2 n \rfloor + 1$$

Of course, for any n, the code string length $l_n = h(n) + 1$. Rissanen [34] has suggested the approximate length function

$$\log^* n \equiv \log_2 n + \log_2 \log_2 n + \log_2 \log_2 \log_2 n + \ldots + C$$

where the series is continued up to the first term which is less than or equal to one, and C is a normalization constant chosen to satisfy

$$\sum_{n=1}^{\infty} 2^{-\log^* n} = 1$$

That is, C is chosen to make the implied probability distribution $\{p_n = 2^{-l_n}\}$ normalized, and the fact that the "string lengths" $\{l_n = \log^* n\}$ are in general not integers is disregarded.

The log* code, and the distributions over the integers implied by it or by the length function $\log^* n$, are in a certain sense universal. The implied distribution has infinite entropy: $\sum_{n=1}^{\infty} l(n)2^{-l(n)}$ is infinite. That is, the probability $p_n = 2^{-l(n)}$ tails off more slowly than does any finite-entropy distribution. In consequence, if the integer n is in fact selected from some finite-entropy distribution $\{q_n : n = 1, 2, \ldots, \}$, the expected string length required to encode n using the log* code is finite. The log* code is feasible for all finite-entropy distributions.

There are other "universal" codes over the integers. All universal codes correspond to probability distributions of infinite entropy. An example is the code derived from the binary prefix code for binary trees discussed in Section 2.1.14. It again has infinite entropy, and is feasible for any finite-entropy distribution. In fact, Rissanen (1983) has shown that in a sense all infinite-entropy distributions which tail off after some finite integer are equivalent. Let p_n and q_n be two such distributions. Suppose an integer is drawn from $\{p_n\}$ and encoded using a code for q_n Define

$$L_N = -\sum_{n=1}^{N} p_n \log q_n \quad , \quad H_N = -\sum_{n=1}^{N} p_n \log p_n$$

Then $\lim\limits_{N \to \infty} L_N / H_N = 1$.

This result proves only a rather weak equivalence among universal codes. The expected string lengths arising when a number from a finite-entropy distribution is encoded using different infinite-entropy codes may differ greatly.

The "binary tree" code and the log* code are actually very similar in their length functions. Numerical calculations show that the lengths in either code of an integer differ by only a few digits for integers at least up to $2^{1000000}$.

Universal codes are of interest in that they allow us to attach a "prior probability" to an unknown integer parameter in some model about data. If we have almost no well-founded prior expectations about the likely value of the integer, use of a "universal" prior still allows us to encode the integer with a string whose expected length is finite, provided we at least have reason to believe that the unknown integer can be represented by a finite string. Similarly, a universal code can be used to define a prior over any infinite countable set. For instance, the set of finite binary strings can be encoded by a string which first gives the length of the given string, then a copy of the given string. Such a code might have length $L + \log^* L$ for a binary string of length L, thus assigning the string a probability

$$2^{-L} \times 2^{-\log^* L}$$

2.2 Algorithmic Complexity

This section introduces a slightly different approach to the measurement of information. We again suppose that we have data recorded in a given binary string, and ask how we might recode it more briefly in another binary string for transmission to a receiver. The measure of "information" in the given data is again the length of the recoded string. We depart from the earlier treatment, in which the receiver was characterized by prior expectations described by probability distributions, by now supposing that the receiver is a specified Turing machine, i.e., a computer.

2.2.1 Turing Machines

For our purposes, a Turing machine is a machine with

(a) A *clock* which synchronizes all its activities. The machine performs one (possibly complicated) action or *step* every time the clock ticks. The period of time from just after one tick up to and including the next tick is called a *clock cycle*.

(b) A finite set of *internal states*. What the machine does in a clock cycle (i.e., what action it performs in synchronism with the tick ending the cycle) depends in part on its internal state. The state remains constant during a cycle, but the machine may change to a different state at the tick. The set of states is indexed by the integers 1 to S, where S is the number of states. We assume the machine starts in state 1.

(c) A binary *work tape*. This is a recording medium like a magnetic tape, divided lengthways into *cells*. Each cell records a binary digit. The action of the machine at a tick depends in part on the value of the digit recorded in the work-tape cell which is under the machine's *work-tape head*. The action may involve changing the digit recorded in this cell, and possibly moving the work-tape one cell to the left or right, so that a different cell will be under the head during the next cycle. The work-tape extends infinitely in both directions. Note every cell of this infinite tape always holds either a "1" or a "0". There is no such thing as a blank cell. For simplicity we may assume the work tape is initially all zeros.

(d) A one-way binary *input tape*. This is the part of the machine which will receive the binary message string, which we imagine to be recorded in the cells of a tape similar to the work tape. The machine has an *input tape head* and its action in a cycle may depend on the binary value recorded in the cell of the input tape currently under this tape head. The action may also involve moving the input tape by one cell to the left, so the next cell can be read. However, the machine cannot move the input tape to the right, and so can never re-read a cell of the input tape. Initially, the first binary digit of the received message is under the input head.

(e) A one-way binary *output tape*. The machine can, as part of its action on a clock tick, write a binary digit into the cell of the output tape under its *output tape head* and move the tape to the left. It cannot read or change what it has written on its output tape. Initially, the first cell of the output tape is under the output tape head, and its contents are undefined.

(f) A list of 4S *instructions*. The action taken at each clock tick depends on the current state, the binary digit under the work-tape head, and the binary digit under the input-tape head. Thus, for a machine of S states there are $S \times 2 \times 2 = 4S$ possible situations. The instruction list contains one instruction for each situation. The instruction dictates the machine's action at the next tick. The instruction has four parts:
Next state: a number in $(1, \ldots, S)$ specifying the next internal state.

Work write: a binary digit to write into the current cell of the work
tape.

Work move: a symbol indicating one of the three choices: move work-
tape left, move work-tape right, or do not move it. The move takes
place after the write action.

Input/Output: two binary digits controlling the input and output
tapes, by selecting one of the four options:

(00) Do nothing

(01) Move input tape one cell to the left

(10) Write "0" on the output and move it one cell to the left

(11) Write "1" on the output and move it one cell to the left

Those familiar with computer design will recognise the instruction list as
equivalent to the microcode of a microcoded computer. In our discussion,
different Turing machines differ only in their numbers of states and their
instruction lists.

When we use a Turing machine (TM) as receiver, we suppose that we
encode the given binary information or data into a binary string which is
written on the input tape. This string is the "transmitted message". The
TM is then started, and it begins to read the input tape, write and read its
work tape, and eventually writes some binary string onto its output tape. We
consider that the TM has correctly decoded the message if the final content
of its output tape is an exact copy of the given binary information or data.
Thus, given a binary string A representing some data or information, we
propose that the amount of information in A (in bits) equals the length of
the shortest input tape which will cause the TM to output a copy of A. This
definition of the information in A is called the *Algorithmic Complexity* of A.
The idea was first proposed by Kolmogorov and later refined by Chaitin [9].
Note that the algorithmic complexity of a binary string A can be entirely
divorced from any interpretation of A as a body of data, propositions or
other sort of meaningful information. The string A is treated simply as an
uninterpreted string of binary digits. However, we will be using the notion of
algorithmic complexity principally for strings which do represent data.

2.2.2 Start and Stop Conditions

For the definition of Algorithmic Complexity to be complete, we need to
impose some technical conditions.

(a) The machine must start in some specified state, say, state 1, with zeroed
work tape.

(b) There are two versions of this condition. Which version we choose to
impose will depend on whether the data string A is regarded as a unique
data object (version (b1)), or a sample of data from some process which
might well produce more data later (version(b2)).

(b1) Immediately after outputting the last digit of A, the machine will stop (by entering a state whose next state is itself regardless of input and work digits, and which performs no tape action).

(b2) Immediately after outputting the last digit of A, the machine will attempt to read input beyond the end of the input I.

(c) No proper prefix of I will cause the machine to output A and also meet condition (b).

If (b1) applies, the conditions effectively require the allowed input sequences for all possible A to form a prefix set. That is, no input I which is allowed as an encoding of A can be the prefix of an input J which is allowed as the encoding of A or any other string.

If instead we require (b2), then if I is an allowed encoding of A, I may be a prefix of an allowed input J which encodes B if and only if A is a prefix of B.

We will say a TM *accepts* an input I if and only if, according to the above rules, I encodes any string. Note that, whichever version of condition (b) is adopted, I must allow the Turing machine to "know" when it has completed the decoding of I and the production of A.

2.2.3 Dependence on the Choice of Turing Machine

Obviously the definition of the algorithmic complexity of a string depends on the choice of the TM. Indeed, if the TM is poorly chosen, there may be no input which will cause it to output a given string. In general, we will be interested only in TMs which are capable of producing any arbitrary string.

There is an obvious correspondence between TMs and efficient coding schemes such as Huffman and Arithmetic codes. For any such code, there is a TM which decodes it. Thus, the Algorithmic Complexity concept embraces all the coding schemes based on probabilities which were mentioned in the discussion of Shannon information. An "optimal" coding scheme which gives the minimum expected message length given certain prior expectations can be decoded by a TM which, in effect, can compute the message probabilities implied by those prior expectations. A coding scheme of the "explanation" type which begins each coded message with an assertion inferred from the data, then codes the data using an optimal code for the probability distribution implied by the assertion, can be decoded by a TM which can interpret the assertion and then compute the implied probability distribution. Hence, it appears that the Algorithmic Complexity formulation is no more than an unduly complicated way of re-formalizing Shannon's approach to information, in which the prior expectations are expressed in the design of a TM rather than directly as probability distributions.

2.2.4 Turing Probability Distributions

If we adopt condition (b1), which requires a TM to stop after producing a finite string A by decoding an acceptable input I, any TM can be regarded as defining a probability distribution over the set of finite binary strings. The definition follows from equating the Shannon measure of the information in a string with the Algorithmic Complexity measure. Let A be a finite binary string and T a TM. We have defined the Algorithmic Complexity (AC) measure of the information in A as the length of the shortest input I which will cause T to output A (assuming the conditions of Section 2.2.2 above). Write the length of I in binary digits as $|I|$. Then, with respect to T,

$$AC(A) = |I|$$

The Shannon measure is defined with respect to a probability distribution $P()$ over strings as

$$\text{Info } (A) = -\log_2 P(A)$$

Equating the measures gives

$$P_T(A) = 2^{-|I|} = 2^{-AC(A)}$$

We can interpret $P_T(A)$ as a probability distribution over strings inherent in the design of T. Loosely, it is that distribution over finite strings for which T decodes an optimal code. Note that $P_T(A)$ is not normalized. The sum over all finite strings of $P_T(A)$ is in general less than one.

If instead we adopt the "stopping rule" (b2), which requires the TM to attempt to read more input after decoding I to A, the AC measure so defined does not define a probability distribution over the set of finite strings, since the set of allowable inputs which decode to finite strings no longer has the prefix property. A string I which decodes to the finite string A may be a proper prefix of a string $I.J$ which decodes to $A.B$. However, when we consider a TM as the "receiver" which must decode an "explanation" of some data string A, the data string A is normally a member of a prefix set of possible data strings, and the receiver may be presumed to have sufficient knowledge of the set to determine when A is complete. No transmission of the data, in "explanation" or any other form, would be possible unless the receiver can tell when the transmission is complete. It is thus reasonable to suppose that the set U of strings A for which the TM is required to decode compressed inputs form a prefix set. If so, then the form of Algorithmic Complexity assuming stopping rule (b2) defines a probability distribution over U again of the form

$$P_T(A) = 2^{-|I|} = 2^{-AC(A)} \quad (A \in U)$$

Thus far, for either stopping rule, there appears to be no useful advantage of the Turing-Machine formulation of information, and its derived probability distributions, over the Shannon formulation which takes probability distributions as given, and derives measures of information from them. The new formulation does, however, have a new feature not normally apparent in the Shannon approach. The new feature arises from the existence of certain special Turing Machines called *Universal Turing Machines*.

2.2.5 Universal Turing Machines

Despite the simplicity of the formal description of a Turing Machine, it in fact enables the description of machines equivalent in most respects to a modern computer. A modern computer differs from the kind of TM we have described only in that it has less restricted input/output facilities, and in particular, by using magnetic discs rather than a tape for its "work-tape", can move much more rapidly between distant cells. Otherwise, the random-access memory, central processing unit and control units of a computer all fit within the TM formalism of a finite set of internal states with a list of fixed rules (the TM instruction list) for moving from state to state and reading and writing cells of input, output and work-tape. The set of internal states of a real computer is usually very large, of the order of $10^{1000000}$, and so could not feasibly be described in the conceptually simple way we have outlined, but a computer is still no more than a TM.

Computers as we know them today have an important property which can also be exhibited by much simpler TMs. Given the right input, a computer can be forced into a state, defined both by its internal state and what it has written onto its work-tape, such that it thereafter behaves like a different TM. That is, the initial input it is given will cause no output, and the output produced in response to any further input will be exactly the same as would be produced by the different TM given that input. The computer may not operate as rapidly as would the TM it imitates: it may require many clock cycles to imitate what would be done in one cycle by the imitated TM, but the final output will be the same. In terms familiar to computer users, we can get computer X to initiate TM Y by inputting to X an "interpreter" program containing a table of constants representing the instruction list of Y. In general, the interpreter program will have to copy this table onto X's work-tape, while leaving room on the tape to imitate Y's work-tape, but this is not difficult. The interpreter program then, in essence, uses some further work-tape cells to remember the number indicating the current state of Y, consults the table to determine what Y would do next, then does it.

A computer or TM which can be *programmed* (i.e., given an input *program*) to imitate any specified TM is called *universal*. The use of a Universal Turing Machine (UTM) as the "receiver" in a communication opens up the possibility of coding schemes not obvious in the Shannon formalism. In particular, the coded message, which is the input to the UTM, can begin with an interpreter program effectively changing the receiver into a different TM. Thus, if the sender of the message determines, after seeing the data, that the "prior expectations" inherent in the design of the receiver TM are inappropriate, she can begin the message with a string effectively redefining the receiver to have different "expectations". The sender may, after inspecting the data to be sent, form a theory that the data conforms (at least probabilistically) to some pattern or regularity. She can devise a coding scheme which would be optimal were the theory true, and begin the coded message with a string programming the receiver to become a decoder for this scheme. The coded message can then continue with the optimally coded form of the data.

The ability to make a UTM imitate any other TM (including any other UTM) using an "interpreter" program of finite length suggests a generality to the Algorithmic Complexity definition of information not apparent in the probability-based definition. In the latter definition, the information in a message is defined in terms of the probability of the events or propositions conveyed by the message. As the probability is subjective, and depends on the prior expectations of the receiver, the numerical value of the information content so defined also depends on the receiver, and the assumption of different receivers can change the value to an arbitrary extent. No obvious bounds can be placed on the ratio between the probabilities assigned to the same message by two different receivers, and hence no bound can be placed on the difference between the two corresponding measures of information, which is the log of the probability ratio. The same arbitrarily large differences in information measure can occur if arbitrary TMs are substituted for the receivers. In particular, some TMs may be incapable of producing certain output strings. In effect, their design assigns a zero probability to these strings, and so leads to an infinite value for their "information content". However, if we consider two UTMs as alternative receivers, the situation is somewhat different.

First, any UTM can, given the right input, produce any output. At worst the input to a UTM can begin with a program which simply copies the remaining input to the output. Thus, any output can be produced encoded as a "copy" program followed by a copy of the desired output.

Secondly, for any string A and any two UTMs, the difference between the lengths of the shortest input I_1 which will cause $UTM1$ to output A and the shortest input I_2 which will cause $UTM2$ to output A is bounded. The length of I_2 need never exceed the length of I_1 by more than the length of a program which will make $UTM2$ imitate $UTM1$. That is, if we know a short input I_1 for $UTM1$, we can make an input I_2 for $UTM2$ which begins with an interpreter program making $UTM2$ imitate $UTM1$, and continues with a copy of I_1. Hence, the difference in the information measures of a string A defined with respect to $UTM1$ or $UTM2$ is bounded, and the bound is independent of A. The bound is the length of a program to make one machine behave like the other. The length of such programs depends on the designs of the two UTMs, but is finite and independent of any string the machines may be later required to produce.

For sufficiently long messages, the differences in information content arising from different choices of receiver UTM become a small fraction of the information measure. This fact suggests that the choice of UTM is relatively unimportant in the Algorithmic Complexity definition of information, at least for long strings having a very high information content. That is, all UTMs have in a sense the same "prior expectations". It is tempting to conclude that we can therefore arrive at a non-subjective definition of information. A non-subjective AC definition of information would then imply a non-subjective definition of the probability of a data string via the relation Information = − log(Probability). Unfortunately, in most practical situations the differences between UTMs are not negligible. The length of the "interpreter" required to make one UTM imitate another can often be large. For example, consider the two UTMs which respectively accept and execute programs written in the two computer languages "C" and "Fortran". These languages are not very different, and the lengths of the two inputs required to make a "C machine" and a "Fortran machine" give the same specified output A are found in practice to be quite similar. That is, whether a program is written in C or Fortran makes little difference to its length, whatever the nature of the computation performed. Hence, the two UTMs must be regarded as being broadly similar, and equivalent to rather similar probability distributions over the set of finite output strings. However, the length of the interpreter (written in C) required to make a "C machine" behave exactly like a "Fortran machine" is many thousands of binary digits. Even after allowing for the facts that the C language is fairly redundant and that practical interpreter programs are usually written to satisfy other requirements as well as brevity, it seems unlikely that the interpreter could be expressed in less than 1000 digits. Hence, the common universality of both UTMs can probably guarantee an equality of C-based and Fortran-based measures of information only to within ±1000 bits. Many data sets from which we might wish to infer theories or estimates might contain only a few thousand bits of information (by any measure), so such a large difference cannot be ignored.

An equivalent way of expressing the importance of the chosen UTM is that the same string will have different probabilities with respect to C and Fortran UTMs, and the universality argument only guarantees that these probabilities will not differ by more than a factor of about 2^{1000}. As we are accustomed to regard probability ratios over 1000:1 as being important, such a guarantee gives no comfort.

2.2.6 Algorithmic Complexity vs. Shannon Information

The algorithmic complexity (AC) of a string is the length of the shortest input required to make a given TM output the string. The Shannon Information of a string is minus the log of its probability in a given distribution over strings. For any given computable distribution over strings P(S), there is a TM such that the AC agrees with the Shannon information for all strings S, at least to within one bit. Informally, the TM is a computer programmed to decode a Huffman or similar optimal code for the distribution P(S). For non-computable distributions, no such agreement can be guaranteed. However, in practice we would expect to deal only with distributions which are either computable or capable of being approximated by computable distributions. Other distributions, by definition, would not allow us to compute the probability of a string, nor to compute its Shannon information. Thus, in practice, the AC model of information can subsume the Shannon model.

If the AC is defined with respect to a given UTM T, and the Shannon information defined with respect to a given probability distribution $P()$, the AC of a string S may exceed $-\log P(S)$, but only by a constant C_{TP} depending on T and P, and independent of S. The constant is the length of the interpreter required to make T imitate a TM which decodes an optimal code for P. Thus, the AC of S is never much more than the Shannon information defined by the given distribution P. However, the AC of S may be considerably less than $-\log P(S)$, since the UTM T is not required to imitate a decoder for an optimum code for P. Instead, the shortest input to T which outputs S may have any form. It may, for instance, comprise an interpreter for a TM based on some other computable distribution Q, followed by S encoded in an optimal code for Q. Thus, the algorithmic complexity of S cannot exceed

$$\min_Q [-\log Q(S) + C_{TQ}]$$

where Q ranges over all computable distributions over strings. This relation can be used to set an upper limit on $AC_T(S)$ by defining $Q(S)$ as follows:

There are 2^L strings of length L. If these are regarded as equiprobable,

$$Q(S) = \Pr_Q(S|L) \times \Pr_Q(L) = \Pr_Q(|S|) \times 2^{-|S|} \quad \text{where } |S| = \text{Length}(S)$$

The Universal "log*" code of Section 2.1.16 defines a computable probability distribution over the strictly positive integers given by $\Pr_{\log *}(N) = 2^{-\log^*(N)}$. Let $\Pr_Q(L) = \Pr_{log*}(L)$. Then

$$\begin{aligned} AC_T(S) &\leq -\log Q(S) + C_{TQ} \\ &\leq -\log \Pr_Q(|S|) - \log 2^{-|S|} + C_{TQ} \\ &\leq +\log^*(|S|) + |S| + C_{TQ} \end{aligned}$$

Hence, for any choice of UTM T, $AC_T(S)$ can exceed $|S| + \log^*(|S|)$ only by a constant independent of S.

If we map the set of finite strings onto the positive integers by the lexographic enumeration of Table 1.1, the integer $N(S)$ corresponding to string S is

$$N(S) = 2^{|S|+1} - 1; \quad |S| \approx \log N(S) - 1$$

Hence, if we regard T as "decoding" an input string I to produce a string S representing the integer $N(S)$, the above bound on $AC_T(S)$ shows that $|I|$ need not exceed

$$\begin{aligned} l(N) &= AC_T(S) \leq |S| + \log^*(|S|) + C_{TQ} \\ &\leq \log N + \log *(\log N) + C_{TQ} \\ &\leq \log^*(N) + C_{TQ} \end{aligned}$$

Hence, the shortest input I causing T to output S representing the integer $N(S)$ defines a Universal code for the integers, albeit a defective one in that not all input strings will represent any integer, as some may fail to give any finite output and many will decode to the same integer.

There are some long, finite strings whose AC will be small for almost any choice of UTM. These are the strings which can be easily generated by a short computer program. Examples are strings of all zeros, strings of alternating zeros and ones, strings representing the first N binary digits of easily computed irrationals such as π, e, $\sqrt{2}$, etc., and strings which exhibit significant departures from the statistics expected in a sequence of tosses of an unbiased coin, such as strings containing many more ones than zeros. Chaitin [8] has proposed that such strings be termed *non-random*. More precisely, a finite string S is non-random (with respect to some UTM T) if $AC(S) < |S| - \delta$, where δ is some fixed value indicating a threshold "significance" or "degree" of non-randomness. Intuitively, this definition regards a string as random if (in our terms) there is no acceptable explanation of the string. Note that the number of strings with $AC_T(S) \leq K$ cannot exceed 2^{K+1}, since each such string must have an input string for T of length $\leq K$, and there are only 2^{K+1} such strings. Thus, of all strings of length L, at most $2^{L+1-\delta}$ can have $AC_T \leq L - \delta$, and hence at most a fraction $2^{-\delta}$ can be non-random. If we set the threshold $\delta = 20$ say, at most one string in a million can be non-random.

This definition of the "randomness" of a finite string may seem strange. Definitions of randomness traditionally relate to processes which produce strings rather than to any finite output of such a process. For instance, we may regard the tossing of an unbiased coin as a process for producing strings of binary digits (1 for a head, 0 for a tail) and regard it as a random process because our (subjective) probability that the next toss will yield a head is (at least in practical terms) 0.5, and is independent of the results of all previous tosses. We regard a binary-output process as somewhat non-random if (possibly using knowledge of the producing process and of previous outputs) we can give probabilities other than a half to subsequent digits, and find that when these probabilities are used to encode the output string, the encoded form is consistently shorter than the raw output. Our judgement of the non-randomness of a process is thus related to the compressibility of its output, suggesting that the AC definition of the randomness of finite strings is in line with the process-oriented definition. The notable difference is that a process will be regarded as (partially) non-random if its output is compressible in the limit of long strings, so the "cost" of specifying the nature of its non-randomness (in effect the length of the compression algorithm used) is immaterial. The AC definition, since it is quite independent of the process producing the finite string and does not regard the string as part of a potentially infinite output, includes the specification of the nature of the non-randomness in the compressed string.

It is important to realize that a random process can produce a string which is non-random in the AC sense, but its probability of doing so is small.

2.3 Information, Inference and Explanation

This section concerns the connexions among Bayesian inference (Section 1.13), the notions of information discussed in Sections 2.1 and 2.2, and the "explanation message" introduced in Section 1.5.2.

Given a body of data represented in a finite binary string D, an "explanation" of the data is a two-part message or binary string encoding the data in a particular format. The first part of the message (the "assertion") states, in some code, an hypothesis or theory about the source of the data. That is, it asserts some general proposition about the data source. The second part (the "detail") states those aspects of the data which cannot be deduced from this assertion and prior knowledge. Section 1.5 suggested that out of all possible theories which might be advanced about the data, the best inference is that theory which leads to the shortest explanation. The measures of information introduced in Sections 2.1 and 2.2 now allow the length of such an explanation to be defined and calculated.

2.3.1 The Second Part of an Explanation

As noted in Sections 1.4 and 1.5, the "theory" asserted of some data is to be read as expressing a relationship or pattern which is expected to hold approximately for data as observed or measured. That is, even if the theory is conceived as an absolute universal and exact relationship (e.g., force = mass × acceleration), the practical application of the theory to real-world data must allow for imprecision, measurement error and (perhaps) outright mistake. Thus, the practical use of a theory requires that it be regarded as expressing an approximate relationship. Such an interpretation of a theory θ, when applied to a body of data x, can well be described by a probability distribution $\Pr(x) = f(x|\theta)$. (Here we assume pro tem that x is discrete.)

That is, $f(x|\theta)$ tells us that if we believe the theory θ, certain values for the data are unsurprising (high probability) and certain other values, if found, should be regarded as very surprising or even flatly unbelievable (probability low or zero).

For example, suppose the data comprised triples (z_i, m_i, a_i) being the forces, masses and accelerations measured in a series of experiments on different objects, all measured to about 1% accuracy. Then the theory "force = mass × acceleration", when applied to such data, really means that in each experiment, we expect the measurement for z_i to be within about 1.7% of the product $m_i a_i$. Of the possible values for z_i (recalling that the value will have been measured and recorded with only a limited number of binary digits), the value closest to $m_i a_i$ will be regarded as the most probable, if we believe the theory, but other values close to $m_i a_i$ will not be regarded as improbable, and even values differing from $m_i a_i$ by 10% or more not impossible. Thus, given two measured values m_i, a_i, the theory is interpreted as implying a probability distribution over the possible values of z_i, having most of the probability concentrated on values close to $m_i a_i$.

Note that this distribution, which might be written as

$$\Pr(z_i) = f(z_i | \theta, m_i, a_i)$$

is tacitly conditional not only on the theory θ and the values m_i and a_i, but also on some "background knowledge" or (in the language of Section 1.5.2) "prior premises". In this case, the background knowledge at least includes knowledge that each data triple comprises a force, a mass and an acceleration, in that order, and that each is subject to about 1% measurement error. Such background knowledge will normally be available with all data sets for which explanations are attempted, and will have some effect on the probability distribution $f(x|\theta)$ implied by theory θ. However, we will not explicitly include it in our notation for the distribution. Instead of writing $f(x|\theta, \text{prior premises})$, we write simply $f(x|\theta)$, since in considering different theories which might be inferred from the data, the same prior premises will obtain for all theories and all possible data values.

Returning to the example, note that for a triple (z_i, m_i, a_i), the distribution $f(z_i|\theta, m_i, a_i)$ is not the full expression for the probability of the data triple $x_i = (z_i, m_i, a_i)$. The distribution $f(z_i|\theta, m_i, a_i)$ encapsulates what the theory θ says about the relationship among z_i, m_i and a_i but the theory implies nothing about the actual masses or accelerations which might be used in the experiments. The full expression for the probability of a data triple assuming theory θ can be written as

$$
\begin{aligned}
\Pr(x_i|\theta) &= \Pr(z_i, m_i, a_i|\theta) \\
&= \Pr(z_i|m_i, a_i, \theta) \Pr(m_i) \Pr(a_i) \\
&= f(z_i|\theta, m_i, a_i) g_m(m_i) g_a(a_i)
\end{aligned}
$$

Here, we are assuming that background knowledge suggests that different masses will occur in the experiments with probability conforming to the distribution $g_m()$, and different accelerations will occur with probability distribution $g_a()$, independently of the masses. For instance, $g_m()$ and $g_a()$ might both be uniform distributions bounded by the largest masses and accelerations our instruments are capable of handling. Since the theory θ (force = mass × acceleration) implies nothing about the experimental selection of masses and accelerations, neither distribution is conditioned on θ. We have also assumed that each triple in the data set is independent: a new mass and acceleration was chosen in each case.

The conclusion to be drawn from the above example is that the probability of a data set x assuming some theory θ, although written simply as the distribution $f(x|\theta)$, will in most cases be conditioned on a large amount of tacitly assumed background knowledge to do with the selection of data, accuracy of measurement and so on. This dependence need not often be made explicit, since we will be concerned with how the distribution changes with differing choices of θ, and not on how it might change with a different choice of experimental protocol or measuring instruments.

With these preliminaries, it is now possible to define the length of the second part of an explanation, which encodes the data assuming the truth of some theory θ. From Section 2.1.2, if θ is assumed to be true, the length of a binary string encoding data x using an optimal code is $\log_2 f(x|\theta)$ binary digits.

If the data are not surprising, given θ and the background knowledge, the second part will be short. If the data are very surprising (assuming θ) the second part will be long. Here, "short" should not be taken to mean very short. Much of the given data may concern matters which θ does not attempt to "explain", and this unexplained detail may require thousands of binary digits to encode. If the (force, mass, acceleration) data set comprised 1000 triples, the encoded values of mass and acceleration might alone require several thousand binary digits. Since θ implies nothing about the distribution of these values, their encoding cannot be shortened by use of the theory. The coding of the forces z_i might also have taken some 7000 digits in the original data string (1000 values $\times$ 7 digits to give each value to a precision of 2^{-7} or $1/128 \approx 1\%$). However, by assuming the theory θ, the second part can now in effect encode a force z_i by encoding just its small difference from the product $m_i a_i$. If the theory that "force = mass $\times$ acceleration" is correct (within the 1% measurement error) then the actual distribution of z_i given m_i and a_i will have most of its probability concentrated on 3 or 4 values close to $m_i a_i$, and only two or three binary digits will be needed to encode z_i. Thus, assumption of θ may shorten the encoding of the data by four or five thousand digits, but this is still only about a third of the original data string length.

2.3.2 The First Part of an Explanation

The first part of an explanation specifies, in some code, the theory θ being asserted of the data. The coding of this part, and the calculation of its length, require careful attention. First, it is usually the case, at least in the limited contexts in which automatic inductive inference is currently feasible, that the set Θ of possible theories to be entertained is heavily restricted. Normally, a good explanation is sought by exploring only a single family of possible theories, all of which have similar structure and mathematical form. The family may be as simple as the family of Normal distributions with unknown mean and variance. It may be rather more complex, such as the family of all finite-state automata or the set of all binary trees, but even in these cases, a single kind of structure is the basis of all members of the family. In discussing the first part of an explanation message, we will assume that these restrictions on the kind of theory which can be asserted are well-known to the receiver of the message, and are accepted as "background premises" which need not be detailed in the explanation, and which will not be modified by the explanation.

For example, suppose that some data x were obtained, and an explanation of x were sought within a restricted set Θ of possible theories comprising just ten theories $\theta_1, \ldots, \theta_{10}$. Here, x might be data collected in investigating an Agatha Christie murder mystery, and the ten theories might correspond to ten suspects who might possibly be the murderer. Assume that "background premises" rule out all other possibilities. An explanation message would then have a first part simply identifying one of the "theories" or suspects, say, $\hat{\theta}$, and a second part encoding x in a code which would be optimal were $\hat{\theta}$ indeed correct. (The length of the second part would be $- \log \Pr(x|\hat{\theta})$.)

In this simple case, a optimal coding of the first part, naming $\hat{\theta}$, need only encode one of ten possibilities, the possible theories $\theta_1, \ldots, \theta_{10}$. An optimal code (in the sense of least expected string length) would, as described in Section 2.1.2, encode theory θ_i using a string of length $- \log \Pr(\theta_i)$. Note that the first part of the explanation naming the inferred theory $\hat{\theta}$ must be decoded by the receiver of the message *before* the receiver knows the data x. The receiver can only discover x by decoding the second part of the explanation, but since the second part uses a code based on the asserted theory $\hat{\theta}$, it cannot be decoded until the receiver knows $\hat{\theta}$, i.e., until after the receiver has decoded the first part. Thus, the form "$\Pr(\theta_i)$" in the expression for the length of the first part means the probability which the receiver places on theory θ_i *before* knowing the data x. This probability is just the prior probability of θ_i occurring in a Bayesian account of the situation (Section 1.13). It is a measure of how probable the theory is thought to be based on considerations other than the present data x. Since "$\Pr(\theta_i)$" can be identified with the Bayesian prior probability of theory θ_i, we will use the notation $h(\theta_i)$ for it, and hence obtain length of first part $= - \log h(\hat{\theta})$ where $\hat{\theta}$ is the theory identified by the first part. To expand on the "murder mystery" analogy, the data x might be a collection of facts obtained by the detective such as:

X1: Edith says she found the front door unlocked at 11.28 pm.
X2: The victim's will was not found in his desk.
X3: The victim was shot no later than midnight.
X4: Thomas was found drunk and unconscious in the kitchen at 7 am
etc.

These are the data to be explained. Background premises, not requiring explanation and not encoded in the message, but assumed known to the receiver might include:

B1: Edith is the victim's mother.
B2: Besides the victim, only Edith, Thomas, ..., etc. were in the house.
B3: Thomas is the butler.

B4: Mabel is the victim's widow.

Background information, together with previous knowledge, might suggest prior probabilities for the theories "Edith did it", "Thomas did it", etc. incorporating such considerations as:

$h(\theta_1 = $ Edith did it): Low. Edith highly reputable person, and mothers rarely murder sons.

$h(\theta_2 = $ Mabel did it): Higher, because a high fraction of murders are committed by spouses, and Mabel known to dislike her husband.

$h(\theta_3 = $ Thomas did it): Very high, by convention.

etc.

These priors depend on background B but not on data X. The conditional probabilities used in encoding the second part of the message are probabilities of the form $\Pr(x|\theta)$ and represent the probability that the data would be as it appears, on the assumption of a particular theory. Factors in these probabilities might be probabilities such as

Pr (Mabel would remove the will if she were the murderer)

Pr (Thomas would get drunk if Edith were the murderer)

etc.

These probabilities are based on our understanding of how the people might have behaved on the assumption that a particular person was the murderer. They are probabilities of getting the data observed, assuming all background premises and a theory premise.

2.3.3 Theory Description Codes as Priors

In the simple case above, where the set of possible theories is discrete and small, it is easy to imagine how background knowledge could lead to an *a priori* probability for, or willingness to believe in, each theory. In more complex situations, there might be no obvious, intuitive way of assigning prior probabilities. Suppose for instance that each possible theory could be represented by a different, finite, binary tree with every non-leaf node having two child nodes, and suppose that each such tree represented a different possible theory. It could well be that with such a complex family of theories, prior experience could give no guidance as to the prior probabilities of the possible theories, except perhaps for a belief that the tree was not likely to be huge. In such a case it might be decided, *faut de mieux*, to encode the first part of an explanation using some code chosen simply on the basis that it had the prefix property, and its strings could be mapped one-to-one onto the set of binary trees by a simple algorithm. That is, the code might be chosen on the basis that its strings could be easily interpreted as straightforward, non-redundant descriptions of binary trees. A possible candidate code is the "binary tree code" of Section 2.1.14. A string of this code gives a direct representation of the structure of a tree, and is coded and decoded by trivial algorithms. The code length for a tree of n nodes is n binary digits. Hence, this code would be optimal if the prior probability distribution over the set of trees gave probability 2^{-n} to a tree of n nodes.

More generally, the adoption in the first part of an explanation of a particular code for the set of possible theories can be regarded as the tacit acceptance of a particular prior probability distribution $h(\theta)$, where for every theory θ, $h(\theta) = 2^{-(\text{length of code for } \theta)}$. Provided the code is non-redundant, i.e., gives only one way of representing each theory, the implied prior is normalized. In practice, a small amount of redundancy in the code may be acceptable, in which case the implied prior is sub-normalized:

$$\sum_{\theta \in \Theta} h(\theta) < 1 \text{ where } \Theta \text{ is the set of possible theories.}$$

The general point being made here is that even when prior knowledge does not lead by any obvious path to a prior probability distribution over the set of possible theories, the choice of a prior may be re-interpreted as the choice of a code for describing theories. If a prefix code can be devised which is non-redundant, and which prior knowledge suggests to be a sensible and not inefficient way of describing the kind of theory likely to be useful in an explanation, then that code can well be adopted for the first part of explanations. In constructing such a code, some thought should be given to the behaviour of the "prior" which is implied by the lengths of the code strings for different theories. If the code turns out to use longer strings for the theories thought a priori to be less plausible, its implied prior may well be an acceptable encapsulation of vague prior beliefs. If however the code assigns long strings to some plausible theories and short strings to implausible ones, it does violence to prior beliefs and should be modified.

In later chapters describing some applications of Minimum Message Length inference, there are several examples of coding schemes for theories which have been constructed as fairly direct representations of the theories, and which lead to intuitively acceptable implied priors over quite complex theory sets.

2.3.4 Universal Codes in Theory Descriptions

In constructing a descriptive code for a complex set of theories, one may encounter a need to devise a code for naming a member of a potentially infinite set. In the example of the preceding section, if the set of possible theories can be regarded as the set of finite binary trees, there may be no obvious upper limit on the number of nodes in the tree, so the adopted code should impose no such limit. Similarly, if the data is a time sequence of values of some random variable observed at regular intervals, one might adopt the set of Markov processes as the set of possible theories, but there may be no obvious prior grounds for limiting the order of the process. If the investigator is reluctant to impose an arbitrary prior on such an infinite set, there is a case for using an appropriate universal code (Section 2.1.16) since such a code guarantees that for any finite-entropy prior over the set, the expected code length will at least be finite. For instance, whatever finite-entropy prior truly represents our prior knowledge about the set of binary-tree theories, the use of the code of Section 2.1.14 at least guarantees that we can expect the length of the first part of an explanation to be finite.

There is a weak objection to the use of universal codes in encoding theories, and in fact to the use of any codes for infinite sets. Given any finite body of data represented as a binary string, we are interested only in explanations which lead to a shorter encoding of the data. Thus, we will never infer a theory whose specification, as the first part of an explanation, is itself longer than the original data. This fact sets an upper limit to the number of binary digits within which the inferred theory can be stated, namely the number of binary digits in the original representation of the data. The number of different theories which can be coded in strings no longer than this upper limit is of course finite, and so in principle the set of theories which can possibly be entertained given some finite data is itself finite. It follows that codes for infinite theory sets are never strictly optimal in framing explanations for finite bodies of data.

The above objection, while valid, is only weak. A universal code for "binary tree" theories allows the representation of trees with millions of nodes, and so cannot be ideal if the limited volume of data available implies that no tree with more than 1000 nodes will be used. However, redesigning the code to eliminate strings for trees with more than 100 nodes, while keeping the relative probabilities of smaller trees unchanged, would make only a small difference to the code lengths for the remaining trees, of order 0.12 of a binary digit. In practice, universal codes can provide simple and convenient codes for integers, trees and other structures, and in some cases their implied probability distributions are not unreasonable reflections of a vague prior preference for the "simpler" members of the set.

2.3.5 Relation to Bayesian Inference

From Section 2.3.2, the length of the first part of an explanation for data x, asserting an inferred theory $\hat{\theta}$ selected from a discrete set of possible theories Θ, is $-\log h(\hat{\theta})$, where $h(\hat{\theta})$ is the Bayesian prior probability of theory $\hat{\theta}$. From Section 2.3.1, the length of the second part, which encodes x using a code which would be optimal if $\hat{\theta}$ were correct, is $-\log f(x|\hat{\theta})$, where $f(x|\hat{\theta})$ is the probability of obtaining data x given that $\hat{\theta}$ is correct. The total explanation length is thus

$$-\log h(\hat{\theta}) - \log f(x|\hat{\theta}) = -\log(h(\hat{\theta})f(x|\hat{\theta}))$$

But, from Bayes' theorem, the posterior probability of $\hat{\theta}$ given data x is

$$\Pr(\hat{\theta}|x) = \frac{h(\hat{\theta})f(x|\hat{\theta})}{\Pr(x)}$$

For given data x, $\Pr(x)$ is constant. Hence, choosing $\hat{\theta}$ to minimize the explanation length is equivalent to choosing $\hat{\theta}$ to maximize $\Pr(\hat{\theta}|x)$, i.e., choosing the theory of highest posterior probability.

If this were the only result of the Minimum Message Length approach, it would represent no real advance over simple Bayesian induction. About all that could be claimed for the MML approach would be that it might assist in the construction of prior probability distributions, by establishing a correspondence between the language or code used to describe a theory and the prior probability of that theory, as in Section 2.3.3. However, the exact equivalence between MML and simple Bayesian induction is apparent only when the set Θ of possible theories is discrete. When the set is or contains a continuum, e.g., when the theories have unknown real-valued parameters, the straightforward equivalence breaks down.

As described in Section 1.13, it is then not possible to assign non-zero prior probabilities to all theories in the set, and $h(\theta)$ becomes a prior probability density rather than a prior probability. Similarly, by direct use of Bayes' theorem, we can no longer obtain posterior probabilities for theories, but only a posterior probability density

$$p(\theta|x) = \frac{h(\theta)f(x|\theta)}{\Pr(x)}$$

As discussed in Section 1.13, it is then not clear that choosing $\hat{\theta}$ to maximize the posterior density $p(\hat{\theta}|x)$ is a sensible or acceptable inductive procedure, since the result depends on exactly how the continuum Θ has been parameterized. The mode of the posterior density is not invariant under nonlinear transformations of the parameters used to specify θ.

As will be fully discussed in later chapters, MML overcomes this problem. Essentially, minimizing the length of the explanation message requires that the first part may specify real-valued parameters to only a limited precision, e.g., to only a limited number of binary or decimal places. The more severely the stated parameter values are rounded off, the fewer binary digits are needed to state the values, thus shortening the first part of the explanation. However, as the stated parameter values are more severely rounded off, they will deviate more and more from the values which would minimize the length of the second part of the message, so the length of the second part will be expected to increase. Minimization of the message length involves a balance between these two effects, and results in the parameters of theories being rounded off to a finite precision. It will be shown that the best precision gives round off errors roughly equal to the expected estimation errors, i.e., to the errors in estimating the parameters likely to arise from random sampling and noise effects in the data.

The fact that an MML explanation message states estimated parameters to limited precision, using only a limited number of binary digits, means that it becomes conceptually possible to attach a non-zero prior probability to the stated values. Again roughly, the prior probability of an estimate value stated as 3.012, using only three decimal places, is given by 0.001 times the prior density $h(3.012)$. It approximates the total prior probability contained in the interval of values 3.0115 to 3.0125, as all values within this interval round to 3.012.

In effect, the MML approach replaces a continuum of possible theories or parameter values by a discrete subset of values, and assigns a non-zero prior probability to each discrete theory or value in the subset. Since each theory now has a non-zero prior probability, a posterior probability can be defined for it, given the data, and the explanation length minimized by choosing the discrete theory or parameter value having the highest posterior probability. Thus, MML reduces the continuum-of-theories problem to a discrete-theories problem, allowing simple Bayesian induction to proceed. Although it is not obvious from this brief account, the resulting MML inferences are invariant with respect to monotonic non-linear transformations of the parameters, as will be shown in Chapter 3.

The non-zero "prior probabilities" which MML gives to a discrete subset of a continuum of theories are clearly not identical to any prior probability appearing in the initial concept of the continuum of possible theories. However, it appears to be useful and legitimate to treat them as prior probabilities, and similarly to treat the resulting posterior probabilities as genuinely indicating the relative merits of competing explanations of the data. That is, if two competing explanations of the same data using two theories θ_1 and θ_2 have lengths l_1 and l_2 binary digits respectively, it is still possible to regard the difference in lengths as indicating the log posterior odds ratio between the theories:

$$l_1 - l_2 = \log_2 \frac{\Pr(\theta_2|x)}{\Pr(\theta_1|x)}$$

$$\frac{\Pr(\theta_1|x)}{\Pr(\theta_2|x)} = 2^{(l_2-l_1)}$$

2.3.6 Explanations and Algorithmic Complexity

Section 2.3.5 has shown a close relation between conventional Bayesian induction and MML induction based on minimizing the length of an explanation. In that section, the treatment of message length was based on Shannon's theory of information. This section discusses the consequences of basing message lengths on the theory of Algorithmic Complexity. As shown in Section 2.2.4, the definition of the Algorithmic Complexity (AC) of a string with respect to a Universal Turing Machine T can be related to Shannon's theory by regarding T as defining a probability distribution over binary strings $P(S)$ such that

$$P_T(S) = 2^{-AC(S)} \quad \text{for all strings } S$$

The fact that $P_T(S)$ as so defined is sub-normalized, i.e., that $\sum_S P_T(S) < 1$, is of no particular concern. This relation between AC and Shannon information suggests that Minimum Message Length induction can be based on AC measures of message length instead of Shannon measures, and indeed this is so.

Given some UTM T and a body of data x represented by a non-null finite binary string D, we will define an "explanation" of D with respect to T as a shorter input I such that I encodes D. That is, when given input I and started in state 1, T reads I and outputs D, then tries to read more input (the formal conditions were stated in Section 2.2.2). As with the Shannon-based treatment, we impose a further condition on the coded message I, namely that it has a two-part format (Sections 1.5.2, 2.3 and 2.3.5). In the Shannon-based treatment of Section 2.3.5, the first part is an (encoded) statement of a theory or estimate $\hat{\theta}$ about the data, and the second part an encoded representation of x using a code which assumes $\hat{\theta}$. However, given a UTM T, a binary representation D of x, and an input string I which causes T to output D, it is not clear how or whether we can interpret I as stating some theory about x and then using that theory to encode D more briefly. The "meaning" of an input string to a UTM can be quite obscure, making it very difficult for a human to determine what part of I, if any, constitutes a "statement of theory". We therefore impose some formal conditions on I to ensure that it can properly be regarded as an explanation. Stating these conditions will be facilitated by some further definitions.

- For any TM T and any input I accepted by T using the second stopping rule (2.2.2), define $O(T, I)$ as the string output by T when given input I. If T does not accept I, $O(T, I)$ is undefined.
- Two TMs T_1 and T_2 are *equivalent* iff, for all inputs I, $O(T_1, I) = O(T_2, I)$. Equivalence is written as $T_1 \equiv T_2$.
- For any TM T and any I accepted by T, define $N(T, I)$ as denoting any one of a set of equivalent TMs such that for all strings J, $O(T, I.J) = O(T, I).O(N(T, I), J)$ where "." denotes concatenation. That is, $N(T, I)$ is a TM which behaves just as T does after T has accepted I. When T accepts I, it produces output $O(T, I)$ and thereafter behaves as if it were the TM $N(T, I)$. We may say that I, when input to T, not only causes output $O(T, I)$ but also programs T to imitate a different TM $N(T, I)$. Note that even if T is universal, $N(T, I)$ may not be, and if T is not universal, $N(T, I)$ is not.

We now define an *explanation* of data string D with respect to a UTM T. A string I is an explanation of D iff

C1: $|I| < |D|$
C2: $O(T, I) = D$

and I has the form $A.B$, where

C3: neither A nor B is null,
C4: $O(T, A) = \Lambda$ (The null string),
C5: $N(T, A) \equiv N(T, I)$, and
C6: A has no proper prefix A_1 such that $A_1.B$ is an explanation of D.

Informally, these conditions require that

- A produces no output, but programs T to behave like $N(T, A)$; B, when input to $N(T, A)$, is accepted and gives output D, but leaves the TM $N(T, A)$ unchanged. That is, $N(T, A)$ in no sense remembers B.
- The division of an explanation I into parts A and B is unique. Suppose it were not. Then I could be divided into three non-null segments. $I = X.Y.Z$ with both divisions $\{A_1 = X, B_1 = Y.Z\}$ and $\{A_2 = X.Y, B_2 = Z\}$ satisfying C1–C6.

Using the second division, C5 $\Rightarrow O(T, X.Y.Z.Z) = D.D$.
Using the first division, C5 $\Rightarrow N(T, X.Y.Z) \equiv N(T, X)$.
Hence, $O(T, X.Z) = D$ contradicting C6.

The intention of these conditions on I is to ensure that it comprises a part A identifiable as a theory about the data, and a part B which encodes the data string D using a code based on that theory. Consider the TM equivalent to $N(T, A)$. Let us call it H. When B is input to H, H accepts B and outputs D, and is not changed by so doing. Thus, $O(H, B) = D$ and $O(H, B.B) = D.D$. We can regard H as a mechanism which decodes the string B to produce D.

2.3.7 The Second Part

Since H is a TM, it defines a (possibly sub-normalized) probability distribution over all data strings S

$$P_H(S) = 2^{-AC_H(S)}$$

where $AC_H(S)$ is the length of the shortest input accepted by H and producing output S. Thus, H defines a probabilistic "theory" about the source of the data. If the data D is probable, given this theory, i.e., if $P_H(D)$ is large, then it is possible to find a short second part B such that $O(H, B) = D$, with $|B| = AC_H(D) = -\log_2 P_H(D)$. However, if D is improbable in the probability distribution $P_H()$, the shortest string B causing H to output D will be long. In the extreme case that $P_H(D) = 0$, no such string B will exist.

Since $H = N(T, A)$ defines a probability distribution $P_H()$ over the possible data, it can be regarded as embodying a probabilistic theory $\hat{\theta}$ about the data which can formally be represented by the conditional probability function

$$f(x|\hat{\theta}) = P_H(D)$$

where string D represents data x, and H embodies $\hat{\theta}$.

2.3.8 The First Part

In an explanation $I = A.B$ using a given TM T, it is the first part (A) which determines H. Thus, A may be interpreted as describing the theory $\hat{\theta}$ to T, or, equivalently, A is an encoding of $\hat{\theta}$ which is decoded by T. Given the close relation between Algorithmic Complexity and Shannon information and the logarithmic relation between Shannon information and probability, we can reasonably define the "probability" of the TM H with respect to T as

$$Q_T(H) = 2^{-|A|}$$

where A is the shortest string such that

$$O(T, A) = \Lambda \quad \text{and} \quad N(T, A) \equiv H$$

(Observe that, if $I = A.B$ is the *shortest* explanation of data string D, and if $N(T, A) = H$, then A must be the shortest string such that $N(T, A) \equiv H$, and B must be the shortest string such that $O(H, B) = D$.)

Thus, just as a TM T defines a probability function $P_T(S)$ over strings, so T also defines a probability function $Q_T(H)$ over TMs. But as with $P_T()$, the normalization of $Q_T()$ requires special mention. We have observed (Section 2.2.4) that, using our preferred (second) stopping rule, $P_T()$ defines a probability distribution over a prefix set of strings rather than over the set of all finite strings, and is (sub-)normalized over the prefix set. If we adopt $Q_T(H)$ as defining a "probability" of TM H with respect to TM T, the function $Q_T()$ is certainly not normalized (or sub-normalized) over the set of all TMs. For instance, $Q_T(T) = 1$, since no input is required to make T behave like T. Rather, we must interpret $Q_T()$ in the same way as we might define and interpret a probability function over a set of propositions or assertions which are not necessarily mutually exclusive nor together exhaustive. If $L = \{w_i : i = 1, 2, \ldots\}$ is such a set of propositions, a probability function $P()$ may be defined over subsets of L so that $P(\Lambda) = 1$ where Λ is the empty subset, $P(w_i)$ represents the probability that w_i is true, $P(w_i \bigwedge w_j)$ is the probability that both w_i and w_j are true, and so on. (Here, "$\bigwedge$" means logical conjunction.) Such a probability function is meaningful even although it obviously does not satisfy either

$$\sum_{w_i \in L} P(w_i) = 1 \quad \text{or} \quad \sum_{\text{subsets of } L} P(\text{subset}) = 1$$

Such a probability function may well be appropriate for expressing subjective, or "prior", probabilities which a person may attach to subsets of a set of non-factual general propositions about the real world such as

> "The U.N. will in future intervene in all international conflicts involving more than one million civilians."
> "World food production will match population growth at least to 2100."
> "Unregulated international trade in goods and services will damage developing nations."
> "Nation states will be irrelevant in 100 years."
> "Vegetarian diets lead to less aggressive populations."
> etc.

In a similar way, $Q_T()$ can be viewed as defining a probability function over a set of Turing Machines which, in a sense, are not mutually exclusive. A TM which can execute FORTRAN programs and which has an extensive matrix-manipulation library is certainly different from a TM which can only execute FORTRAN, but in a useful sense the latter is a subset of the former. The latter "knows" the propositions defining FORTRAN, the former "knows" these and also the propositions defining matrix arithmetic.

When we consider explanation messages intended to be decoded by some TM T, the first part of the explanation $I = A.B$ is both a description of the TM $H = N(T, A)$ and an encoding of a "theory" about the data. In conventional Bayesian induction, the set of possible theories is normally required to be a set of mutually exclusive and together exhaustive models, so the prior probability function or density over the set is properly normalized. However, in more general scientific induction, we do not necessarily regard theories as mutually exclusive. Rather, some theories may be regarded as additions to other theories, or completely independent. The theory of electromagnetic phenomena expressed by Maxwell's equations is somewhat dependent on Newton's theory of motion, but is an addition to Newton's theory rather than an alternative. A reasonable 19th-century assessment of prior probabilities would regard the prior probability of (Newton.Maxwell) as smaller than prior (Newton), but would not treat (Newton.Maxwell) as an alternative to (Newton). Rather, it would give priors related by

$$\text{Prior(Newton)} \quad = \quad \text{Prior(Newton.Maxwell)}$$
$$+ \text{Prior(Newton.All other electromagnetic theories)}$$

Further, given Newton's theory of motion, his theory of gravity would be regarded as having a prior probability almost independent of the prior probability of Maxwell's theory, as the two deal with different phenomena. I say "almost" because the occurrence in both the gravitational and electromagnetic theories of an inverse-square law would suggest to most people that believing the truth of one theory would incline one favourably to the other, but on the other hand, the assumption in Maxwell's theory of interactions' being mediated by local effects propagating with finite speed would, if accepted, cast doubt on Newton's model of instantaneous gravitational action at a distance.

The above discussion suggests that, if we regard the length of the first part of an explanation $I = A.B$ as indicating a prior probability

$$2^{-|A|} = Q_T(H)$$

for the "theory" embodied in the TM $H = N(T, A)$, the fact that $Q_T()$ is not normalized should be of no concern. The "theory" embodied in H may be the conjunction of several "theories", each individually expressible in some TM. As theories are not necessarily mutually exclusive, the sum of their (prior) probabilities is meaningless. We can properly regard $Q_T(H)$ as being the prior probability, with respect to T, of the theory embodied by H. $Q_T()$ cannot as it stands be regarded as a probability distribution over all TMs.

2.3.9 An Alternative Construction

The definition of an AC-based explanation given in Section 2.3.6 leads, as shown above (Section 2.3.8), to the conclusion that the first part of the explanation does not necessarily nominate one of a set of mutually exclusive models or theories. Rather, it may encode a set of mutually compatible theories, in effect asserting that all of these theories are true of the data. This interpretation goes beyond the usual Bayesian assumption that the models to be entertained are mutually exclusive, and may be seen as a useful enrichment of the usual framework. However, a relatively minor re-definition of a TM-based explanation message can be made which narrows the difference between the two formalisms. Given a TM T and a data string D, the alternative definition of an explanation string $I = A.B$ is:

C1: $|I| < |D|$
C2': $O(T, I) = $ '0'.D
C3: Neither A nor B is null.
C4': $O(T, A) = $ '0'
C5: $N(T, A) = N(T, I)$

In the above, only C2' and C4' differ from the original conditions of Section 2.3.6, and the original C6 is no longer necessary. The effect of the change is that T, on accepting A, must output a single binary zero and immediately begin to read B. With this change, the set of possible first-parts A clearly form a prefix set, since the output of the initial zero signals the end of A. Hence, the probability function over the set of Turing machines $H : H = N(T, A), O(T, A) = $ '0' defined by $Q_T(H) = 2^{-|A|}$ is (sub-)normalized, and $Q_T(H)$ may be regarded as a conventional prior probability for H, or the "theory" which it embodies.

With this definition, it is still possible, given a sufficiently powerful T, to choose A in ways which result in H "knowing" different combinations of Newton's theory of motion, his theory of gravity, and/or Maxwell's electromagnetic theory. The only difference is that now the first part of the explanation is required explicitly to announce its own completion. In general, if one defines some input string J for a given TM by requiring J to cause some specified output $O(T, J)$ or final state $N(T, J)$, imposing on J the further requirement that it be a member of a prefix set has the effect of requiring some lengthening of the input. Without the additional requirement, J need encode only the specified output string or final TM. With the requirement, J must also in effect encode its own length. Thus, if J_1 satisfies the original definition, and J_2 satisfies also the prefix condition, we in general expect

$$|J_2| \approx |J_1| + \log^* |J_1|$$

where the log* term represents the additional information needed to specify $|J|$. However, in changing the definition of an explanation as outlined above, we may expect little change in the length of the explanation. The change now requires part A to determine its own length, which will in general require part A to be slightly longer than with the original definition of Section 2.3.6. But in that definition, the second part B of the explanation had to convey to the TM that the TM was to make no permanent change of state dependent on B, but rather to decode B and forget it. With the alternative definition of this section, B need not convey this fact to the TM, since the ending of A is already known to the TM. Overall, it is difficult to see that the two different ways of defining an explanation would lead to significantly different explanation lengths or inferred "theories".

2.3.10 Universal Turing Machines as Priors

In the Algorithmic Complexity formalism, we assume a Turing Machine T is to be the receiver of the explanation message. The choice of T is equivalent to the adoption of prior probability distribution $Q_T()$ over the possible inferred models which might be used in the explanation. For most induction problems which can feasibly be automated, the set of possible models entertained is severely limited, usually to a relatively sample family of parameterized data probability functions $\{f(x|\theta) : \theta \in \Theta\}$. In such cases, the Turing Machine which embodies a reasonable prior over Θ will not need to be very powerful. It need only have sufficient power to decode the "theory description code" (Section 2.3.3) used to nominate the inferred theory $\hat{\theta}$, and then to compute the resulting data probability function $f(x|\hat{\theta})$. For this purpose, a finite state machine or stack machine will usually suffice, and when the explanation-decoding process is simulated on a conventional computer, the processing and memory demands are small. Further, the theory description code will usually be quite simple, and be such that the end of the theory description is obvious.

That is, the possible first-parts of explanations will form a prefix set, so the distinction between the alternative explanation definitions of Sections 2.3.6 and 2.3.9 becomes irrelevant. The implied prior probability function $Q_T()$ is usually properly normalized.

Although the above simple situation obtains in most currently feasible applications of MML induction, the possibility of using a Universal Turing Machine (UTM) as the receiver is of some interest for the philosophy of induction.

When a UTM is the receiver, the implied prior probability function over possible theories cannot be normalized. It is well known that there is no algorithm which can always determine whether a given program for a given UTM will ever cause output. Hence, however we design the UTM, there will be input strings which never cause output, and so do not correspond to any meaningful assertion, and we can have no general method for weeding them out.

Although it is not normalized, the "prior probability distribution" over possible assertions or theories about the data which is implied by a UTM has a unique virtue. It allows the assertion of *any* theory giving rise to a computable probabilistic model of the data. This scope arguably includes all theories at present seriously considered in scientific enquiry. That is, it can be argued that the scientific community does not accept any theory which does not lead to probabilistic expectations about what data might be observed, or such that the expectations implied by the theory cannot be computed at least in principle. We admit that such an argument cannot at present be compelling. There are current theories which cannot as yet be expressed as computable probability distributions over the set of possible raw data which an investigator might observe, and yet these theories are widely accepted as meaningful, if not necessarily correct. For example:

> "In recent years the personal language of symbolic forms with which he invests his constructions has been simplified and clarified, his compositions are more controlled, and his colour has become richer, denser and more flexible."
>
> (Bernard Smith, writing about the artist Leonard French, in *Australian Painting 1788-1970*, 2nd edition, OUP Melbourne, 1971, p. 311.)

This passage asserts a general proposition about a change over time in the observable characteristics of a time-series of coloured images. It is intended to be, and probably is, meaningful to anyone with a cursory knowledge of modern art, and would lead the reader to have different expectations about earlier and later images in the series. Most readers would be able to decide, after seeing the images, whether the data conformed to these expectations or not. Thus, we must accept the passage as making an assertion about the data series which implies a kind of pattern or regularity capable of verification (or falsification). However, the science of cognition and the art of computing are still nowhere near being able to express the assertion as a computable probability distribution over a time-series of digitized colour images.

To give another example of more "scientific" character, a theory proposing an evolutionary tree for a group of plant species might purport to explain observed characteristics of their leaves, flowers, seeds, etc. At present, reliable identification and measurement of parts like stamens and anthers from raw photographic data is probably beyond the competence of computerized pattern recognition. Hence, such a theory cannot at present be held to make a computable assertion about photographic data.

As there is yet no compelling evidence to the contrary, we may reasonably believe that theories using terms like "flexible colour" and "anther" may eventually be shown to have computable implications about raw image data. However, at the present stage of computer and software technology, it must be admitted that there are many important and respected inductive theories which can be formally represented as probabilistic models of data only if we accept as "data" the interpretations and inferences of human observers. We cannot restrict the idea of explanations to messages conveying only raw data about visual fields, instrument readings, air pressure waves, etc. The informed human must be accepted as an indispensable instrument for translating such data into terms admitting formal analysis. Thus, in applying our theory of "explanations" to data concerning visual images, sounds and other phenomena at present beyond full automated analysis, we will suppose that the data string to be encoded as an explanation has already been processed by human interpretation. The data string presented for an "explanation" of the grammatical forms of spoken English will not be the sampled pressure-wave values as recorded on a compact disc, but rather the sequence of spoken words written in the usual 26-letter alphabet. The data presented for an "explanation" of the evolutionary relationships among a family of plant species will not be a binary string representing digitized photographs of the various plants, but rather a binary string representing the "characters" and measurements of the plants as determined by a competent botanist. At present, any theory about natural language grammars, evolutionary trees and the like can be represented as a probability distribution or TM decoder only for such pre-processed data expressed in terms whose interrelationships are more or less understood. Although, as humans, we are obviously capable of extracting word sequences from sounds, and formal character-lists of plants from visual images, we do not as yet fully understand how we do this preprocessing, and cannot incorporate an account of the preprocessing into a formal theory.

The above proviso applies equally whether theories are framed in terms of probability distributions or Turing machines or any other way admitting formal analysis. Thus, while the proviso limits the scope of the approach to inductive inference presented here, it seems to us to limit equally the scope of all other approaches known to the author. Henceforth we assume all data to be available in a pre-processed, or "interpreted", form. Thus, we will assume that any scientific theory about the data is expressible as a computable probability distribution over the set of possible data strings.

With this assumption, the use of a UTM as receiver allows an explanation message to assert and use any scientifically meaningful theory whatsoever about the data. Equivalently, we may say that any given UTM implies an (un-normalized) prior over possible theories which gives a non-zero prior probability to every meaningful theory and estimate. Further, the results of computability theory show that no receiver which is less powerful than a UTM can accept all computable probability distributions. Equivalently, no computable prior can assign a non-zero prior probability to every possible theory unless it corresponds to the prior implied by some UTM.

In less abstract terms, if we choose to frame explanations of data in such a way that any meaningful theory can be asserted, then the explanations will be decodable only by receivers which are at least as powerful as a UTM. The code used to assert the inferred theory will necessarily be redundant, i.e., capable of asserting nonsense theories, and will therefore imply an unnormalized prior over the set of meaningful theories. The lack of normalization is unfortunate, but is an inevitable consequence of the fact that it is not in general possible to compute whether a given input string for a given UTM will ever result in output. That is, because of the "halting problem", there is no computable method of excluding "meaningless" theories from the code used to assert theories.

Strictly speaking, the UTM is *too* powerful to serve as a realistic model of the receiver of an explanation. A UTM has an unbounded work-tape, which serves as an unbounded memory capable of storing any finite number of binary digits. In the real world, no computer or human has access to such an unbounded memory, nor has the luxury of the unbounded time needed to use it. Any computation or act of inference performed by any person or computer has been performed using limited memory and time. When bounds are placed on the memory and/or computation time of a TM, it cannot be universal. In fact, such a bounded machine is reduced in principle to a much more limited kind of machine called a *finite-state machine* (FSM). In the theory of computability, a FSM is the simplest and most lowly of computing machines. It can compute only a limited set of functions, called *regular expressions*, and can at best be made to imitate only a small range of other FSMs, all simpler than itself. If the real world only offers us FSMs as possible receivers of explanations, it may seem curious to base an account of induction and explanation on a model using UTMs which do not exist, and which have very different theoretical properties. This objection to our approach must be conceded as valid. If we assume a UTM as the receiver of an explanation, we allow in principle that the explanation may assert and use in encoding the data a theory whose consequences could never be computed by any present human or computer. That is, we allow in principle explanations which cannot be decoded within feasible limits of memory and time. A fully developed account of inductive inference should incorporate such limits. Although there is a growing body of theory, called the theory of Computational Complexity, which characterizes what can and cannot be computed within specified limits of memory and time, we have not attempted here to extend our account to allow for such limits. There are arguments in our defense. Firstly, the limits on available memory and computation effort are (for machines) determined by current techniques, and are improving rapidly. Thus, it would be hard to fix on specific values for the limits. Secondly, Computational Complexity is not as yet able to characterize the resource needs of computations very well. For many computations of interest in inductive inference, it is not known whether the needs grow exponentially or only as a polynomial function with increasing volumes of data. Thus, the boundaries of the "feasible" are not well defined. Thirdly, while any one human or computer has only finite memory and computational power, scientific inference is not really an activity of a single agent. Rather, scientific investigation is an activity carried out by a culture, and the memory and information processing power available to the culture grow as time goes by. In accounting for the history of the inductive inferences formed in such a culture, it is not unreasonable to suppose that the culture has been able to record in its "work-tape" of libraries and human skulls whatever information it has needed to record. That is, the unbounded memory of a UTM, while clearly not available to an individual or single computer, is in some sense available to a lasting culture. Finally, while an individual or computer may be only a FSM in the terms of Computability Theory, a sufficiently large and powerful FSM can imitate a UTM within the limits of its resources. Unless the computational resources required to decode an explanation exceed these limits, a powerful FSM can decode the explanation in the same way as a UTM. In practice, working out how some given data can best be encoded as an explanation appears to be much more difficult than decoding the resulting explanation. Informally, it seems harder to make a good inductive inference than it is to compute its consequences. Our account of inductive inference does not directly concern how inferences are formed, but rather attempts merely to characterize "good" inferences as ones which allow a concise explanation of observed data. If induction is computationally harder than the deduction involved in decoding an explanation, it is reasonable to expect that the receiver of a real-world explanation will need less computational power than the sender who had to make the induction used in the explanation. Hence, we might expect that any explanation framed by a human could be decoded by a human. We do not expect in the real world that a receiver will often encounter explanations whose decoding is beyond the limits of the receiver's computational resources.

Some theories current in science might appear to be counter-examples to the above argument. For instance, it is widely believed that quantum mechanics gives an accurate theory of the electron motions in atoms, and how these interact to bind atoms together as molecules. Yet the theory is computationally so difficult that no exact solutions of its equations have yet been computed for any non-trivial molecule. (Here, we mean by an exact solution not an algebraic solution in closed form, which may well not exist, but a computational algorithm capable of providing a numerical solution to an accuracy comparable with the accuracy of observed data.) But in this case, practical explanations of observed data on, say, the absorption spectra of molecules, do not directly encode the data in terms of quantum mechanics. Rather, the explanations assert and use approximate "theories" derived from quantum mechanics but using grossly simplified terms and variables whose behaviour is computationally simpler than that of the variables actually appearing in the original theory. For example, the explanation of some data concerning lithium atoms, which have three electrons, may describe these electrons as occupying distinct "orbits" around the nucleus of the atom. The notion of an "orbit" or state for an electron derives from exact solutions to the quantum-mechanical equations describing a *single* electron orbiting a nucleus. These solutions show that such a single electron must exist in one of a discrete set of states, or orbits. To assume that the three electrons of a lithium atom will occupy three such states is a distortion and simplification of the theory. Strictly, the theory only allows the definition of complex "states" involving all three electrons: the trio of electrons can exist in one such complex state out of a set of states. Describing such a complex state as being the result of three separate electrons individually occupying three distinct single-electron states amounts to replacing quantum mechanics proper with a simplified but different "theory" which, fortunately, happens to work quite well. The conclusion of this digression is this: there certainly are current theories whose consequences cannot feasibly be computed as yet. However, explanations of data purporting to use these theories will usually be found to encode the data using approximate, derived theories rather than the pure theory. The derived theories are not feasibly deducible from the pure theory: if they were, the explanation could begin with the pure theory. Typically, the derived theory will introduce terms having no exact definition in terms of the pure theory, and its "laws" are not deduced from the pure theory. Rather, the terms and laws of the derived theory are themselves inductive inferences based on observed regularities in the data. They may be suggested or inspired by the pure theory, or may even have been formed before the pure theory, but their acceptance is based on their own explanatory power as much as on their in-principle relationship to the pure theory. For example, the theory of atomic interactions uses terms such as "orbitals" having no strict interpretation in quantum dynamics. The thermodynamics of gasses uses terms such as "collision", "viscosity" and "temperature" having no strict definition in atomic interaction theory. Gas flow dynamics uses terms such as "turbulence" not directly definable in thermodynamics. Meteorology uses "cold front" and "cyclone", terms not exactly definable in flow dynamics. Thus, while we may argue that meteorological phenomena are ultimately explicable in terms of quantum mechanics, no such explanation (as we have defined the term) has ever been given, or is likely in the near future. Assertion of the laws of quantum mechanics is in practical terms useless in attempting a brief encoding of weather data, even if we have reason to believe that encoding based on quantum mechanics should be possible, and would be decodable by a UTM.

To summarize this discussion, the assumption of a UTM as receiver allows an in-principle extension of our account to cover any theory with computable consequences. It offers a formal definition of a prior over all computable theories. However in practical terms, we are unable to use theories in an explanation if the computational effort becomes excessive. The limits on available computing or reasoning resources are more likely to be reached in trying to frame explanations than in decoding them. While the UTM is a useful abstract model, real explanations are unlikely to tax the computing power of modern computers when used as receivers, even though these computers are only FSMs. The limits on what is computationally feasible provide one limit on how far reductionism can be pushed in explanations.

2.3.11 Differences among UTMs

We have noted above that one of the attractions of proposing UTMs as receivers of messages, and basing measures of information on the lengths of messages decodable by a UTM, is that the ability of one UTM to imitate another bounds the differences between information measures based on different UTMs. Because the bound depends only on the UTMs, and not on the content or length of the message, some writers have concluded that the differences among UTMs are essentially negligible. For very long messages, the differences in message length arising from different choices of receiver UTM become a very small fraction of the lengths. However, as we have also noted, the differences in absolute terms are not necessarily small, and can easily be several thousand binary digits. Recall that the difference in length between two explanations corresponds to the difference in the logarithms of their posterior probabilities, so an absolute difference in length of 100 digits corresponds to a huge probability ratio, even for explanations of millions of digits. It is particularly important to note that the differences can never be neglected when we are considering explanation messages.

When we frame an explanation of some data for a given UTM receiver, the explanation message begins with an assertion of a theory inductively inferred from the data. This assertion may be regarded as an interpreter program which programs the receiver to imitate a different TM (also perhaps universal). We will call the new TM the "Educated Turing machine" (ETM) since it now "understands" the asserted theory. The ETM is designed to decode the rest of the explanation message, which conveys the data in a code which is optimal if the asserted theory is true. That is, the ETM expects the data to be coded assuming the truth of the theory.

In comparing the merits of competing theories about some given data, our approach is to take each theory in turn, and use it to frame an explanation of the data. We then prefer the theory which gave the shortest explanation. The same receiver UTM is assumed for all theories. Clearly, the length of the first part of each explanation, which asserts the theory, is an important component of the explanation length, and cannot be neglected. If we were to neglect it, and so assess theories purely on the basis of the lengths of the second part of the corresponding explanations, we would always be able to find a "theory" (i.e., an ETM) such that the given data was an inevitable consequence of the theory, and the length of the second part was zero. That is, we could always find an explanation whose first part transformed the receiver UTM into an ETM designed to output the given data without reading any input. Our preferred inference would then be "the world is such that the data has to be <given data string>". That is, the data would be built into the "theory". It is only the inclusion of the length of the first part of the message, asserting the theory, in the assessment of explanations which prevents such nonsense, and provides a balance between the complexity of the theory and its ability to account for the data. Hence, in comparing explanation lengths, we cannot neglect the first part.

However, when a UTM is used as receiver, the first part is precisely an interpreter which transforms the UTM into the ETM. Its length is the difference in length between an encoding of the data for one UTM (the assumed receiver) and for another (the ETM). If, as shown above, this difference is vital, we must conclude that, at least for explanation messages, the differences in length arising from different choices of receiver are not in general negligible. It is just such a difference which prevents the inference of absurdly complex theories which have all the data built in.

Having shown that differences between receiver UTMs are not in general negligible, we must now ask how should the receiver UTM be chosen. Exactly the same considerations apply as in the specification of a prior in Bayesian inference. The length of the interpreter required to make the receiver UTM imitate the ETM which accepts an optimal data code given some theory may be equated to the negative log prior probability of that theory. That is, the design of the receiver UTM embodies and defines a prior probability distribution over the set of possible theories. We should choose our receiver UTM so that the prior it defines accords well with our prior knowledge. The only real difference between choosing a receiver UTM and choosing a prior in a conventional Bayesian analysis is that the UTM "prior" admits all computable theories, and so is not normalized.

There is, however, a useful cosmetic change in the way we may think about the choice of prior. If we use the conventional Bayesian approach, we will probably think of the different possible theories as parameterized members of different families of data probability distributions, which of course they are. However, this line of thought does not seem to lead in any obvious way to attaching prior probabilities to the different families or their members, except when we can draw on a history of many previous similar data sets, each of which has been somehow identified as coming from a particular family. In this probably rare situation, we might well equate the prior probability of a family to the fraction of previous data sets identified as coming from this family. More generally, it might be difficult to translate our prior knowledge into a prior probability distribution over possible theories.

Regarding the receiver as a UTM gives us another way of thinking about priors which can be helpful. For example, suppose our data concerns the observed vibrations of various points on a building which is subjected to an impact at some point. We might be interested in inferring from the data a theory about the internal structure of the building and the mass and stiffness of its structural members. Assuming the receiver to be a UTM, we realize that our explanation will begin with an inferred description of the structure, and that the UTM will have to compute from this description things such as the natural modes of vibration of the structure, their frequencies, amplitudes and phases at each point in the building, and the rates at which the vibrations decay. Before we even think of prior probabilities over different structural arrangements, it is clear that we expect the UTM to have to deal with vectors, sine waves, exponential decay functions, forces, positions, masses and so on. If the UTM is to embody our prior knowledge of the form the theory is likely to take, it should be a computer already equipped to perform high-precision arithmetic, to calculate trigonometric and exponential functions, to handle matrices and eigenvalue calculations, etc. If it is so equipped, we can now concentrate on designing it to accept structural descriptions in some code matched to the kinds of structure that prior knowledge leads us to expect. The code might encode horizontal and vertical members more briefly than oblique ones, since we expect the former to be more common. It might build in expectations that the structure will fill only a small fraction of the volume of the building, that the members will be more massive low in the building, that slab members will more often be horizontal than vertical, and so on. But if we consider the length of input required to bring an initially naive UTM up to the specifications we want in our receiver, we will probably find that most of the input is concerned with the basic mathematical functions we want the UTM to perform, rather than the expectations we have about specific structural details. That is, we may find that the important parts of our prior (in terms of determining the length of our encoded theory) are not to do with just what kind of building we expect to infer, but rather with the expectation that the explanation will use the kind of mathematics previously found to be applicable to structures in general. In other words, thinking about the receiver as a UTM makes us realize that, in specifying a prior, a very important part is specifying a *language* in which we can hope to describe our inferences concisely. The choice of language for the first part of the explanation can greatly affect its length, and should be made on the basis of our prior experience in stating propositions of the kind we expect to infer from the data. Of course, any further and more specific prior expectations can be incorporated in the UTM design: our expectations about the kind of structure to be found in an office block, in a small domestic residence, or in an aircraft hangar are rather different.

When the set of possible theories is sufficiently restricted, a UTM may be unnecessarily powerful as the receiver of an explanation. Rather, the explanation may be decoded by a non-universal TM or FSM. In such a case, rather than considering the detailed design of the receiver as embodying our prior, it is usually more convenient to think in terms of the code or language which will be used to assert the inferred theory in an explanation. This code must, of course, be decoded or "understood" by the receiver machine, and in fact, specifying the language to be accepted by the receiver is just another way of specifying its machine design. In later chapters we give examples of explanations and discuss the choice of theory-description language in each example. Generally, if the receiver is not universal, it will be possible to specify a non-redundant theory-description language, corresponding to a normalized prior over theories. It appears in practice that thinking about what kind of theory-description language should be efficient, given our prior expectations, is less confusing and more "natural" than attempting to specify the prior directly.

2.3.12 The Origins of Priors Revisited

We have argued in Section 1.15 that conventional Bayesian statistical reasoning gives no satisfactory account of how we might come to have a prior probability distribution over possible theories or estimates. It does show how an original prior is modified by data to give a posterior, and that this posterior is an appropriate prior for the analysis of further data, but fails to give any line of reasoning leading to the original prior. We now argue that, by considering the prior to be inherent in the choice of a receiver UTM, we can reason soundly about the original prior.

We saw that an attempt to choose a conventional Bayesian prior expressing "complete ignorance" founders when the set of possible theories is or includes a continuum. What might be regarded as expressing prior ignorance about one parameterization of the continuum (e.g., a uniform or minimum-entropy density) is not colourless in a different parameterization of the same continuum. However, if we consider the explanation receiver to be characterized by a UTM design rather than by a prior, the picture is different. The TMs we have described are specified by an instruction list, containing four instructions for every internal state of the TM. Hence, the complexity of a TM is monotone increasing with its number of states. An overly simple TM, say one with only one or two states, cannot be universal. There must be, and is, a simplest UTM, or a small set of simplest UTMs. Adoption of such a UTM as receiver can reasonably be regarded as expressing no expectation

about the theory or estimate to be inferred, save that it will be computable
(i.e., imply a computable probability distribution over possible data strings.)
Adoption of any UTM (or TM) with more states seems necessarily to assume
something extra, i.e., to adopt a "less ignorant" prior. We therefore suggest
that the only prior expressing total ignorance is that implied by a simplest
UTM.

This definition of total ignorance avoids the objections raised to maximum-
entropy priors and other uninformative priors over continua of theories. Even
if we choose to regard a theory as a member of a parameterized continuum,
to be identified by one or more real-valued parameters, the theory must be
specified to the receiver UTM by an assertion of finite length. Since the set of
finite binary strings is countable, the set of theories which can ever be used in
the explanation of any data is also countable, i.e., discrete. For the simplest
UTM (or indeed any specified UTM), the theory to be asserted will have a
shortest assertion, and its prior probability (as implied by the UTM) is just

$$2^{-\text{length of this assertion}}$$

Questions of non-linear transformation of parameters simply do not arise.
When we choose to regard a theory as being identified by parameter values,
i.e., co-ordinates in a continuum, the continuum and its parameters are ar-
tifacts of our own devising. For a UTM receiver, every theory which can be
asserted is a discrete entity having a shortest assertion and hence a non-zero
prior probability. Whether we choose to regard two computable theories as
near or distant neighbours in some continuum is of no consequence, and has
no direct effect on their relative prior probabilities as defined by the UTM.

(Note, however, that in an explanation of the kinematic behaviour of many
bodies, the theory may well involve hypothesising a "mass" for each body,
and it to be expected that examination of the UTM "program" asserting
this theory will be found to deal with these quantities using representations
recognizable as masses rather than logs, arctans or cubes of masses. That
is, the "theory" may effectively assert a "natural" parameterization of what
humans would recognize as concepts involved in the theory.)

The definition of the "simplest" UTM which we have offered could be
questioned, since it assumes a particular way of describing a UTM, viz., its
instruction list. Other methods of description are possible, and would lead
to a different scale of simplicity. For instance, real-world machines represent
their current internal state as a pattern of binary values called *state variables*
held in electrical circuits. Our form of description treats this pattern simply as a binary number indexing the entries in the instruction list. However,
in most real computers, at least some of these state variables have specific
meanings which are relatively independent of the values of other state variables. That is, the state variables of the machine are such that the values
they will assume in the next cycle, and the actions taken by the machine,
can be expressed as relatively simple Boolean functions, each involving only

a few of the state variables and perhaps the input and work-tape digits. Machines are often designed in this way because it is cheaper or faster to build circuits to compute these simple Boolean functions than to store and refer to an arbitrary instruction list. If this style of design is followed rather than the instruction-list style, the machine may well require more states to do the same job, but less actual hardware. Hence, it is at least arguable that the complexity of a machine should be judged by the number of Boolean logic circuits needed to build it rather than by its number of states.

We concede the uncertainty in the notion of "simplest UTM" resulting from alternative design styles and the possibility of ties on whatever criterion is used. However, it seems plausible that a UTM accepted as "simplest" by any reasonable criterion will incorporate no significant prior knowledge about any sort of data. Any such machine, and its implied prior, should be acceptable as an expression of prior ignorance. It would, for instance, require an input of many digits simply to give it the ability to perform addition and multiplication of integers.

The model we propose for a wholly uninformative prior avoids the objections raised to the conventional Bayesian proposals. It is primitive in an absolute sense which cannot directly be applied to priors expressed as probability densities or distributions over enumerable sets of theories. An aspect of its primitive nature is that it excludes no computable theory about data strings. In a conventional Bayesian analysis, considerable care may be given to choosing an uninformative prior density for the parameters of a model family, yet the usual restriction of the analysis to a single family of models is equivalent to the assumption of an extremely "informative" prior which eliminates all other potentially applicable families. However, our proposed primitive prior is too primitive to be usable in most explanations, because in fact we have some prior knowledge about virtually all data sets we encounter. We offer the model only to show that, if the measurement of information is based on a UTM, the existence of a simplest UTM provides a basis on which Bayesian reasoning can build without having to resort to leaps of faith or arbitrary choices of parameterization.

2.3.13 The Evolution of Priors

It is of some interest to consider how Bayesian reasoning or inductive inference of the form we propose could lead in the real world from a most-primitive prior to the development of scientific theories as we know them today. We will attempt to sketch such a development based on our explanation model of induction.

Suppose we start with a primitive UTM, say, $UTM0$, and obtain some data. Since $UTM0$ requires a long string to make it imitate an ETM of any significant complexity, it is likely that the shortest explanation of the data will assert only a simple inferred theory about it. Indeed it is possible that no explanation will be possible, i.e., that no input to $UTM0$ which is shorter than the data string can cause $UTM0$ to output the data string. However, as more data accumulates, an explanation will eventually become possible if the data exhibits any computable pattern. Suppose this first body of explicable data $D1$ has a shortest explanation comprising an interpreter for an "educated" TM, $ETM1$ say, and a second part or "detail" encoding $D1$ as an input to $ETM1$. In other words, the detail encodes $D1$ using a code which would be optimal were the theory embodied in $ETM1$ true. If we like to translate into Bayesian terms, the length of the assertion (the interpreter) is the negative log of the prior probability of the theory in the prior distribution defined by $UTM0$. The length of the detail is the negative log of the probability of the data, given the asserted theory. The total explanation length is the negative log of the joint probability of theory and data, as shown in Figure 2.14.

$$
\begin{aligned}
\text{length of assertion} \quad &= \quad -\log \Pr(\text{Theory } ETM1) \\
\text{length of detail} \quad &= \quad -\log \Pr(\text{Data } D1 | ETM1) \\
\text{length of explanation} \quad &= \quad -\log \Pr(ETM1)\Pr(D1|ETM1) \\
&= \quad -\log \Pr(ETM1, D1) \\
&= \quad -\log \Pr(D1) - \log \Pr(ETM1|D1)
\end{aligned}
$$

Fig. 2.14. Length of explanation.

Since $\Pr(D1)$ does not depend on the theory, choosing $ETM1$ to minimize the explanation length is equivalent to choosing the theory of highest posterior probability given data $D1$.

In a purely Bayesian inference, we would at least in principle carry forward the entire posterior distribution $\Pr(\text{Theory}|D1)$ into the analysis of further data. We suggest that a more realistic model of scientific enquiry is to suppose that what is carried forward is not the whole of the posterior, but rather the Educated Turing Machine $ETM1$ which is effectively built by reading the assertion of, or interpreter for, $ETM1$ into the primitive $UTM0$. This suggestion requires that we impose a new condition on $ETM1$, namely that it be universal. The condition is not onerous, and need not involve any significant increase in the length of the interpreter for $ETM1$. In its simplest implementation it requires only that, after reading the detail and outputting $D1$, the interpreter for $ETM1$ should go to the initial internal state of the primitive $UTM0$, but leaving a copy of itself on the $UTM0$ work-tape. A more sophisticated implementation would design $ETM1$ to recognize a special bit sequence on the input tape, different from any sequence used for encoding data in the detail, and signalling that the machine should accept any following input digits as defining a new interpreter or modifying the $ETM1$ interpreter. Reserving such an *escape sequence* in the code decoded by $ETM1$ necessarily increases slightly the length of the detail encoding a data string, since the detail must avoid this sequence. However, by using a sufficiently long escape sequence, the increase in detail length can be made as small as we please.

We suggest, then, that when the best explanation of data $D1$ is found, using theory $ETM1$, the receiver Turing machine is left as a UTM, but is now $ETM1$ rather than the primitive $UTM0$. New data $D2$ will be analysed using $ETM1$ as the receiver rather than $UTM0$. Thus, rather than carrying forward a posterior derived from data $D1$, in our suggested model we carry forward only the theory inferred from $D1$. In analysing new data $D2$, we will seek the shortest explanation which, when input to $ETM1$, will cause it to output $D2$. If $D2$ comes from exactly the same source as $D1$, but is an independent and less voluminous sampling from that source, we may well find that the shortest explanation of $D2$ has no assertion part, but consists simply of a detail using the code expected by $ETM1$. That is, $D2$ gives no reason to change or elaborate the theory $ETM1$ already inferred from $D1$. More generally, $D2$ may have a shortest explanation whose assertion part slightly modifies the machine $ETM1$, e.g., by refining some parameter estimate embodied in $ETM1$. If $D2$ is much more voluminous than $D1$, and is drawn from a wider and/or different range of phenomena, the explanation of $D2$ may assert a quite new theory $ETM2$, which however uses some concepts and laws already embodied in $ETM1$ as subroutines. In this case, while $ETM2$ may be a new theory, its assertion to $ETM1$ may be much shorter than the assertion required to program $UTM0$ to imitate $ETM2$, since $ETM1$ already contains subroutines useful in imitating $ETM2$. Note that this economy in the description of $ETM2$ may be found even when the data $D2$ concerns phenomena totally different from the source of $D1$. For instance, suppose $D1$ was an extensive list of (mass, force, acceleration) triples observed in experiments applying forces to particles. No matter how the data is presented in $D1$, it seems inevitable that the inferred theory $ETM1$ would involve some form of addition and multiplication operations. The detail of the explanation of $D1$ could then condense each triple into the form (mass, acceleration, force $-$ mass $\times$ acceleration) where the third component of each triple, being typically very small, could be encoded briefly. That is, we can confidently expect that the "computer" $ETM1$ would contain routines for addition and multiplication. Suppose then that a second body of data $D2$ is obtained concerning the length, width, soil type, rainfall and harvest yield of a set of paddocks. An explanation of $D2$ might assert some subtle theory about the interaction of soil type and rainfall, but it would surely also relate the yield of a paddock to its area, i.e., length $\times$ width. The presence in $ETM1$ of a routine for multiplication would make the interpreter for $ETM2$ shorter when the receiver is $ETM1$ than when the receiver is $UTM0$, since $ETM1$ would not have to be instructed how to multiply. It is quite possible that $D2$ would be inexplicable to $UTM0$, because the shortest input encoding $D2$ for $UTM0$ was longer than $D2$, yet $D2$ would be explicable to $ETM1$.

It is noteworthy that our model permits data about particle dynamics to modify the "prior" used in analysing crop yields. Such an accumulation of prior knowledge is difficult to account for in the conventional Bayesian framework, but emerges naturally when probability distributions are modelled by Turing machines and their inputs. In principle, it appears that our model of induction provides for the inductive inference of all mathematics useful in the explanation of real-world data.

The idea that nothing is retained from the analysis of previous data save the *ETM* used in its explanation is over-simplified. The retention and reconsideration of some previous data clearly play a role in scientific enquiry. To consider an extreme situation, if data is acquired in small parcels, the first parcel may have no acceptable explanation for receiver $UTM0$, so consideration of the parcel leaves $UTM0$ unchanged. The same may apply to all subsequent parcels. No theory will ever be formed unless sufficient parcels are retained and encoded in a single explanation. Even when the first body of data $D1$ permits an explanation asserting $ETM1$ to $UTM0$, and a second body $D2$ permits an explanation asserting $ETM2$ to receiver $ETM1$, it may well be that the shortest explanation of the combined data $(D1, D2)$ which can be decoded by $UTM0$ asserts a theory $ETM3$ different from $ETM2$. In that case we should prefer $ETM3$ to $ETM2$, as giving the shorter explanation of the whole data. In a Bayesian analysis, it would make no difference whether we considered the whole data sequentially, first $D1$ then $D2$, or as a single body of data $(D1, D2)$. In the sequential analysis, the complete posterior distribution $\Pr(\text{Theory}|D1)$ would be taken as the prior in analysing $D2$, and no relevant information in $D1$ would be lost. However, in the model we propose, the educated machine $ETM1$ resulting from the analysis of $D1$ does not capture quite all of the relevant information in $D1$. Hence, when it is used as the prior, or receiver machine in analysing $D2$, the final theory $ETM2$ cannot in general be expected to be the best possible given all the data. Moreover, the final theory $ETM2$ may depend on which of the two data sets $D1$ and $D2$ was analysed first. Ideally, a process of scientific enquiry based on our model should not discard data once it has been used to infer a theory, but rather accumulate all data. Periodically, a revision of current theory should be performed to see whether, if all the data is considered at once and encoded for transmission to the original primitive $UTM0$, some theory may be found yielding a shorter explanation than that given by the currently accepted, and incrementally developed, $ETMn$. This ideal would bring our model back into line with the ideal Bayesian analysis, in that no information would be lost, and the order of presentation of the data would not matter.

The actual practice of scientific enquiry seems to fall somewhere between the extremes of, on the one hand, a purely incremental approach where only the current "best" theory is carried over into the explanation of new data, and on the other hand, retention of all data and requiring any new theory to be inferred from the entire data set without reference to previous theory. Some old data is certainly retained and used as evidence to be explained by new theory, especially when the new theory involves discarding a previous theory based on the old data. But much data which has led to scientific theories is never re-examined after the theory has been accepted. A proponent of a new theory of gravitation would feel little need to demonstrate that the raw data used by Kepler could be explained by the new theory (if the data are still available). Rather, old bodies of data are often treated as being fully summarized by the theories inferred from them, with perhaps some anomalies noted. A new theory is then regarded as explaining the old data if it can be shown deductively that any data explicable by the old theory necessarily admits of an equally brief detail when encoded using the new theory. The new theory may also, one hopes, be capable of explaining the data regarded as anomalous under the old theory.

In the physical sciences, the pattern of enquiry seems in the twentieth century to have followed the incremental model rather well. Few of the new concepts and patterns enunciated in theories accepted in the twentieth century seem to have been discarded in later, more successful theories. This assertion is just an impression based on a modest education in the physical sciences, and might well be disputed by those better informed. However, we can at least observe that a great number of concepts and patterns used in 19th century physics are still to be found in present theory. In Turing Machine terms, our present Educated Turing Machine uses many of the important subroutines of its 19th century predecessors.

Lastly, a slightly different form of incremental theory development may, and I think does, occur. Suppose, as above, that some data $D1$ admits of an explanation $I1$ accepted by the primitive $UTM0$, and that $I1$ comprises an assertion part $A1$ followed by a detail part $X1$. Let $ETM1$ denote the UTM $N(UTM0, A1)$, i.e., the Educated Turing Machine which results when $UTM0$ is fed the theory $A1$. More packets of data $D2$, $D3$, $D4$, etc. are collected, and it is found that each admits of explanation by the theory $A1$. That is, detail packets $X2$, $X3$, $X4$, etc. can be found, each shorter than the corresponding data packet, which when fed to $ETM1$ cause it to output the data packets. It may be that none of the additional data packets is sufficient to justify scrapping assertion $A1$ and replacing it by a new theory. That is, none of the new data packets admits of an explanation to $UTM0$, but each admits of a null-assertion explanation to $ETM1$. Now suppose that, although the original data packets are not carried forward into new analysis, the detail packets are. It may then be noted that the concatenation of the details $X2.X3.X4\ldots$ seems to have some regularities. Then this string may admit of a two-part explanation to $ETM1$. That is, there may exist an input string $I2$ which begins with an assertion part $A2$ and which, when input to $ETM1$, causes it to output $X2.X3.X4\ldots$.

Now, if a program for a UTM causes it to read an input, process it and output a results string, the program can easily be modified to make the UTM remember the results string instead of outputting it, and then behave as if it had been given this string as input. Thus, if there exists a string

$I2 = A2.Y2$ which when input to $ETM1$ makes it output $X2.X3.X4\ldots$
there exists a trivially different string $J2$ which when input to $ETM1$ makes
$ETM1$ compute and remember the string $X2.X3.X4\ldots$ then "read it back"
and produce the output string $D2.D3.D4\ldots$. The final situation then is that
a string has been found with the structure

$$I3 = A1.J2 = A1.A2.Y2$$

which when input to $UTM0$ makes it produce the string $D2.D3.D4\ldots$.

What we are suggesting is that, after some theory A1 has been inferred
from data $D1$, and found to explain further data $D2$, $D3$, etc. fairly well, it
may be noticed that the 'details' required by these additional data, when considered in toto, exhibit significant regularities. A further theory A2 is inferred
from these regularities, allowing an acceptable explanation of these details to
be framed for input to a UTM which already uses theory A1. That is, the
original theory A1 is not rejected or even substantially modified. Rather, a
further theory A2 is inferred, which succeeds in more briefly explaining the
new data, *using the concepts and regularities already embodied in A1.*

An example may clarify. The "wave" theory of optics successfully explained numerous properties of light, including refraction, diffraction, colours,
the resolution limits of telescopes, etc. but "details" in these explanations
needed to provide unexplained values for things like the speeds of light in
different media and the dispersion of different colours. Maxwell's electromagnetic theory used the wave-motion theory, but allowed many of the details
unexplained in the original theory to be explained in terms of the dielectric
and magnetic properties of different media. The modern quantum theory uses
all the terms of wave motion and Maxwell, but allows explanation of much
that Maxwell's theory left as unexplained detail (besides accounting for phenomena such as the photo-electric effect which Maxwell's theory could not
account for quantitatively). In this history, later theories did not really su-
percede the earlier ones, which remain valid theories. Rather, they built upon
their predecessors by finding regularities in what for the earlier theory were
unexplained "details".

An alternative (but I think equivalent) account of this form of evolution
arises from regarding a theory as the definition of a new language, or at least
a tightening of the "grammar" of an existing one. Data observed must be
recorded in some "observational" code or language, say, $L0$. When a UTM
is used as the receiver of data, it defines a new language $L1$: the set of
input strings which will it can translate into meaningful $L0$ strings. In the
simplest situation, $L1$ differs little from $L0$, but when the UTM is given an
"explanation" message with a non-trivial assertion part, its reading of the
assertion changes it into a different (educated) UTM which will now accept
a language $L1$ which (if the explanation is acceptable) allows a more concise
representation of data strings actually observed, and which it can translate
back into the original language of observation $L0$. If regularities are found

in the sentences of $L1$ which encode real data, the UTM may be further "educated" by a new assertion to understand a yet more concise language $L2$ which it can now translate first into $L1$ and then into $L0$. The new assertion does not replace the old one. In fact, the new assertion is couched in the language $L1$ defined by the old assertion, and in general would not even be acceptable to the original uneducated UTM.

The above is an account in mechanical terms of what seems to have happened during the evolution of scientific theories. Old theories, if they had explanatory power, are not usually wholly discarded. They introduced new language and concepts in which their "detail" accounts of observations were expressed, and this new language was often used to express new theories. Most modern accounts of the physical world still use terms such as mass, energy, wavelength, density, cause, etc. with pretty much their old meanings, and we would probably be reluctant to scrap our current scientific language in favour of one which made no use of them.

3. Strict Minimum Message Length (SMML)

This chapter begins the development of methods for constructing short explanation messages, and hence obtaining statistical or inductive inferences from data. Actually, the detailed construction of an explanation as a sequence of binary digits will not be developed. Rather, as our interest lies in the inferences to be drawn, we will need only to develop methods for computing the *lengths* of explanations based on different possible inferences. The preferred inference is that giving the shortest explanation message, and the difference in length between two explanations based on different hypotheses can be used as a measure of the relative posterior probabilities of the hypotheses. The actual digit sequences need not be computed.

The discussion will be based on Shannon's theory of information, using the relation

$$\text{information} = -\log(\text{probability})$$

and assuming use of coding techniques such as Huffman and Arithmetic codes which give message lengths closely approximating the ideal

$$\text{length in binary digits} = \text{information in bits} = -\log_2(\text{probability})$$

The Algorithmic Complexity theory of information will not be needed.

The chapter begins by defining the assumptions and functions which are taken as describing the induction problem to be addressed. Essentially, these are the assumptions and functions conventionally used in describing a statistical estimation problem in a Bayesian framework, but the conventional view of parameter spaces is slightly generalized so that our treatment encompasses model selection or "hypothesis testing" as well as estimation in the same framework.

The main result of the chapter is an exact formal characterization of estimator functions which exactly minimize the expected length of explanations, and derivation of certain relations which must be satisfied by the estimators. Such estimators are called "Strict Minimum Message Length" (SMML). Unfortunately, these relations are in general insufficient to allow calculation of the SMML estimator, and in fact the estimator function has been obtained for only a few very simple problems. Some of these simple cases are used as illustrative examples, as despite their simplicity, they give some insight into the general nature of MML estimators.

Later chapters will develop approximations to SMML estimators which can feasibly be calculated for a wide range of problems.

3.1 Problem Definition

Our aim is to construct a brief (ideally the briefest possible) explanation of given data, where an explanation is a message conveying first, a general assertion about the source of the data, and second, the data itself encoded in a code which would be optimal were the assertion true. We now consider what is involved in the construction of such a message.

The assertion inferred from the data, and stated in the first part of the message, will in general imply a probabilistic model for the data, i.e., probability distribution over the set of all data values which might be observed if the assertion were true. Even when the inference embodied in the assertion is not inherently probabilistic, it will still imply such a probability distribution in most cases. For instance, if the data comprises many independent triples, each comprising a voltage, current and resistance observed in an experiment on direct current electricity, the basic inference conveyed in the first part might be voltage = current × resistance, which asserts a deterministic rather than a probabilistic relationship. However, we would not expect the triple of numbers representing the measured voltage, current and resistance exactly to satisfy this relation. Inevitable measurement error, and the finite precision of the numbers representing the measurements, would cause some variation around the ideal relation. Thus, in practice the assertion would have to take the form of a probabilistic relation, e.g., "The measured voltage is a random variate from a Gaussian distribution with mean V = measured current *times* measured resistance and standard deviation $0.001 \times V + 0.01$ volt". Although later we shall extend our discussion to include the possibility of truly deterministic assertions, we shall now assume that the inferred assertion always implies a probability distribution over the set of possible data values. We will write this distribution as the function $f(x|\hat{\theta})$, where x is a data value and $\hat{\theta}$ denotes an inference. Note that the data value x will usually be a set of numbers (or other types of value such as Boolean values). For instance, x could be a set of 100 triples, each triple being a voltage, a current and a resistance expressed in suitable units.

3.1.1 The Set X of Possible Data

Initially, we shall depart from the usual treatment of data in statistics by assuming that the set of possible values which the data x might take is finite. That is, we regard x as a discrete variable, not as continuously variable. (At this point we admit to a confusion of notation: in some contexts we use x to mean a data value, and in other contexts, e.g., in the distribution function

$f(x|\hat{\theta})$, we use x to denote a random variable, rather than a value which might be taken by that variable. The intent of each use of x should be clear enough from the context.)

The assumption of discrete data values may appear restrictive: surely we must be able to deal with data about quantities such as length, temperature and mass, which can take any value in a continuum. However, we are concerned with the construction of a message which conveys observed data; and any observed data presented for analysis will be presented in some form, be it a set of numbers, a picture, a sequence of sounds, etc., which can be represented by or translated into a finite string of symbols from a finite alphabet. Certainly, any observation which can be conveyed from one person to another via telecommunication media is often so represented during its transmission. One may argue perhaps that data presented for analysis by the analyst's own direct sense impressions need not be so representable, but even in this case, it is plausible that the sequence of nerve impulses which carry data from sense organs to the analyst's brain can be represented without loss of information by a sequence of numbers representing, with finite frequency and precision, the time-varying action potentials of the nerve cells involved. Thus, if we assume that any data value must be representable by a finite string in a finite alphabet, we are actually making our model of the inference problem more rather than less realistic.

If every possible data value can be represented as a finite string in a finite alphabet, it can be represented as a finite string in a binary alphabet, i.e., as a finite string of "0s" and "1s", using standard coding techniques. Since the set of finite binary strings is countable, it follows that the set X of all possible data values which may arise in a given inference problem is countable. We could, indeed, argue that X is not only countable but finite. There are experimental procedures which could in principle yield any one of an infinite, countable set of data strings. For instance, we could decide to record as our data the sequence of head or tail outcomes in a coin-tossing experiment, and to continue tossing until we produce either a run of 10 heads or a run of 10 tails. As there is an infinite number of binary strings containing no group of 10 consecutive 0s or 10 consecutive 1s, the set of data strings which could in principle be obtained is infinite. But in the real world, we may be sure that no analyst will ever be given an infinite data string, since transmitting it would take an infinite time, and if none of the really possible data strings is infinite, the set X of all really possible strings must be finite.

While the finiteness of X is assured in the real world, we do not need this property, but will assume countability. Since the set X of possible data values which can arise in any given inference problem is countable, we may index its members. Normally, we will use the index i, and write

$$X = \{x_i : i = 1, 2, 3, \ldots\}$$

3.1.2 The Probabilistic Model of Data

For any inference $\hat{\theta}$ asserted in the first part of the explanation message, there must be, as we have noted, an implied probability distribution function $f(x|\hat{\theta})$.

For given $\hat{\theta}$, $f(x|\hat{\theta})$ must be defined for all $x \in X$ and satisfy $0 \leq f(x|\hat{\theta})$ and $\sum_{x \in X} f(x|\hat{\theta}) = 1$.

We assume that the statement of $\hat{\theta}$ in the first part of the explanation is sufficient to define the corresponding function $f(x|\hat{\theta})$ to the receiver of the message. In the simplest problems of statistical estimation, the mathematical form of $f()$ may be assumed to be given, and already known to the receiver. In such cases, the assertion $\hat{\theta}$ need only give the estimated value(s) of any free parameter(s) of the distribution function. For example, the data may consist of the outcomes of a sequence of 100 tosses of a coin, and it may be accepted without question that these outcomes are independent Bernoulli trials, i.e., that each outcome has probability p of being a head, and probability $(1 - p)$ of being a tail, where p is some unknown probability characteristic of the coin and the tossing apparatus.

All this is accepted as given, is not subject to further inference, and is assumed already known to the receiver of the explanation. Thus, both sender and receiver of the message know that the probability of obtaining a particular data string containing n heads and $(100 - n)$ tails has the form

$$p^n (1 - p)^{(100-n)}$$

What is not known is the value p of the probability of heads. The framer of the explanation must form some estimate of p, assert that estimate in the first part of the explanation, then encode the data sequence using a code based on that estimate. Thus, in this case, the assertion or inference $\hat{\theta}$ is simply a single number $\hat{p}$, the estimate of p, and this is sufficient, given what is already known to the receiver, to define the probability distribution $f(x|\hat{\theta})$ as

$$\hat{p}^n (1 - \hat{p})^{(100-n)}$$

where $\hat{p} = \hat{\theta}$ and n is the number of heads in the sequence x.

Our early development and examples will concentrate on just such simple cases, where $\hat{\theta}$ is no more than a set of estimated parameter values. Later, we will consider more complex problems, where the assertion $\hat{\theta}$ asserts one of a number of possible forms for the function $f()$, as well as any parameters for the chosen form. Ultimately, we consider inferences where $\hat{\theta}$ comprises, in effect, an arbitrary computer program for computing $f()$.

3.1.3 Coding of the Data

The first part of an explanation message asserts some inference or estimate $\hat{\theta}$ which implies a probability distribution $f(x|\hat{\theta})$ over the set X of possible data values. Having chosen $\hat{\theta}$ and constructed the first part, the transmitter of the message can now construct an optimum code for the distribution $f(x|\hat{\theta})$, and hence encode the actual data value as the second part of the explanation. The explanation message is readily decodable by any receiver. Having received and decoded the first part (the coding of which is discussed later), the receiver can compute the distribution function $f(x|\hat{\theta})$ and hence construct the same optimum code over X as was used by the sender. Knowing this code, the receiver can now decode the second part of the explanation, and so recover the data value x.

Since the code used in the second part is chosen to be optimal for the distribution $f(x|\hat{\theta})$, the coding probability for data x is $f(x|\hat{\theta})$, and the length (in nits) of the second part is $-\log f(x|\hat{\theta})$.

This simple result will be used in all our calculations of explanation lengths. Once a probabilistic model has been specified by the inference $\hat{\theta}$, there is no further complication in principle to the construction and use of an optimal code for the data. We now turn to the problem of coding the first part of the message, which asserts $\hat{\theta}$. This problem is not so straightforward, and various approaches to it will be the main concern of this work.

3.1.4 The Set of Possible Inferences

The value $\hat{\theta}$ asserted in the first part of the message specifies a particular probabilistic model $f(x|\hat{\theta})$ for the data. In any one inference problem, the range of models considered possible before the data is seen will usually be quite restricted. In simple estimation problems, it will have been decided *a priori* that the only models to be considered are those within a parameterized family of fixed mathematical form. For instance, it may be accepted *a priori* that the data will be modelled as a set of 100 values drawn independently from a Normal distribution, so the set of possible models is just the family of distributions spanned by some possible range of values for the mean and standard deviation of the Normal. In this case, $\hat{\theta}$ need only specify the esti-mated values of the mean and standard deviation: the sample size (100), the assumption of independence, and the mathematical form of the Normal dis-tribution are not in question, and are assumed already known to the receiver of the explanation.

In an even simpler case, the set of models considered may be discrete, with no free real-valued parameters. For instance, where the data is a coin-toss sequence, knowledge of the circumstances may dictate that the only models to be considered are (a) that the sequence is a Bernoulli sequence with $\Pr(\text{head}) = 1/2$, and (b) that it is a Bernoulli sequence with $\Pr(\text{head}) = 1/3$. In this case, $\hat{\theta}$ need only have two possible values, naming options (a) and (b) respectively. The form of the probability distributions, and their two $\Pr(\text{head})$ values, are not in question and need not be included in the first part of the message.

More complex problems may involve a much more complex set of possible models for the data. For instance, it may be believed that the data, a sequence of numeric values, are independently drawn from a probability distribution which is a mixture of a number of different Normal distributions, but the number, means, standard deviations and relative abundances (or mixing proportions) of these Normals are all unknown. Then $\hat{\theta}$ must specify the number of component Normals inferred from the data, and estimates of the parameter values of each component. The set of possible models is the union of several sets. The set of one-component models is a two-dimensional continuum spanned by the mean and standard deviation of the one Normal. The set of two-component models is a continuum of five dimensions, the means, the SDs and the relative abundance of the two Normals. The set of three-component models is a continuum of eight dimensions, and so on.

Whatever its structure, we will denote the set of possible inferences (or data models) by Θ, and use θ to mean a variable taking values in Θ, and also to denote a specific value in Θ.

3.1.5 Coding the Inference $\hat{\theta}$

The first part of the message encodes an inference or estimate $\hat{\theta}$ which, together with assumptions already known to the receiver, specifies a probabilistic model for the data. If we aim to produce the shortest possible explanation, the code used to encode $\hat{\theta}$ must be chosen with some care. We have seen that, to name one of a countable set of possibilities each having a known probability of occurrence, techniques exist for constructing a code which gives minimal expected message length, and that such codes have the property that the message asserting some possibility has a length of minus the logarithm of the probability of that possibility. However, in attempting to devise an optimum code for the set Θ, two difficulties emerge.

Assume *pro tem* that Θ is discrete. The construction of an optimal code for Θ requires that we know the probability $\Pr(\theta)$ for every $\theta \in \Theta$. However, it is not obvious what the meaning of $\Pr(\theta)$ is, nor how the analyst might come to know this probability if it indeed exists. We have mentioned an example, in which the inference θ was (a form of) Newton's second law of motion, supposedly inferred from measurements of the accelerations of various particles. What sense can we apply to the notion of the "probability" of such an inference? A modern physicist would consider the assertion false (probability approaching zero), as there is abundant evidence that the assertion only approximates the truth when the particles' speeds are small, and that at high speeds, relativistic "laws" are more accurate. More reasonably, consider the situation in Newton's own time, when Newton or a follower might have made the assertion in an explanation of experimental data. The sender of the explanation likely considered the assertion to be very probably true (Prob →1) but, as we have seen, the coding probabilities used in the construction of a code must reflect the expectations of the receiver. What probability might have been assigned to the assertion of Newton's second law by an intelligent person prior to Newton, assuming that person to be already familiar with the terms mass, force and acceleration, and with simple mathematics?

We might argue that such a receiver could conceive of an infinite number of possible relations, e.g., $F = m + a$, $F = ma^{1.03}$, $F = ma$ if motion upwards but $F = \frac{1}{2}ma$ if motion downwards, etc. Given an infinite number of possible relations, why should the receiver assign a non-zero probability to any particular relation? This argument would suggest that the receiver's "probability" would be zero for any arbitrary relation.

In fact the difficulties raised by such arguments are, in this context, only apparent. We are concerned with assigning probabilities to possible inferences only for the purpose of devising a code capable of efficiently encoding the assertion of an inference. We may imagine this code to be negotiated and agreed between the sender and the receiver of the explanation *before* the data is acquired and an inference made from it. It is only necessary that we define a code which, in the light of the rational expectations of the receiver based on knowledge of the circumstances of the data collection and any other relevant prior knowledge, cannot be materially improved. In other words, we turn the question on its head and ask, not what is the probability of every possible inference given the receiver's prior knowledge, but rather what code for inferences will the receiver accept as efficient? We may then, if we wish, regard an inference coded with a string of length l binary digits as having a "prior probability" of 2^{-l}. In practice, the question of how inferences might efficiently be encoded appears to raise fewer philosophical arguments and objections than the question of assigning probabilities to inferences. Our view is that the questions are in fact equivalent, but we accept that treating the "probability" of an assertion as being reflected in the code or language a rational person would use in making the assertion is at odds with some interpretations of "probability". To avoid confusion with situations where prior knowledge allows the assignment of prior probability having a conventional frequentist interpretation, we will use the term "coding probability" for the quantity 2^{-l}, or more generally for the probabilities implied by adoption of a particular code.

In the "Newton's Law" example, for instance, we need only ask, what kind of language or code might Newton's peers have accepted as likely to be efficient for specifying a universal relation among the quantities involved in particle dynamics. It is not unreasonable to suppose that the language of mathematics, in which the operations of addition, multiplication and exponentiation are coded very concisely, reflected quite well the prior expectations of Newton's contemporaries as to what kinds of quantitative relations would prove to be useful in explaining the physical world. That is, we may reasonably suppose the relations concisely represented in the mathematical notation of the time to be those then regarded as having a high probability of appearing in the assertion of inferred "natural laws". Thus, even though the set of conceivable relations among the quantities of dynamics might have been infinite, the language conventionally used in discussing the subject corresponded to a probability distribution over the set placing most probability on relations involving elementary mathematics.

Given a code accepted as efficient for naming one of a discrete set of possible inferences, the probability implied by the length of the string encoding an inference will be termed the "coding probability" of that inference. Alternatively, in situations where we have more conventional statistical evidence, e.g., a history of previous similar situations, for the probability of a particular inference being true, we will regard that probability as the "prior probability" of the inference, and adopt a code for the first part of the explanation which encodes the inference with a string of length

$$- \log \text{ (Prior Probability)}$$

That is, we will use a coding probability equal to the prior probability. An example of the latter situation is the diagnosis of a disease $\hat{\theta}$ given some patient symptom data x. The relative frequency of occurrence of the disease $\hat{\theta}$ in the population may be taken as the prior probability of the disease $\hat{\theta}$, being in fact the probability that a randomly selected member of the population would have the disease. (Here, "population" should be defined as the population of persons presenting for diagnosis, rather than the entire population.)

3.1.6 Prior Probability Density

Often the set of possible inferences considered will be or include a continuum spanned by one or more "parameters". An example is the set of all Normal distributions, which is a continuum with two parameters, the mean and standard deviation. Since such a set is not discrete, it is not possible to associate non-zero prior probabilities with its members. Equivalently, an arbitrary member of the set can be specified only by a message giving its parameter values to infinite precision, which in general would require an infinitely long message.

Instead of a prior probability for each possible inference, we must consider a Prior Probability density, which we will write as $h(\theta)$ ($\theta \in \Theta$). The meaning of $h(\theta)$ is that, for any subset t of Θ, the prior probability that θ lies in t is

$$\int_{\theta \in t} h(\theta) \, d\theta$$

We will use the density $h(\theta)$ in a generalized sense to include problems where the set Θ includes continua of different dimensionality. For instance, the set of models for some data x comprising 100 scalar values might be that the values are independently and randomly drawn from a population which either has a single Normal distribution with mean μ_0, SD σ_0, or is the sum of two equally abundant subpopulations having different Normal distributions with parameters respectively (μ_1, σ_1) and (μ_2, σ_2), the two possibilities being equally likely.

Then Θ is the union of two continua, a two-dimensional continuum Θ for the one-Normal case and a four-dimensional continuum Θ_2 for the two-Normal case. When $\theta \in \Theta_1$, the element $d\theta = d\mu_0 d\sigma_0$; when $\theta \in \Theta_2$, $d\theta = d\mu_1 d\sigma_1 d\mu_2 d\sigma_2$. Since the two cases are equally likely *a priori*, we would have

$$\int_{\theta \in \Theta_1} h(\theta) d\theta = \int_{\theta \in \Theta_2} h(\theta) d\theta = \frac{1}{2}$$

Although formally this generalized definition of $h(\theta)$ allows us to deal with unions of continua in a single expression, in practice it is usually expedient to deal with the various continua in a complex Θ separately.

For the next several sections it will be sufficient to consider Θ to be a single continuum spanned by some vector parameter θ of fixed dimensionality. For discrete sets Θ we regard the assignment of prior probabilities to members of the set as equivalent to choosing an efficient code for the set. The equivalence does not extend directly to a continuum set Θ because neither a non-zero probability nor a finite code string can be determined for arbitrary points in a continuum.

However, if we consider some partition of Θ into a countable set S of regions $\{s_j : j = 1, 2, \ldots\}$, a probability $p_j = \int_{\theta \in s_j} h(\theta) \, d\theta$ can be associated with each region, and, given that the receiver knows the details of the partition, an efficient code for naming regions will encode region s_j with a string of length $-\log p_j$. We may therefore regard as equivalent the choice of a prior probability density $h(\theta)$ and the choice of a family of codes for naming the regions in any agreed partition of Θ. A family of codes is involved since different partitions require different codes.

In the usual practical case for Θ with one dimension and a scalar parameter θ, the equivalence is very simple.

Suppose we wish to encode values of θ as integer multiples of some small unit Δ. (For instance, we may wish to encode values of a temperature to the nearest $0.1°C$, i.e., as integer multiples of a tenth-degree unit.) Then, if we consider θ to have a prior density $h(\theta)$, we should accept as an efficient code one which encodes value θ with a string of length approximately $-\log(h(\theta)\Delta)$. In this context, a string encoding some value $\hat{\theta}$ would be interpreted as asserting that θ lay in the interval $\hat{\theta} - \Delta/2 \leq \theta < \hat{\theta} + \Delta/2$.

A later chapter will discuss the questions involved in selection of a prior distribution or density, especially when there is little prior knowledge about the possible models. For the time being, we take the prior density as given as part of the inference problem specification.

3.1.7 Meaning of the Assertion

An explanation consists of two parts: a string encoding the assertion of a model for the data, and a string encoding the data using a code which would be optimal were the assertion true. Henceforth, we will often use "assertion" to refer to the first string, and "detail" to refer to the second.

If it is agreed *a priori* to restrict the possible models to a continuum Θ, then the detail will encode data x using a code based on some model distribution $f(x|\hat{\theta})$, $\hat{\theta} \in \Theta$, and the assertion will encode the chosen model or parameter $\hat{\theta}$.

Given that an assertion of finite length cannot name an arbitrary $\hat{\theta} \in \Theta$, it is tempting to interpret an assertion which encodes $\hat{\theta}$ to some finite precision Δ (e.g., to one decimal place if $\Delta = 0.1$) as meaning only that the data conforms to some model in the range $\hat{\theta} \pm \Delta/2$. But if we interpret the assertion in this way, as specifying an interval rather than a value for θ, we can no longer suppose that the distribution $f(x|\hat{\theta})$ defines an optimal code for the data values expected if the assertion is true. Given only that $\hat{\theta} - \Delta/2 \leq \theta < \hat{\theta} + \Delta/2$, the probability of obtaining data value x is not $f(x|\hat{\theta})$ but rather

$$\Pr(x|\hat{\theta} \pm \Delta/2) = \frac{\int_{\hat{\theta}-\Delta/2}^{\hat{\theta}+\Delta/2} h(\theta) f(x|\theta) d\theta}{\int_{\hat{\theta}-\Delta/2}^{\hat{\theta}+\Delta/2} h(\theta) d\theta}$$

We will write the numerator as $\Pr(x, \hat{\theta} \pm \Delta/2)$. It is the joint probability that θ lies in $\hat{\theta} \pm \Delta/2$ and the data x has value x. The denominator, $\Pr(\hat{\theta} \pm \Delta/2)$ is the prior probability that $\theta \in \hat{\theta} \pm \Delta/2$. Using this interpretation of the assertion $\hat{\theta}$, the length of the assertion is $-\log \Pr(\hat{\theta} \pm \Delta/2)$, and the length of the detail is $-\log \Pr(x|\hat{\theta} \pm \Delta/2)$, giving a total message length

$$
\begin{aligned}
&- \log \left(\Pr(\hat{\theta} \pm \Delta/2) \Pr(x|\hat{\theta} \pm \Delta/2) \right) \\
=\ &- \log \Pr(x, \hat{\theta} \pm \Delta/2) \\
=\ &- \log \int_{\hat{\theta}-\Delta/2}^{\hat{\theta}+\Delta/2} h(\theta) f(x|\theta) d\theta
\end{aligned}
$$

For fixed x and $\hat{\theta}$, this expression is a monotonically decreasing function of Δ. Thus, our aim of obtaining a short encoding of the data would lead us to choose Δ as large as possible, i.e., to choose the interval $\hat{\theta} \pm \Delta/2$ to cover the entire set Θ. The "assertion" would then mean only "$\theta \in \Theta$". It would have zero length, and would tell us nothing not already known *a priori*. The message length for data x would be

$$
\begin{aligned}
&- \log \int h(\theta) f(x|\theta) d\theta \\
=\ &- \log r(x), \quad \text{say.}
\end{aligned}
$$

There is nothing wrong with this mathematics. Indeed, a code for X chosen with no objective other than to minimize the expected message length would encode x with a string of the above length, since $r(x)$ is just the probability of obtaining data x from any model in Θ. The expected length of such an optimally coded representation of the data is a quantity of some interest, which we write as

$$ I_0 = -\sum_X r(x) \log r(x) $$

It gives a lower bound for the expected length of any code for X.

The above argument shows that, if we treat the assertion as specifying an interval Δ of models for the data, conciseness gives no basis for choosing Δ, and no basis for inferring any non-trivial assertion about the source of the data. We therefore insist that the assertion in an explanation be interpreted as asserting a single model $\hat{\theta}$ for the data, and that the detail encode the data using this single model $f(x|\hat{\theta})$, rather than some mixture of models centred on $\hat{\theta}$.

3.2 The Strict Minimum Message Length Explanation for Discrete Data

We are now in a position to define the construction of explanations of minimal expected length, called SMML explanations, but first rehearse our assumptions.

An inference problem is specified by:

X a discrete set of possible data values.

x a variable taking values in X.

x the actual data value.

Θ a set of possible probabilistic models for the data. We assume initially that Θ is a simple continuum of some known dimension D.

θ a variable taking values in Θ. The values in Θ are probabilistic models of known mathematical form. Hence, Θ may be regarded as the continuum of a D-component vector parameter θ.

$f(x|\theta)$ the function embodying the mathematical form of the data models. For a model characterized by parameter value θ, $f(x|\theta)$ gives the probability distribution over X implied by that model.

$h(\theta)$ a prior probability density over Θ.

X, Θ and the functions $h(\theta), f(x|\theta)$ are assumed to be known *a priori* to both sender and receiver of the explanation message. Before the data is acquired, the sender and receiver agree on a code for X, using knowledge only of X, Θ, $h(\theta)$ and $f(x|\theta)$. Messages in the code are required to be "explanations". That is, a message begins with an "assertion" naming an inferred model $\hat{\theta} \in \Theta$, then continues with a "detail" encoding data x using the model $f(x|\hat{\theta})$. We will seek that code giving least expected explanation length.

Useful quantities derived from the problem specification include:

$$r(x) = \int h(\theta) f(x|\theta) d\theta$$

the marginal prior probability of data x. We usually call it simply the marginal probability. We sometimes write r_i for $r(x_i)$.

$$I_0(x) = -\log r(x)$$

the length of a string encoding data x in the optimal, non-explanation, code for X.

$$I_0 = -\sum_{x \in X} r(x) \log r(x)$$

the expectation of $I_0(x)$.

The assertion, being a finite string, can name at most a countable subset of Θ. Let $\Theta^* = \{\theta_j : j = 1, 2, 3, \ldots\}$ be this subset. The code for Θ^* may encode the different possible assertions with strings of different length.

The choice of the code for Θ^* is equivalent to adopting a coding probability distribution over $\Theta^* : \Pr(\theta_j) = q_j > 0 : j = 1, 2, 3, \ldots$, with $\sum_j q_j = 1$. (Summation over index j will be taken to be over the set $\{j : \theta_j \in \Theta^*\}$.) That is, the length of the string encoding assertion θ_j is $-\log q_j$. We will sometimes write $q(\theta_j)$ instead of q_j.

The choice of Θ^* and $\{q_j\}$ determine the code for the assertion part of an explanation. The code for the detail is determined by the data model $f(x|\hat{\theta})$. If some data value $x_i \in X$ is "explained" using assertion $\hat{\theta} = \theta_j$, the length of the detail is $-\log f(x_i|\theta_j)$, and the length of the explanation is $-\log q_j - \log f(x_i|\theta_j)$. These codes are agreed between sender and receiver before the actual data is known.

The agreed explanation coding scheme will in general allow a data value $x_i \in X$ to be encoded in any of several ways, i.e., by making any of several assertions. For data x_i, and any $\theta_j \in \Theta^*$ for which $f(x_i|\theta_j) > 0$, an explanation asserting θ_j will have finite length and will correctly encode x_i. Since the sender of the explanation is trying to make the explanation short, we assume that the sender will, given data x_i, choose to make the assertion which minimizes the explanation length. This choice can be described by an "estimator function" $m(x)$ mapping X into Θ^* so that

$$-\log q(m(x_i)) - \log f(x_i|m(x_i)) \leq -\log q_j - \log f(x_i|\theta_j)$$
$$\forall j : \theta_j \in \Theta^*, \theta_j \neq m(x_i)$$

We will later show the inequality to be strict, so there is no ambiguity in this definition of $m()$.

Then the length of the explanation for data x is

$$I_1(x) = -\log q(m(x)) - \log f(x|m(x))$$

and the expected length is

$$I_1 = -\sum_{x \in X} r(x) \left[\log q(m(x)) + \log f(x|m(x))\right]$$

We now consider how Θ^* and the coding distribution $\{q_j : \theta_j \in \Theta^*\}$ should be chosen to minimize I_1, i.e., to allow the shortest possible explanation on average.

Define $t_j = \{x : m(x) = \theta_j\}$. That is, t_j is the set of data values any of which results in assertion θ_j being used in the explanation. Then I_1 can be written as

$$
\begin{aligned}
I_1 &= -\sum_{\theta_j \in \Theta^*} \sum_{x_i \in t_j} r_i \left[\log q(m(x_i)) + \log f(x_i|m(x_i))\right] \\
&= -\sum_{\theta_j \in \Theta^*} \sum_{x_i \in t_j} r_i \left[\log q_j + \log f(x_i|\theta_j)\right] \\
&= -\sum_{\theta_j \in \Theta^*} \left(\sum_{x_i \in t_j} r_i\right) \log q_j - \sum_{\theta_j \in \Theta^*} \sum_{x_i \in t_j} r_i \log f(x_i|\theta_j)
\end{aligned}
$$

The first term gives the expected length of the assertion, and the second gives the expected length of the detail. The first term is minimized, subject to $\sum_j q_j = 1$, by choosing coding probabilities

$$q_j = \sum_{x_i \in t_j} r_i$$

That is, the coding probability assigned to assertion or estimate θ_j is the sum of the marginal probabilities of the data values resulting in estimate θ_j. It is the probability that estimate θ_j will actually be used in the explanation.

The second term is the expected length of the detail encoding x using the inferred model. If we consider the contribution to this expectation from the set t_j, i.e.,

$$-\sum_{x_i \in t_j} r_i \log f(x_i|\theta_j)$$

this is the only contribution to I_1 which depends on the value θ_j, and I_1 is minimized by choosing θ_j to minimize this expression. It will be recognized that θ_j then maximizes the logarithmic average of the likelihood function $f(x_i|\theta_j)$ over $x_i \in t_j$.

3.2.1 Discrete Hypothesis Sets

Even when the set Θ of possible models is discrete, each member having a non-infinitesimal prior probability, it may well happen that the minimum expected message length for some set X of possible data is achieved using a set Θ^* of assertable hypotheses which is a proper subset of Θ. This outcome is likely if the information expected to be obtained from the data is insufficient to distinguish all members of Θ with much reliability. By omitting some hypotheses from Θ^*, the coding probability of the assertable hypotheses is increased, leading to a shorter assertion, and the increase in detail length resulting from a more limited choice of models for the data may on average be a less important effect.

Note that in such a case, the "coding probability" q_j used for some model $\theta_j \in \Theta^*$ will not in general equal the discrete prior probability $h(\theta_j)$. In fact, even if $\Theta^* = \Theta$, in general $q_j \neq h(\theta_j)$ for some or all hypotheses.

3.2.2 Minimizing Relations for SMML

We have found above three necessary relations which must be satisfied by SMML explanations.

Obeyed by the estimator, $m()$:

$$\text{R1: } q(m(x))f(x|m(x)) \geq q_j f(x|\theta_j) \text{ for all } x \text{ and all } \theta_j \neq m(x)$$

The estimator $m()$ "explains" data x using the assertion of highest posterior probability.

Obeyed by the code for assertions, represented by $q()$:

$$\text{R2: } q(\theta_j) = \sum_{x:m(x)=\theta_j} r(x) = \sum_{x \in t_j} r(x)$$

The code for assertions is optimal for the probability distribution over Θ^* expressing the probability that an assertion will occur in an explanation.

Obeyed by the assertions or estimates $\{\theta_j\}$:

$$\text{R3: } \theta_j \text{ maximizes } \sum_{x \in t_j} r(x) \log f(x|\theta_j)$$

These three relations are unfortunately insufficient fully to define the SMML explanation code, as will appear in the first example below.

Using R2, we can write

$$I_1 = - \sum_{\theta_j \in \Theta^*} q_j \log q_j - \sum_{\theta_j \in \Theta^*} \sum_{x_i \in t_j} r_i \log f(x_i|\theta_j)$$

but in general there will be many choices of Θ^* and $m()$ which satisfy R1 and R3, yet do not minimize I_1.

For many non-optimal choices of Θ^* and $m()$, it may be that no re-assignment of a single data value x to a different estimate will decrease I_1, so R1 is satisfied; no adjustment of any one coding probability q_j will decrease I_1, so R2 is satisfied; and no adjustment of any one estimate value θ_j will decrease I_1, so R3 is satisfied. Reduction of I_1 may only be possible by a simultaneous change to many assignments of data to estimates, and to many coding probabilities and estimates.

3.2.3 Binomial Example

The data consist of the outcomes of 100 ordered trials, each giving success (S) or failure (F). Thus, the data x can be written as a string of 100 S or F symbols and X is the set of binary strings of length 100. It is believed that the outcomes are independent of one another, and that the probability of success in a trial is some fixed value θ, which is unknown. The family of possible data models is therefore

$$f(x|\theta) = \theta^n (1-\theta)^{100-n}$$

where n is the number of successes in the string x.

The unknown parameter θ is considered equally likely *a priori* to have any value in Θ, the interval 0 to 1, so $h(\theta) = 1$.

An explanation will begin with an assertion asserting some estimated value $\hat{\theta}$ for θ. The detail, encoding x, can consist of 100 segments, each encoding one outcome. A success will be encoded by a segment of length $-\log \hat{\theta}$, a failure by a segment of length $-\log(1-\hat{\theta})$. The length of the detail is therefore

$$-n\log \hat{\theta} - (100-n)\log(1-\hat{\theta}) = -\log f(x|\hat{\theta})$$

The marginal probability of obtaining a particular data vector x with n successes is

$$
\begin{aligned}
r(x) &= \int_0^1 f(x|\theta)h(\theta)d\theta \\
&= \int_0^1 \theta^n (1-\theta)^{100-n}d\theta \\
&= \frac{n!(100-n)!}{101!} \\
&= \frac{1}{101\binom{100}{n}}
\end{aligned}
$$

The minimum expected non-explanation message length $I_0 = -\sum_x r(x)\log r(x) = 51.900$ nits.

We now consider the construction of an explanation code for X. We might first consider the naive choice defined by the estimator $\hat{\theta} = n/100$.

For this estimator, the set of possible estimates $\Theta^* = \{\theta_j : \theta_j = j/100, j = 0, 1, 2, \ldots, 100\}$ has 101 members. The set t_j of data values x resulting in estimate θ_j is just $\{x : j \text{ successes }\}$, so

$$
\begin{aligned}
q_j &= \sum_{x \in t_j} r(x) \\
&= \binom{100}{j} \frac{1}{101 \binom{100}{j}} \\
&= \frac{1}{101} \qquad \text{for all } j
\end{aligned}
$$

The explanation length for data x with n successes is:

$$
\begin{aligned}
I_1(x) &= -\log q_n - \log f(x|\hat{\theta}) \\
&= -\log q_n - \log\left[(n/100)^n (1 - n/100)^{100-n}\right]
\end{aligned}
$$

and the expected length is

$$
\begin{aligned}
I_1 &= \sum_{n=0}^{100} \frac{1}{101}\left[\log 101 - \log\left(\left(\frac{n}{100}\right)^n \left(1 - \frac{n}{100}\right)^{100-n}\right)\right] \\
&= 54.108 \text{ nits}
\end{aligned}
$$

This choice of Θ^* and $m()$ satisfies all relations R1–R3, but it does not minimize I_1. It is substantially less efficient than the non-explanation code: $I_1 - I_0 = 2.208$ nits. Consider for example the two sets of data strings with 20 and 21 success counts respectively. The explanation lengths of data strings with 20 or 21 success counts are respectively:

n	Assertion	Detail	Total length
20	$\log 101 = 4.615$	$-20 \log 0.20 - 80 \log 0.80 = 50.040$	54.655
21	$\log 101 = 4.615$	$-21 \log 0.21 - 79 \log 0.79 = 51.396$	56.011

Suppose now we modify Θ^*, replacing $\theta_{20} = 0.20$ and $\theta_{21} = 0.21$ by a single estimate $\theta_{20.5} = 0.205$. We also modify the estimator so that $m(x : n = 20) = m(x : n = 21) = \theta_{20.5}$ That is, $t_{20.5}$, the set of data strings resulting in estimate $\theta_{20.5}$, includes all strings with 20 or 21 successes, and $q_{20.5} = 2/101$. The rest of Θ^* and $m()$ is left unaltered, leaving the explanation lengths of all other data strings unchanged. The explanation lengths for strings with $n = 20$ or 21 are now:

n	Assertion	Detail	Total length
20	$-\log(2/101) = 3.922$	$-20 \log 0.205 - 80 \log 0.795 = 50.045$	53.970
21	$-\log(2/101) = 3.922$	$-21 \log 0.205 - 79 \log 0.795 = 51.403$	55.325

The new choice of Θ^* and $m()$ gives reduced lengths for $n = 20$ or 21. For each set of strings, the detail has become slightly longer, because the new estimate $\theta_{20.5}$ is not exactly optimal for either set. However, the new estimate has coding probability 2/101 rather than 1/101, since it will occur as the assertion whenever the data shows either 20 or 21 successes. The length of the assertion is thus reduced by $\log 2$ for these data strings, outweighing the increased detail length. Since the new code gives unchanged explanations for other strings, on average it gives shorter explanations than the original code, i.e., a lower value of I_1, and hence is preferable.

The grouping together of data strings with different but similar success counts into the same t_j set can be carried further with advantage. If carried too far, however, the estimates available in Θ^* for use in assertions become so few, and so widely separated in value, that no good detail code may be found for some values of n. As an extreme case, consider the code with only one member of Θ^*, $\theta_{50} = 0.50$. As this estimate will perforce be used for all data strings, it has coding probability 1.0, giving a zero assertion length. However, the detail length, which will be the same for all strings, is

$$-n \log 0.5 - (100 - n) \log(1.0 - 0.5) = -100 \log 0.5 = 69.3 \text{ nits}$$

This value for I_1 is much worse than for even the naive code ($I_1 = 54.108$).

The optimum explanation code represents a compromise between a large number of possible estimates (which gives shorter details but requires long assertions) and a small number (which gives short assertions but longer details). The optimum for the simple Binomial SMML problem can be found by an algorithm due to Graham Farr, which will be described in Section 3.2.8. The solution is not unique, as discussed later in Section 3.2.5.

Table 3.1 shows one of the optimal solutions. There are 10 possible estimates in Θ^*. The column headed t_j shows the range of success counts in the

Table 3.1. Success count ranges and estimates.

j	t_j	θ_j
1	0	0
2	1–6	0.035
3	7–17	0.12
4	18–32	0.25
5	33–49	0.41
6	50–66	0.58
7	67–81	0.74
8	82–93	0.875
9	94–99	0.965
10	100	1.0

strings resulting in estimate θ_j. These ranges may seem surprisingly wide. In fact, they are not unreasonable, and do not lead to implausible explanations of the data. For instance, if the data results in the assertion "$\hat{\theta} = 0.41$", it would not be surprising to find anything from 33 to 49 successes. The probability that 100 trials with $\theta = 0.41$ would yield exactly 41 successes is 0.0809, and the probabilities of 33 and 49 successes are 0.0218 and 0.0216 respectively, which are over a quarter that of the most probable value 41. We will see later that the spacing of the estimates $\{\theta_j\}$ in Θ reflects the expected error in estimating θ.

The solution of Table 3.1 gives $I_1 = 52.068$, which is very little more than the length of the optimal non-explanation code: $I_1 - I_0 = 0.168$. Recall that for the naive code with 101 possible estimates, $I_1 - I_0 = 2.208$. Moreover, for increasing sample size, i.e., more trials, $I_1 - I_0$ for the naive code increases roughly as $\frac{1}{2} \log N$, whereas $I_1 - I_0$ for the SMML code approaches a limiting value of $\frac{1}{2} \log(\pi e/6) = 0.176\ldots$.

3.2.4 Significance of $I_1 - I_0$

For any data value x resulting in SMML estimate $\hat{\theta} = m(x)$, the explanation length is

$$I_1(x) = -\log(q(\hat{\theta})f(x|\hat{\theta}))$$

For the same data x, the length of its encoding in the optimal non-explanation code is

$$I_0(x) = -\log r(x)$$

The difference

$$I_1(x) - I_0(x) = -\log \frac{q(\hat{\theta})f(x|\hat{\theta})}{r(x)}$$

is formally identical to the negative log of the Bayes posterior probability of estimate $\hat{\theta}$ given data x. If the set of possible models were discrete, with $\Theta = \Theta^*$ and $q(\theta)$ the prior probability of model θ, the correspondence would be exact. However, when Θ is a continuum and Θ^* a discrete subset of $\Theta, q(\theta \in \Theta^*)$ is not the prior probability that "θ is true": indeed no non-zero probability could be attached to such a statement. Nonetheless, the difference can play the role of a negative log posterior probability, and its expectation $(I_1 - I_0)$ is a good measure of the "believability" of the estimates.

I_1 exceeds I_0 because an explanation conveys not only the data but also something not logically deducible from the data, viz., the estimate. The explanation code permits the encoding of x using any model or estimate in Θ^* for which $f(x|\theta) > 0$. The explanation chosen for the data tells the receiver the data, and also which of these possible estimates was chosen. It is therefore necessarily longer than a message which encoded the data and nothing more.

In the binomial example, a string of outcomes with $n = 8$ say is optimally explained using $\hat{\theta} = 0.12$, with an explanation length 30.940 (vs. $I_0(x) = 30.56$). The same string could also be encoded in the same agreed code using $\hat{\theta} = 0.035$ with a length of 32.766, or using $\hat{\theta} = 0.25$ with a length 39.46, or even using $\hat{\theta} = 0.965$ for a length 311.53. Any of these "explanations" of the string conveys the choice of estimate as well as the string itself, and the choice of estimate is not dictated by the code.

3.2.5 Non-Uniqueness of Θ^*

The optimum choice of Θ^* is not necessarily unique. For instance, in the binomial example, the inference problem is exactly symmetric with respect to success and failure. However, the Θ^* shown in Table 3.1 is not symmetric. Indeed, no partition of the 101 possible success counts into ten sets $t_1, \ldots, t_{10}$ could be symmetric. Thus, the "mirror image" of the SMML estimator shown in this table gives a different but equally efficient explanation code. However, it is easily shown that for any given optimal choice of Θ^*, the estimator function is unique. Equality in relation R1, which would allow an alternative optimal $m()$, cannot occur.

Different optimal choices of Θ^* will, of course, give different estimator functions. The non-uniqueness of SMML estimators has little practical consequence. The different optimal estimators will give different estimates for some data values, but the different estimates are always plausible.

3.2.6 Sufficient Statistics

Usually, the data value x is a vector comprising many component numbers and/or discrete values. For some families of model distribution $f(x|\theta)$ it is possible to find a function $s(x)$ such that the model distribution function $f(x|\theta)$ can be factored into the form

$$f(x|\theta) = g(s(x)|\theta)v(x)$$

where $g(s|\theta) = \sum_{x:s(x)=s} f(x|\theta)$ is the probability of obtaining data yielding function value s, and where the function $v(x)$ does not depend on θ. Such a function $s()$ is called a "sufficient statistic", and its value for some x contains all the information about θ which can be recovered from x. In particular, if θ_1 and θ_2 are different estimates which might be asserted,

$$\frac{f(x|\theta_1)}{f(x|\theta_2)} = \frac{g(s(x)|\theta_1)}{g(s(x)|\theta_2)}$$

The data x itself is of course always a sufficient statistic, but for some model families there exists a sufficient statistic $s(x)$ having fewer components than x. The value $s(x)$ is then a more compact representation of all the information in x relevant to θ.

For any choice of Θ^* and $q(\theta)$, where $\theta \in \Theta^*$, the estimator $m()$ is determined by relation R1, which may be written as

$$\frac{q(m(x))f(x|m(x))}{q(\theta)f(x|\theta)} > 1$$

for all $\theta \in \Theta^*$, $\theta \neq m(x)$.

If $s = s(x)$ is a sufficient statistic, relation R1 becomes

$$\text{R1:} \qquad \frac{q(m(x))g(s|m(x))}{q(\theta)g(s|\theta)} > 1$$

whence it follows that $m(x)$ depends on x only via the sufficient statistic s. Further, since all data values x having the same value s for $s(x)$ clearly result in the same estimate, we can write the estimator as

$$m(x) = m_s(s(x))$$

or simply $m_s(s)$, and we can define for all $\theta \in \Theta^*$

$$t_s(\theta) = \{s : m_s(s) = \theta\}$$

and

$$r_s(s) = \sum_{x:s(x)=s} r(x)$$

Clearly, the coding probability $q(\theta)$ is the same whether we work with the raw data x or the sufficient statistic s:

$$q(\theta) = \sum_{s:m_s(s)=\theta} r_s(s) = \sum_{x:m(x)=\theta} r(x)$$

Defining S as the set of possible values of s,

$$S = \{s : s = s(x), x \in X\}$$

the entire estimation problem, and its SMML solution, can be expressed in terms of Θ , $h(\theta)$, S and $g(s|\theta)$ ($s \in S, \theta \in \Theta$). That is, we need not consider the set of all possible data values at all, only the set of all possible sufficient statistic values. We then choose $\Theta^*, m_s(s)$ and $q(\theta)$ ($\theta \in \Theta^*$) to minimize

$$I_{1s} = -\sum_{s \in S} r_s(s) \left[\log q(m_s(s)) + \log g(s|m_s(s))\right]$$

Of course, $I_{1s} \neq I_1$. I_{1s} is the expected length of a message encoding s only, not x. In fact

$$I_1 = I_{1s} - \sum_{x \in X} r(x) \log v(x)$$

where the second term is the expected length of a message segment encoding x for a receiver who already knows $s(x)$. Since $v(x)$ does not depend on θ, this length does not depend on the choice of Θ^* or the estimator and so can be ignored in constructing the SMML estimator.

In message-coding terms, the use of a sufficient statistic s suggests that an explanation of x can be encoded in three parts. The assertion of the estimate

$\hat{\theta}$ remains as previously discussed. The detail now comprises a part encoding s, of length $-\log g(s|\hat{\theta})$, and a part encoding x using a code which assumes knowledge of $s(x)$, of length $-\log v(x)$. This third part is independent of $\hat{\theta}$, and can be ignored in choosing the assertion. However, its contribution to $I_1(x)$ cannot be ignored if we wish for instance to compare the length of an explanation of x assuming the model family $f()$ with explanations assuming some other model family. A function $s(x)$ which is a sufficient statistic for the family $f()$ may not be a sufficient statistic if some other model family is assumed.

A sufficient statistic is particularly useful if it has no more components than θ. Such a statistic can be called a "minimal sufficient statistic". In some of the literature, the word "sufficient" seems to be reserved for minimal sufficient statistics. When a minimal sufficient statistic exists, it is a function of the data no more complex than the unknown parameter, but gives all the information about the parameter which can be extracted from the data.

3.2.7 Binomial Example Using a Sufficient Statistic

The binomial example, where the data consist of an ordered list of outcomes from 100 trials, each independently having $\Pr(\text{success}) = \theta$, can be reframed to use a sufficient statistic. If x denotes the binary string representing the outcomes,

$$f(x|\theta) = \theta^n (1-\theta)^{100-n}$$

where n is the number of successes in the string x. The success count n is clearly a (minimal) sufficient statistic, since we can write

$$f(x|\theta) = \left[\binom{100}{n} \theta^n (1-\theta)^{100-n} \right] \cdot \left[\frac{1}{\binom{100}{n}} \right]$$

The first factor is $g(n|\theta)$, the probability of obtaining a success count n regardless of the order of the n successes and $(100-n)$ failures, and the second factor $v(x)$ is independent of θ. (In this simple problem, $v(x)$ happens to depend on x only via n, but this is not in general the case for other models.)

The set S is just the set of integers $\{0, \ldots, 100\}$,

$$r_s(n) = \sum_{x:n \text{ successes}} r(x) = \frac{1}{101}$$

and

$$I_{1s} = -\sum_{n=0}^{100} \frac{1}{101} \left[\log q(m_s(n)) + \log g(n|m_s(n)) \right]$$

Minimization of I_{1s} by choice of Θ^* and $m_s()$ gives exactly the SMML estimator of Table 3.1, with $I_{1s} = 4.783$ nits.

A form of message encoding the full outcome sequence x can easily be developed using the sufficient statistic. The message would have three parts: an assertion of an estimate $\hat{\theta}$, a statement of the success count n, and an encoding of x assuming knowledge of n. The first two parts have a combined expected length of I_{1s}. The third part, encoding x using knowledge of n, can be done as follows:

Define n_k as the number of successes in the first k outcomes, with $n_0 = 0$ and $n_{100} = n$. Then encode each outcome in turn. Encode a success on trial k with a string of length $-\log((n - n_{k-1})/(101 - k))$ and a failure with length $-\log(1 - (n - n_{k-1})/(101 - k))$. It is easily seen that the length of this part, for a string with n successes, is $\log\binom{100}{n}$, i.e., $-\log v(x)$. Also the length of the second and third parts together exactly equals the length of the original detail, which was encoded without use of the sufficient statistic.

3.2.8 An Exact Algorithm for the Binomial Problem

The binomial problem is one of the very few for which an exact, computationally feasible algorithm is known for finding an optimal SMML code. The following algorithm is due to Graham Farr [15]. Suppose N trials are conducted, yielding n successes. We take the datum to be just the sufficient statistic n, as in Section 3.2.7. Assume some prior is given on the unknown success probability, and that this leads to the marginal success count probabilities $\{r(n) : n = 0, \ldots, N\}$ Assume that the optimum code results from a partition of the success counts into a number of groups, each group containing a set of consecutive values of n. Let such a group be denoted by $[k, m]$, meaning the group

$$\{n : k \leq n \leq m\} \quad (0 \leq k \leq m \leq N)$$

For this group, the coding probability is given by

$$q_{k,m} = \sum_{n=k}^{m} r(n)$$

The SMML estimate is

$$\theta_{k,m} = \frac{1}{Nq_{k,m}} \sum_{n=k}^{m} n\, r(n)$$

and the contribution of the group to I_1 is

$$G(k, m) = -q_{k,m} \log(q_{k,m}) - \sum_{n=k}^{m} r(n) \log\binom{N}{n}$$
$$-Nq_{k,m}\left[\theta_{k,m} \log(\theta_{k,m}) + (1 - \theta_{k,m}) \log(1 - \theta_{k,m})\right]$$

The expected message length I_1 is then the sum over all groups in the partition of $G(,)$. The algorithm is based on the observation that, if the partition has a group starting at $n = m$, then the contribution to I_1 of all success counts in the range $m \leq n \leq N$, and hence the optimum sub-partition of this range, is independent of how success counts less than m are partitioned. Let $Q[m]$ denote an optimum partition of the range $m \leq n \leq N$, let $T(m)$ denote its cost, i.e., the contribution to I_1 of counts in this range, and let $Q[N]$ be the singleton group $[N, N]$. Then it follows from the above observation that for all $m < N$, $Q[m]$ consists of a group $[m, k]$ followed by the optimum partition $Q[k+1]$ for some k in the range $m \leq k \leq N$, where we define $Q[N+1]$ as void, $T(N + 1) = 0$, and its cost $T(m) = G(m, k) + T(k + 1)$.

Hence, the successive partitions $Q[m]$ can be determined in order of decreasing m. Each determination requires only a linear search for the value k_m which minimizes $T(m)$. In fact it can be shown that for all m, $k_m \leq k_{m-1}$, which limits the number of values which need to be tried. As this number is of order $N^{\frac{1}{2}}$, and as successive values of $G(m, k)$ can be found in constant time, the overall complexity of the algorithm is of order $N^{\frac{3}{2}}$.

The final SMML solution is given by $Q[0]$, and I_1 by $T(0)$. $Q[0]$ is easily reconstructed from the values of k_m. Unfortunately, the algorithm does not generalize readily to trinomial and higher-order distributions.

3.2.9 A Solution for the Trinomial Distribution

The Farr algorithm used to find the SMML estimator for Binomial distributions is applicable to any problem where there is a scalar sufficient statistic with a finite range. There is as yet no known feasible algorithm for the exact SMML solution for problems outside this very limited family. The trinomial distribution, where the data is a finite sequence of N symbols which can take three rather than two possible values, say, 1, 2 and 3, and the log likelihood is

$$LL = \sum_{i=1}^{3} n_i \log p_i$$

where the sequence contains n_i occurrences of symbol i, and $\sum_i p_i = 1$ has no known exact SMML solution.

Figure 3.1 shows an approximate solution for $N = 60$ and a Uniform prior over the two-dimensional space of probabilities. The sufficient statistic (n_1, n_2, n_3) is represented by a point in a triangular grid, with the apex point representing $(60, 0, 0)$, the left bottom corner $(0, 60, 0)$ and the central point $(20, 20, 20)$. The figure shows a partition of the sufficient-statistic set found by attempting to minimize I_1 by simulated annealing. Data points in the same data group are marked with the same mark and would result in the same estimate (which suitably scaled would correspond to the centroid of its group). There are 37 groups, so Θ^* comprises 37 possible estimates. The groups are generally hexagonal near the centre of the diagram, but become distorted near the edges. The partition is no doubt not an optimal solution, but must be quite close, as it leads to $I_1 - I_0 \approx 0.294$ nit, rather less than the large-sample "lower bound" shown for two-parameter estimators in Table 3.4.

Fig. 3.1. Partition for the Trinomial Distribution, $N = 60$.

3.3 The SMML Explanation for Continuous Data

We have noted, perhaps pedantically, that all data available for analysis are necessarily discrete. A model of a real-valued quantity conventionally treats the quantity as continuously variable, and uses a probability density rather than a discrete probability distribution. In such a case, we think of the discrete value appearing in actual data as the result of a rounding-off or quantization of the underlying real-valued model quantity. For instance, when we measure a temperature, we usually conceive of it as being able to take values in a continuous range. We will treat a temperature datum "297.2°K, measured to the nearest 0.1°K as meaning that the temperature lies somewhere in the interval 297.15°K to 297.25°K. If our probabilistic data model implies a probability density $\rho(T)$, then we consider the (discrete) probability of obtaining the datum 297.2 to be

$$\int_{297.15}^{297.25} \rho(T) \, dT$$

Provided the rounding-off quantum Δ (in this case 0.1°K) is small compared with the width of any peak or other feature of the density $\rho(T)$, we may adequately approximate the above integral by $\rho(297.2)\Delta$. More generally, provided the rounding quantum is independent of the measured value, the discrete probability for any measured value T can be taken as $\rho(T)\Delta$.

Assuming a small, constant quantum Δ, the SMML approach can be recast to use, instead of a discrete model distribution $f(x|\theta)$ over a discrete set X of possible data values, a model probability density over a continuum of possible data values. We will re-use the symbolic forms $f(x|\theta)$ for the density of a real-valued (possibly vector) data variable x over a continuum X. Similarly, we redefine

$$r(x) = \int_{\Theta} h(\theta) f(x|\theta) \, d\theta$$

as the marginal probability density of x. As before, we seek an estimator $m(x)$, $x \in X$, mapping data values into Θ^*, a discrete subset of Θ. The set of possible estimates Θ^* must remain discrete and countable, so that the chosen estimate can be encoded in an assertion of finite length. Thus, we still associate a non-zero coding probability $q(\theta)$ with each possible estimate. The length of the explanation of some rounded-off data value x is then approximately

$$-\log q(m(x)) - \log [f(x|m(x))\Delta]$$
$$= -\log q(m(x)) - \log f(x|m(x)) - \log \Delta$$

If the marginal probability of the rounded-off data value x is approximated by $r(x)\Delta$, the expected message length becomes

$$-\sum_{\substack{\text{rounded-off} \\ \text{data values}}} \Delta \, r(x) \left[\log q(m(x)) + \log f(x|m(x)) + \log \Delta\right]$$

For sufficiently small Δ, this sum over the discrete set of actual possible rounded-off data values can be approximated by an integral over the continuum of unrounded values:

$$I_1 = -\int_X r(x) \left[\log q(m(x)) + \log f(x|m(x)) + \log \Delta\right] \, dx$$

$$= -\log \Delta - \int_X r(x) \left[\log q(m(x)) + \log f(x|m(x))\right] \, dx$$

Similarly, the length of the optimum non-explanation code for the set of rounded values can be approximated by

$$I_0 = -\log \Delta - \int_X r(x) \log r(x) \, dx$$

and the difference by

$$I_1 - I_0 = -\int_X r(x) \log \frac{q(m(x)) f(x|m(x))}{r(x)} \, dx$$

The difference $(I_1 - I_0)$ does not depend on the rounding-off quantum Δ, so if Δ is small enough to make the above approximations valid, its actual value can be ignored and the SMML estimator constructed by minimizing $(I_1 - I_0)$. In dealing with problems involving continuous data, we will often omit mention of Δ altogether, and write simply

$$I_0 = -\int_X r(x) \log r(x) \, dx$$

$$I_1 = -\int_X r(x) \log(q(m(x)) f(x|m(x))) \, dx$$

These expressions no longer give expected message lengths, but as we will usually be concerned only with the differences in explanation lengths using different models or estimates, no confusion should arise.

Note that the difference $(I_1 - I_0)$ for continuous data retains its formal correspondence with the expected negative log of a posterior probability.

For small Δ, the set t_j of rounded-off values resulting in some estimate $\theta_j \in \Theta^*$ may be regarded rather as a region of the continuum X, and the sum

$$\sum_{x:m(x)=\theta_j} r(x) = \sum_{x \in t_j} r(x)$$

in the minimizing relation R2 replaced by an integral:

$$\text{R2:} \quad q(\theta_j) = \int_{x \in t_j} r(x) \, dx$$

The other minimizing relations become

R1: $q(m(x))f(x|m(x)) > q_j f(x|\theta_j)$ for $\theta_j \neq m(x)$

R3: θ_j maximizes $\displaystyle\int_{x \in t_j} r(x) \log f(x|\theta_j)\, dx$

where now $f(x|\theta)$ is a probability density on X.

3.3.1 Mean of a Normal

The data x are an ordered set of N real scalar values $x = (y_1, y_2, \ldots, y_n, \ldots, y_N)$ which are believed to be independently drawn from a Normal density of un-known mean θ and known standard deviation σ, so

$$f(x|\theta) = \prod_{n=1}^{N} \frac{1}{\sqrt{2\pi}\sigma} e^{-(y_n-\theta)^2/2\sigma^2}$$

$$-\log f(x|\theta) = \frac{N}{2}\log(2\pi) + N\log\sigma + \frac{1}{2\sigma^2}\sum_n (y_n - \theta)^2$$

Define the sample mean $\bar{y} = \frac{1}{N}\sum_n y_n$. Then

$$\sum_n (y_n - \theta)^2 = N(\bar{y} - \theta)^2 + \sum_n (y_n - \bar{y})^2$$

Hence,

$$-\log f(x|\theta) = \frac{N}{2}\log(2\pi) + N\log\sigma + \frac{N(\bar{y}-\theta)^2 + \sum_n (y_n - \bar{y})^2}{2\sigma^2}$$

and $f(x|\theta)$ can be written as

$$\left[\frac{1}{\sqrt{2\pi}} \frac{\sqrt{N}}{\sigma} e^{-(\bar{y}-\theta)^2 N/2\sigma^2} \right] \cdot \left[\left(\frac{1}{\sqrt{2\pi}\sigma} \right)^{N-1} \frac{1}{\sqrt{N}} e^{-\sum (y_n - \bar{y})^2/2\sigma^2} \right]$$

The first factor is the density of $\bar{y}$ given θ, $g(\bar{y}|\theta)$, and is a Normal density with mean θ and standard deviation $\sigma/\sqrt{N}$. The second factor does not depend on θ, so $\bar{y}$ is a (minimal) sufficient statistic.

Purely to keep this example simple, we assume $h(\theta)$ has a uniform density $1/L$ over some range $L \gg \sigma/\sqrt{N}$. Then the marginal density $r_s(\bar{y})$ is also uniformly $1/L$ except for smooth "shoulders" near the ends of the range, which we shall ignore. The uniformity of these assumed forms over the range L implies that, except near the ends of the range, Θ^* will be a uniformly spaced set of values within the range. Let the separation between adjacent members $\hat{\theta}_j$ and $\hat{\theta}_{j+1}$ of Θ^* be w, and ignore possible variation of w near the ends of the range. Then the set $t_s(\hat{\theta})$ of $\bar{y}$ values resulting in estimate $\hat{\theta} \in \Theta^*$ is just

$$t_s(\hat{\theta}) = \{\bar{y} : \hat{\theta} - w/2 < \bar{y} < \hat{\theta} + w/2\}$$

and $q(\hat{\theta}) = w/L$.

This choice of Θ^* clearly satisfies the relations R1–R3.

Since all members of Θ^* and all sets $t_s(\hat{\theta})$ are similar, the expected explanation length I_{1s} is just the average length for $\bar{y}$ values within one t_s set. Consider the set $t_s(\hat{\theta})$ for some $\hat{\theta} \in \Theta^*$. Then

$$
\begin{aligned}
I_{1s} &= - \int_{\hat{\theta}-w/2}^{\hat{\theta}+w/2} \frac{1}{w} \left[\log(w/L) + \log \left\{ \frac{1}{\sqrt{2\pi}} \frac{\sqrt{N}}{\sigma} e^{-N(\bar{y}-\hat{\theta})^2/2\sigma^2} \right\} \right] d\bar{y} \\
&= -\log \frac{w}{L} - \log \left(\frac{1}{\sqrt{2\pi}} \frac{\sqrt{N}}{\sigma} \right) + \frac{1}{w} \int_{\hat{\theta}-w/2}^{\hat{\theta}+w/2} \left[N(\bar{y}-\hat{\theta})^2/2\sigma^2 \right] d\bar{y} \\
&= -\log \frac{w}{L} + \frac{1}{2} \log(2\pi\sigma^2/N) + \frac{Nw^2}{24\sigma^2}
\end{aligned}
$$

Choosing w to minimize I_{1s} gives $-1/w + Nw/(12\sigma^2) = 0$, $w^2 = 12\sigma^2/N$,

$$I_{1s} = \log L + \frac{1}{2} \log \frac{2\pi}{12} + \frac{1}{2} = \log L + \frac{1}{2} \log(\pi e/6)$$

The optimum non-explanation code for $\bar{y}$ gives

$$I_{0s} = \log L$$

since $r_s(\bar{y})$ is uniformly $1/L$.

Hence, $I_{1s} - I_{0s} = I_1 - I_0 = \frac{1}{2} \log(\pi e/6) = 0.176 \ldots$ nit $= 0.254 \ldots$ bit.

This example was chosen for its extreme simplicity, but it shows some general features which appear in many SMML estimators.

First, the spacing w between adjacent estimates in Θ^* is perhaps surprisingly large, and the $t_s(\hat{\theta})$ sets correspondingly wide. The estimate $\hat{\theta} = m_s(\bar{y})$ may differ from the sample mean by as much as $w/2 = 1.73\sigma/\sqrt{N}$. The exact difference $(\hat{\theta} - \bar{y})$ will depend on the exact location of the possible estimates in Θ, which will be dictated in this example by the fall-off of $r_s(\bar{y})$ near the ends of the range. For most values of $\bar{y}$, far from the ends of the range, it is more instructive to consider the average value of $(\hat{\theta} - \bar{y})^2$. The difference $(\hat{\theta} - \bar{y})$ will be equally likely to take any value in $\pm w/2$, so the average squared difference is $w^2/12$. But if the N data values were drawn from a Normal with mean θ, the expected squared difference between θ and the sample mean $\bar{y}$ is σ^2/N. The SMML code chooses $w^2/12 = \sigma^2/N$. Thus, as measured by average squared difference, the SMML estimate will differ from the sample mean no more than will the true mean θ. This is an example of a general property of SMML estimators, which will be further discussed later: on average, the relation between the SMML estimate $\hat{\theta}$ and the data closely mimics the relation between true parameter θ and the data.

Second, the spacing w does not depend much on the value of the prior probability density: in this simple case, not at all. The spacing is determined by a trade-off between two terms in I_1 (or I_{1s}). The term $(-\log w)$ favours large w, because large w results in a short assertion. The term $(Nw^2/24\sigma^2)$ favours small w, and arises from the increase in the average detail length as a large spacing forces some data to be encoded with an estimate which does not fit the data well. The factor (N/σ^2) arising in this simple problem shows how rapidly the detail length $-\log g(\bar{y}|\hat{\theta})$ increases as $\hat{\theta}$ is displaced from its minimizing value $\bar{y}$.

3.3.2 A Boundary Rule for Growing Data Groups

The argument used in the preceding section to find the optimum size for a group t_j of data (or sufficient statistic) values mapping into a single estimate $\hat{\theta}_j$ can be generalized. The resulting construction does not in general give a strictly optimal explanation code, but usually comes close.

Let $\hat{\theta}$ be some estimate in Θ^*. We can attempt to form its associated data group $t = \{x : m(x) = \hat{\theta}\}$ so as to minimize the average value in the group of $I_1(x) - I_0(x)$, i.e., to maximize

$$
\begin{aligned}
A &= \frac{1}{q}\sum_{x \in t} r(x)\log[qf(x|\hat{\theta})] - \frac{1}{q}\sum_{x \in t} r(x)\log r(x) \\
&= \frac{1}{q}\sum_{x \in t} r(x)\log[qf(x|\hat{\theta})/r(x)]
\end{aligned}
$$

where $q = \sum_{x \in t} r(x)$.

$$
A = \log q + \left(\frac{1}{q}\right)\sum_{x \in t} r(x)\log[f(x|\hat{\theta})/r(x)]
$$

Consider the effect on A of adding to t some data value y not at present in t. We consider $\hat{\theta}$ to be fixed.

The new average is

$$
A_1 = \log q_1 + \left(\frac{1}{q_1}\right)\sum_{x \in t_1} r(x)\log[f(x|\hat{\theta})/r(x)]
$$

where $t_1 = t \cup \{y\}$,

$$
q_1 = q + r(y)
$$

Writing $r(y) = \delta$, and assuming $\delta \ll q$, the difference $A_1 - A$ is given to first order in δ by

$$
\begin{aligned}
A_1 - A &\approx \frac{\delta}{q} - \frac{\delta}{q^2}\sum_{x \in t} r(x)\log[f(x|\hat{\theta})/r(x)] + \frac{\delta}{q}\log[f(y|\hat{\theta})/r(y)] \\
&\approx \frac{\delta}{q}\left\{1 - \frac{1}{q}\sum_{x \in t} r(x)\log[f(x|\hat{\theta})/r(x)] + \log[f(y|\hat{\theta})/r(y)]\right\}
\end{aligned}
$$

So, $A_1 > A$ if

$$\log[f(y|\hat{\theta})/r(y)] > \frac{1}{q}\sum_{x\in t} r(x)\log[f(x|\hat{\theta})/r(x)] - 1 > A - \log q - 1$$

i.e., if $\log[qf(y|\hat{\theta})/r(y)] > A - 1$.

Recalling the definition of A, this condition can also be written as

$$I_1(y) - I_0(y) < \mathrm{Av}_{x\in t}[I_1(x) - I_0(x)] + 1$$

Thus, it improves the average of $[I_1(x) - I_0(x)]$ to add to the group any data value y for which $[I_1(y) - I_0(y)]$ exceeds the group average by less than 1.

We will term this the "boundary rule".

The binomial explanation code of Table 3.1 was exactly reproduced by growing a group, starting with $n = 0$, until a further addition of an n value would not reduce the average $I_1(n) - I_0(n)$ in the group. As the added values of n had $r(n) = 1/101$, which is not small compared to the q value of the group, each new average A_1 was computed exactly, with a revised estimate value, rather than using the boundary rule.

In general, the boundary rule works well for the binomial problem. The exact solution, as given by Farr's algorithm (Section 3.2.8), is found in most cases, and when it is not, the boundary rule gives a solution with I_1 exceeding the optimum by only 0.01 or less.

When the data (or sufficient statistic) set is continuous rather than discrete, the marginal probability of an added value is infinitesimal, so the boundary rule may be used. It is easy to show that, for the Normal problem of Section 3.3.1, the test is exactly satisfied by the code derived in that section. All sufficient statistic sets have the form

$$\hat{\theta} - w/2 \leq \bar{y} < \hat{\theta} + w/2$$

where $w^2 = 12\sigma^2/N$.

For some $\hat{\theta} \in \Theta^*$, the average value of $I_1 - I_0$ in the set $t(\hat{\theta})$ was shown to be

$$A = \frac{1}{2}\log(\pi e/6)$$

For any value $\bar{y}$ in the set, we have

$$\begin{aligned} I_{1s}(\bar{y}) &= -\log q(\hat{\theta}) - \log g(\bar{y}|\hat{\theta}) \\ &= -\log\frac{w}{L} + \frac{1}{2}\log(2\pi\sigma^2/N) + N(\bar{y} - \hat{\theta})^2/2\sigma^2 \end{aligned}$$

and $I_{0s}(\bar{y}) = \log L$.

Substituting $w^2 = 12\sigma^2/N$ gives

$$\begin{aligned} I_{1s}(\bar{y}) - I_{0s}(\bar{y}) &= \frac{1}{2}\log(\pi e/6) - \frac{1}{2} + N(\bar{y} - \hat{\theta})^2/2\sigma^2 \\ &= A - \frac{1}{2} + N(\bar{y} - \hat{\theta})^2/2\sigma^2 \end{aligned}$$

For $\bar{y}$ a data value at the upper or lower boundary of a set, $|\bar{y} - \hat{\theta}| = w/2$, hence

$$
\begin{aligned}
I_{1s}(\bar{y}) - I_{0s}(\bar{y}) &= A - \frac{1}{2} + N(w^2/4)/2\sigma^2 \\
&= A - \frac{1}{2} + N(12\sigma^2/4N)/2\sigma^2 \\
&= A + 1
\end{aligned}
$$

Thus, the boundary rule is exactly satisfied by the SMML code.

This boundary rule is only a rule of thumb, and in general will not lead to a strictly optimal explanation code. If the data groups are grown one after another, choosing the latest group to minimize $\mathrm{Av}(I_1(x) - I_0(x))$ within it is a rule which has no regard to the consequences for groups not yet grown, and so cannot be expected to achieve a global minimization of $I_1 - I_0$. However, the rule of thumb gives some insight into the size of data group to be expected in an optimal code. It can also be used to construct codes which, while not optimal, are often very close.

3.3.3 Estimation of Normal Mean with Normal Prior

The data are N independent values drawn from a Normal with unknown mean θ and known standard deviation σ. The prior density $h(\theta)$ is also of Normal form, with mean zero and S.D. α.

$$
f(x|\theta) = \prod_{n=1}^{N} \frac{1}{\sqrt{2\pi}\sigma} e^{-(y_n - \theta)^2/2\sigma^2}
$$

$$
h(\theta) = \frac{1}{\sqrt{2\pi}\alpha} e^{-\theta^2/2\alpha^2}
$$

As before, the sample mean $\bar{y} = \frac{1}{N}\sum y_n$ is a sufficient statistic, so we deal with it alone, replacing $f(x|\theta)$ by

$$
g(\bar{y}|\theta) = \frac{1}{\sqrt{2\pi\sigma^2/N}} e^{-N(\bar{y} - \theta)^2/2\sigma^2}
$$

For simplicity of notation, we will treat $\bar{y}$ as the data, and write $r_s(\bar{y})$ as $r(\bar{y})$, $t_s(\hat{\theta})$ as

$$
t(\hat{\theta}) = \{\bar{y} : m(\bar{y}) = \hat{\theta}\}
$$

and so on.

Without loss of generality, we scale the problem so that $\sigma^2/N = 1$.

The marginal density of data is

$$
r(\bar{y}) = \frac{1}{\sqrt{2\pi}\beta} e^{-\bar{y}^2/2\beta^2}
$$

where $\beta^2 = \alpha^2 + 1$

Consider some estimate $\hat{\theta} \in \Theta^*$ and its associated data group $t(\hat{\theta})$, which will be an interval $a < \bar{y} < b$.

Define

$$q = \int_a^b r(\bar{y}) \, d\bar{y}$$

$$s^2 = \int_a^b \bar{y}^2 r(\bar{y}) \, d\bar{y}$$

Then, within the group $t(\hat{\theta})$

$$\text{Av}(\bar{y} - \hat{\theta})^2 = s^2/q - \hat{\theta}^2$$
$$\text{Av}(\bar{y}^2) = s^2/q$$

so

$$\text{Av}(I_1(\bar{y})) = -\log q + \frac{1}{2}\log(2\pi) + (s^2/q - \hat{\theta}^2)/2$$

$$\text{Av}(I_0(\bar{y})) = \frac{1}{2}\log(2\pi\beta^2) + (s^2/q)/2\beta^2$$

$$\text{Av}(I_1(\bar{y}) - I_0(\bar{y})) = (s^2/2q)(1 - \frac{1}{\beta^2}) - \log\beta - \log q - \hat{\theta}^2/2$$

At the lower end of the data group, $\bar{y} = a$, and

$$I_1(a) = (\hat{\theta} - a)^2/2 + \frac{1}{2}\log(2\pi) - \log q$$

$$I_0(a) = a^2/2\beta^2 + \frac{1}{2}\log(2\pi\beta^2)$$

$$I_1(a) - I_0(a) = (\hat{\theta} - a)^2/2 - a^2/2\beta^2 - \log\beta - \log q$$

Applying the "boundary rule" rule of thumb from Section 3.3.2 at a gives a "boundary condition" BCa:

$$I_1(a) - I_0(a) = \text{Av}(I_1(\bar{y}) - I_0(\bar{y})) + 1$$
$$BCa : (\hat{\theta} - a)^2/2 - a^2/2\beta^2 = (s^2/2q)(1 - 1/\beta^2) - \hat{\theta}^2/2 + 1$$

Similarly applying the rule at the upper boundary $\bar{y} = b$:

$$BCb : (b - \hat{\theta})^2/2 - b^2/2\beta^2 = (s^2/2q)(1 - 1/\beta^2) - \hat{\theta}^2/2 + 1$$

We also have the minimizing relation R3, which must be obeyed by any optimal code:

$$\hat{\theta} \quad \text{maximizes} \quad \int_a^b r(\bar{y}) \log g(\bar{y}|\hat{\theta}) \, d\bar{y}$$

which in this case reduces to $\hat{\theta} = \text{Av}(\bar{y})$:

$$\text{R3:} \quad \hat{\theta} = \int_a^b \bar{y} r(\bar{y}) \, d\bar{y}/q$$

Whereas, in the similar example with $h(\theta)$ constant, the optimum code satisfied the group boundary conditions as well as R1–R3, we now find that the three equations BCa, BCb and R3 are incompatible, except for a single data group with $\hat{\theta} = 0$, $a = -b$. If we wish to use the boundary condition rule of thumb to construct a good, but not exactly SMML estimator for this problem, we can use any two of the three equations for growing groups, but these groups will not in general satisfy the third equation. We show below the estimators constructed by group growth using two of the possible choices of equations. In both cases, the construction began with the central data group $(j = 0)$ satisfying all three equations:

$$\hat{\theta}_0 = 0; \quad a_0 = -b_0;$$

$$\begin{aligned} b_0^2 &= s_0^2/q_0 + 2\beta^2/(\beta^2 - 1) \quad \text{(from BCb)} \\ &= s_0^2/q_0 + 2\beta^2/\alpha^2 \end{aligned}$$

The next group, $j = 1$, has lower limit $a_1 = b_0$ fixed, and its estimate value $\hat{\theta}$ and upper limit b_1 obtained by simultaneous solution of either BCb and R3, or BCa and BCb. Then the group $j = 2$ is grown and so on. The complete estimator is symmetric about $\bar{y} = 0$. For the calculations leading to the estimators shown below, the marginal distribution $r(\bar{y})$ was truncated beyond $\pm 30\beta$, i.e., thirty standard deviations.

The third pair of equations, BCa and R3, leads to the growth of groups which are well-sized near the middle of the marginal distribution of x, but which become over-large, eventually unbounded, for $|\bar{y}| > 2\beta$. This failure is not surprising, as growing a new group from lower bound a to satisfy BCa and R3 is seeking to choose an upper bound b so that the lower boundary a (which is fixed) should not be moved!

The results shown are for $\sigma^2/N = 1$, $\alpha = 2$. That is, the width of the prior density peak is only twice the width of the distribution of $\bar{y}$ about the true mean θ. The prior density is thus far from constant, and changes by large factors over the width of a single data group.

As well as showing the estimators grown by the rule of thumb, Table 3.2 also shows an estimator very close to the exact SMML solution. It was formed by iterative relaxation of one of the rule-of-thumb estimators. Holding the group boundaries $\{b_j\}$ constant, the estimates $\{\hat{\theta}_j\}$ and their prior probabilities $\{q_j\}$ were calculated using minimizing relations R3 and R2. Then, holding the $\{\theta_j\}$ and $\{q_j\}$ constant, the boundaries were recomputed using relation R1. The process was repeated until convergence.

Table 3.2. Estimators for Normal mean with $N(0, 4)$ prior grown by boundary rule, and exact SMML.

	Using R3,BCb		Using BCa,BCb		Exact SMML	
j	$\hat{\theta}_j$	b_j	$\hat{\theta}_j$	b_j	$\hat{\theta}_j$	b_j
0	0	1.893	0	1.893	0	1.920
1	3.056	5.797	3.074	5.792	3.091	5.980
2	6.499	10.482	6.503	10.465	6.667	10.884
3	10.923	16.838	10.909	16.808	11.311	17.544
4	17.125	25.979	17.097	25.934	17.821	27.113
5	26.169	39.444	26.125	39.377	27.295	41.196
6	39.570	59.466	39.503	59.381	41.317	62.145
7	59.550	67.082	50.585	67.082	62.225	67.082
	$I_1 - I_0 = 0.17559$		$I_1 - I_0 = 0.17567$		$I_1 - I_0 = 0.17531$	

Table 3.3. Estimate separations and data group sizes for boundary rule and SMML.

	Using BCa,BCb		Exact SMML	
j	$\hat{\theta}_j - \hat{\theta}_{j-1}$	$b_j - a_j$	$\hat{\theta}_j - \hat{\theta}_{j-1}$	$b_j - a_j$
0	3.074	3.787	3.092	3.841
1	3.074	3.899	3.092	4.059
2	3.429	4.673	3.576	4.904
3	4.407	6.344	4.644	6.660
4	6.188	9.126	6.510	9.569
5	9.027	13.443	9.474	14.083
6	13.379	20.004	14.022	20.948

All estimates shown have the same highest boundary, $b_7 = 67.082$. This is fixed by the truncation of the marginal density $r(\bar{y})$ to 30 standard deviations, and so the $j = 7$ entries are not correct. However, the effect of this error on estimates and boundaries for $j < 7$ is minute. Changing the truncation of $r(\bar{y})$ from 30 to 32 standard deviations changes b_6 by less than 0.001. The most obvious feature of Table 3.2 is that the "boundary rule" constructions give expected message lengths exceeding the optimum by only a negligible 0.0004 nit or less.

While Table 3.2 shows that, even with the strongly non-uniform prior of this example, the boundary rule construction gives an estimator close to the SMML estimator, we are more directly interested in its ability to approximate the optimum sizes of data groups. Table 3.3 compares the widths $(b_j - a_j)$ of the data groups, and the spacing $(\hat{\theta}_j - \hat{\theta}_{j-1})$ between members of Θ^*, for the (BCa, BCb) approximation and the SMML estimator. It will be seen that the boundary rule gives data group sizes and estimate spacing quite close to those of the SMML estimator, and varying with $\bar{y}$ in much the same way.

Table 3.3 shows how the very non-uniform prior affects the SMML code as compared to the case of a flat prior. In the latter, the width of each data group, and the spacing between adjacent members of Θ^*, are constant with value $\sqrt{12} \approx 3.464$. For the tight Normal prior, the estimates in Θ^* are a little closer together near the peak of the prior, with spacing about 3.1 but become more widely separated in the tail of the prior. The data groups are a little wider near the peak and much wider in the tail. I suspect that a high slope in the prior induces large data groups, and that, to the extent that data group enlargement leaves possible, a negative second derivative in the log-likelihood function induces closer spacing of estimate values, but have not studied the matter.

The most notable effect of the non-uniform prior is the shift of estimates towards higher-prior values. Whereas for a flat prior, the estimate for a data group lies midway between the upper and lower $\bar{y}$ limits of the group, the estimate $\hat{\theta}_4$ in Table 3.2 is 17.82, much closer to the lower boundary $a_4 = b_3 = 17.54$ than to the upper limit $b_4 = 27.11$.

3.3.4 Mean of a Multivariate Normal

The Normal example with uniform prior density can be extended to cover the multi-dimensional Normal distribution. Now, the data are an ordered set of N real vector values $\underline{x} = (\underline{y}_1, \underline{y}_2, \ldots, \underline{y}_N)$ with mean $\bar{y}$, believed to be drawn from a multivariate Normal with unknown mean $\underline{\theta}$ and known covariance matrix $\mathbf{M}$. The dimension of data and mean is D. Henceforth we will drop the vector indication and write data values, sample mean and population mean simply as $y_n, \bar{y}, \theta$. Then the model density is

$$f(x|\theta) = \left(\frac{1}{\sqrt{2\pi}}\right)^{ND} \left(\frac{1}{\sqrt{|\mathbf{M}|}}\right)^N \prod_{n=1}^{N} e^{-0.5(y_n-\theta)^T \mathbf{M}^{-1}(y_n-\theta)}$$

It is convenient to define a non-singular linear transformation of the data and θ spaces

$$z = \mathbf{A}^{-1}y, \ \phi = \mathbf{A}^{-1}\theta$$

such that $\mathbf{A}^T \mathbf{M}^{-1} \mathbf{A} = \mathbf{I}_D$, the $D \times D$ identity matrix. Then the model density of the transformed data $x_A = \{z_1, \ldots, z_N\}$ is

$$f_A(x_A|\phi) = \left(\frac{1}{\sqrt{2\pi}}\right)^{ND} \prod_{n=1}^{N} e^{-(z_n-\phi)^2/2}$$

and the model density of the sufficient statistic $\bar{z} = \frac{1}{N}\sum_{n=1}^{N} z_n$ is

$$g(\bar{z}|\phi) = \left(\sqrt{\frac{N}{2\pi}}\right)^D e^{-N(\bar{z}-\phi)^2/2}$$

For simplicity, we assume the prior density of θ to be uniform in some large region. Hence, the prior density of ϕ and the marginal density $r(\bar{z})$ of $\bar{z}$ are

$$h(\phi) = r(\bar{z}) = \frac{1}{V}$$

As before, we are neglecting "edge effects" near the boundary of V.

In the one-dimensional case, the set Θ^* of possible estimates was an evenly spaced set of values, and the data group $t(\hat{\theta})$ associated with an estimate $\hat{\theta}$ was an interval of sample mean values, all intervals having the same width w. We assume without proof that in D dimensions, the set Φ^* of possible $\hat{\phi}$ estimates is a regular lattice of points in Φ, the space of ϕ, and that every data group $t(\hat{\phi}) = \{\bar{z} : m(\bar{z}) = \hat{\phi}\}$ has the same size and shape.

Let w be the volume of a data group $t(\hat{\phi})$. Then the total marginal probability of $\bar{z}$ values within $t(\hat{\phi})$ is w/V, and by R2, the coding probability $q(\hat{\phi}) = w/V$ for all $\hat{\phi} \in \Phi^*$.

As all data groups are similar, the average value I_{1s} equals the average value of $I_{1s}(\bar{z})$ over any one data group. Hence,

$$
\begin{aligned}
I_{1s} &= -\log q(\hat{\phi}) - \mathrm{Av}_{\bar{z} \in t(\hat{\phi})} \log g(\bar{z}|\hat{\phi}) \qquad \text{for some } \hat{\phi} \in \Phi^* \\
&= -\log(w/V) + (D/2)\log(2\pi/N) + (N/2)\,\mathrm{Av}_{\bar{z} \in t(\hat{\phi})}\,(\bar{z} - \hat{\phi})^2
\end{aligned}
$$

But $I_{0s} = -\mathrm{Av}\,\log r(\bar{z}) = \log V$

$$I_{1s} - I_{0s} = -\log w + (D/2)\log(2\pi/N) + (N/2)\,\mathrm{Av}_{\bar{z} \in t(\hat{\theta})}\,(\bar{z} - \hat{\phi})^2$$

The quantity $\mathrm{Av}_{\bar{z} \in t(\hat{\phi})}\,(\bar{z} - \hat{\phi})^2$ depends on both the volume w and the shape of the group $t(\hat{\phi})$. Because $q(\hat{\phi})$ is the same for all groups, it is clear that $t(\hat{\phi})$ is the Voronoi region of $\hat{\phi}$, i.e., the set of values closer to $\hat{\phi}$ than to any other estimate in Φ^*, where closeness is measured by Euclidean distance. The shape of a group is thus determined by the geometry of the lattice Φ^*. If the lattice geometry and group shape are held constant, and the volume w of the groups is changed, the average squared distance of points within a group from its centre $\hat{\phi}$ will vary with w as $\mathrm{Av}_{\bar{z} \in t(\hat{\phi})}\,(\bar{z} - \hat{\phi})^2 = D\kappa_D w^{2/D}$ where κ_D depends on the lattice geometry

$$I_{1s} - I_{0s} = -\log w + (D/2)\log(2\pi/N) + (N/2)D\kappa_D w^{2/D}$$

which is minimized when $w = (N\kappa_D)^{-D/2}$, $(N/2)D\kappa_D w^{2/D} = D/2$, and $\mathrm{Av}_{\bar{z} \in t(\hat{\phi})}\,(\bar{z} - \hat{\phi})^2 = D/N$.

$$
\begin{aligned}
I_1 - I_0 &= I_{1s} - I_{0s} \\
&= (D/2)\log(2\pi/N) + (D/2)\log(N\kappa_D) + D/2 \\
&= (D/2)\{1 + \log(2\pi\kappa_D)\}
\end{aligned}
$$

The best choice of lattice geometry will minimize κ_D. A lattice with minimal κ_D is called an optimum quantizing lattice. Quantizing lattices have been discussed by Conway and Sloane [11]. For general D, the optimum lattice is unknown. However, following Zador [61], we can establish bounds on κ_D.

Suppose $w = 1$. The volume of a hypersphere of radius r is $r^D \, \pi^{D/2}/(D/2)!$ where by $n!$ we mean, for non-integral n, $\Gamma(n+1)$. Then the radius of a sphere with unit volume $(w = 1)$ is

$$r = \{(D/2)!\}^{1/D}/\sqrt{\pi}$$

and the average squared distance of points within the sphere from its centre is

$$r^2 D/(D + 2) = D\frac{\{(D/2)!\}^{2/D}}{(D + 2)\pi}$$

Of all regions with unit volume, the hypersphere has the smallest average squared distance. But the Voronoi region of a lattice cannot be spherical. Hence,

$$\kappa_D > \frac{\{(D/2)!\}^{2/D}}{(D + 2)\pi}$$

The upper bound is obtained by considering Φ^* to be not a lattice but a random selection of points in Φ with unit average density. For some $\bar{z}$, let a be the distance from $\bar{z}$ to the nearest point in Φ^*, i.e., $a^2 = (\bar{z} - \hat{\phi})^2$. The probability that $a > b$ is the probability that there is no member of Φ^* within a sphere of radius b and centred on $\bar{z}$.

$$
\begin{aligned}
\Pr(a > b) &= \exp(-\pi^{D/2} b^D/(D/2)!) \\
\text{Density}(a) &= \frac{\pi^{D/2}}{(D/2)!} D\, a^{D-1} \exp(-\pi^D/2a^D/(D/2)!) \\
\text{Av}(a^2) &= ((D/2)!)^{2/D}(2/D)! \, / \, \pi
\end{aligned}
$$

With probability one, a random choice of Φ^* can be improved by iterative application of relations R1–R3. Hence,

$$\kappa_D < ((D/2)!)^{2/D}(2/D)! \, / \, (\pi D)$$

As D increases, both bounds approach $1/(2\pi e)$ from above. Using Stirling's approximation in the form

$$\log(n!) = (n + 1/2)\log n - n + (1/2)\log(2\pi) + O(1/n)$$

the lower bound gives

$$
\begin{aligned}
\log \kappa_D \; > \; & \frac{2}{D}\left[\frac{D+1}{2}\log\frac{D}{2} - \frac{D}{2} + \frac{1}{2}\log(2\pi)\right] - \log(D + 2) \\
& - \log\pi + O(1/D^2) \\
> \; & (\log(D\pi) - 2)/D - \log(2\pi e) + O(1/D^2)
\end{aligned}
$$

Similarly, and using $\log(\epsilon!) = -\gamma\epsilon + O(\epsilon^2)$ for small $\epsilon > 0$, where γ is Euler's constant, the upper bound gives

$$\log \kappa_D < (\log(D\pi) - 2\gamma)/D - \log(2\pi e) + O(1/D^2)$$

Substituting in $I_1 - I_0 = (D/2)\{1 + \log(2\pi\kappa_D)\}$:

$$\frac{1}{2}\log(D\pi) - \gamma > (I_1 - I_0) > \frac{1}{2}\log(D\pi) - 1 \quad \text{to order } 1/D$$

Table 3.4 gives upper and lower bounds on κ_D for a few values of D, and, where available, the best known value κ_D. It also gives for the multivariate Normal the lower bound on $(I_1 - I_0)$ implied by the lower bound on κ_D, and the approximate form $(I_1 - I_0) \approx \frac{1}{2}\log(2\pi D) - \gamma$. For $D > 1$, this approximation falls between the upper and lower bounds, approaching the upper bound for large D. It is above the value achieved by the known lattices for $D = 2$ and 3. It may therefore be regarded as a slightly pessimistic approximation for $D > 1$.

Table 3.4. Bounds and approximations for Lattice Constant κ_D.

D	Lattice Constant κ_D			$I_1 - I_0$	
	Upper Bound	Known	Lower Bound	Lower Bound	Approx
1	0.5000	0.08333	0.08333	0.17649	-0.00485
2	0.15916	0.08019	0.07958	0.30685	0.34172
3	0.11580		0.07697	0.41025	0.54446
10	0.07614		0.06973	0.82869	1.14644
100	0.06133		0.06080	1.88649	2.29773
1000	0.05896		0.05891	3.02741	3.44903
10000	0.05860		0.05860	4.17765	4.60032

This is a remarkable result. The problem of estimating the mean θ of a multivariate Normal with known covariance can be transformed (as was done above) to one of estimating the mean ϕ of a multivariate Normal with covariance matrix $\mathbf{I}_D$. For this problem, the model density $g(\bar{z}|\phi)$ can be factored as

$$g(\bar{z}|\phi) = \prod_{k=1}^{D} \sqrt{N/(2\pi)} \, \exp(-N(\bar{z}_k - \phi_k)^2/2)$$

where $\bar{z}_k$ and ϕ_k are the kth components of the D-dimensional vectors $\bar{z}$ and ϕ. This separation into independent factors shows that the problem can be seen as D independent problems. Each is a simple univariate problem: to estimate ϕ_k given a sample with mean $\bar{z}_k$. Since we have assumed $h(\phi)$ to be uniform, the D different problems do not interact through either the prior or the likelihood: they are entirely separate. Conversely, if we are presented with D independent problems, each concerned with estimating the mean of a univariate Normal with known standard deviation, we can, if we choose, bundle them all together as a single multivariate problem.

If each component problem were considered separately, we would obtain D separate explanation messages. Each would assert one component of $\hat{\phi}$, and encode one component of the N z-vectors. Each explanation would have $I_1 - I_0 = \frac{1}{2}\log(2\pi e)$, and so if the explanations were concatenated to provide a single message encoding all the data, the expected length of that message would be $(D/2)\log(2\pi e)$ longer than the shortest non-explanation message for all the data. That is, we would have, for the whole message encoding all the data,

$$I_1 - I_0 = \frac{1}{2} D \log(2\pi e)$$

If all the data is encoded in one explanation, we would obtain the same $(I_1 - I_0)$ difference, growing linearly in D, if we chose a rectangular lattice for the set Φ^*. However, the rectangular lattice is not an optimum quantizing lattice for $D > 1$, having $\kappa_D = 1/12$ for all D. Use of an optimum lattice, or failing that a random selection of Φ^*, gives a smaller $I_1 - I_0$ growing only logarithmically with D.

Even if D univariate mean estimation problems are completely independent, we do better to encode their data in a single explanation. Closely parallel results are well known in Information Theory. If several analogue quantities are to be digitized for storage or transmission, the digitizing process, by representing each quantity to only a fixed number of digits, inevitably introduces a "quantization noise". By quantizing several quantities at once, treating them as forming a vector and choosing their digital representation from a quantizing lattice, the mean quantizing noise is reduced without use of extra digits.

A curious consequence of combining independent problems in this way is that now the SMML estimate for one problem depends on data from other, unrelated problems. Consider a possible set of $\hat{\phi}$ values for $D = 2$. For two dimensions, an optimum quantizing lattice consists of hexagonal $t(\hat{\phi})$ Voronoi regions. Note that there is no reason for the lattice to be aligned with the component axes. Supposing the two components of the problem are independent, the model distribution for the first component $\bar{z}_1$ is Normal with mean ϕ_1, S.D. $1/\sqrt{N}$. The second component $\bar{z}_2$ has a Normal distribution, with mean ϕ_2, S.D. $1/\sqrt{N}$.

If the data for problem 1 happened to yield a sample mean $\bar{z}_1 = a$, the SMML estimate $\hat{\phi} = (\hat{\phi}_1, \hat{\phi}_2)$ could take any of a range of values. Which value would be estimated for ϕ_1 would depend on the sample mean $\bar{z}_2$ obtained in the unrelated second problem.

This behaviour of SMML estimators may appear bizarre. Note, however, that all possible values for the estimate $\hat{\phi}_1$, given $\bar{z}_1 = a$, are believable. Their average squared difference from the sample mean a is $1/N$, exactly equal to the expected squared difference of the true mean ϕ_1 from the sample mean. For any D, we have

$$\text{Av}\,(\bar{z} - \hat{\phi})^2 = D/N = \sum_{k=1}^{D} \text{Av}\,(\bar{z}_k - \hat{\phi}_k)^2$$

As there is no distinction among the D components, each component should contribute equally to the average. Hence, $\text{Av}\,(\bar{z}_k - \hat{\phi}_k)^2 = 1/N$ for any component k.

Although the shape of the optimum Voronoi region for large D is unknown, the close approximation of the optimum lattice to the behaviour of spherical regions suggests that the optimum SMML data group t is nearly spherical in this problem for large D. Making this assumption, and using for the optimum volume w, gives a region of radius $r = \sqrt{(D+2)/N}$. Consider one component $(\hat{\phi}_k - \bar{z}_k)$ of the difference between sample mean and estimate. As noted above, this may take many different values, depending on the values of all components of $\bar{z}$. Indeed, without knowledge of the other components of $\bar{z}$, all that can be said of the vector difference $(\hat{\phi} - \bar{z})$ is that it lies in a sphere centred on the origin and radius r. Thus, if the values of other components of $\bar{z}$ are thought of as random variables selected from their respective (uniform) marginal distributions, the value of $(\hat{\phi}_k - \bar{z}_k)$ for given $\bar{z}_k$ can be regarded as a random variable generated by projecting a random point within this sphere onto the k axis. The density of $(\hat{\phi}_k - \bar{z}_k)$ is then proportional to

$$\{r^2 - (\hat{\phi}_k - \bar{z}_k)^2\}^{(D-1)/2} = \{(D+2)/N - (\hat{\phi}_k - \bar{z}_k)^2\}^{(D-1)/2}$$

For any D, this density has mean zero and standard deviation $1/N$, as already noted. For large D, it closely approximates a Normal shape. Hence, the density of the SMML estimate $\hat{\phi}_k$ is (approximately) Normal$(\bar{z}_k, 1/N)$.

But consider the ordinary Bayes posterior density of $\hat{\phi}_k$ given $\bar{z}_k$,

$$\text{Post}(\hat{\phi}_k | \bar{z}_k) = h_k(\hat{\phi}_k) \sqrt{\frac{N}{2\pi}}\, \exp\{-N(\bar{z}_k - \hat{\phi}_k)^2/2\} \,/\, r_k(\bar{z}_k)$$

As $h_k(\hat{\phi}_k)$ and $r_k(\bar{z}_k)$ are both uniform, this is also Normal$(\bar{z}_k, 1/N)$. Thus, for this problem, the SMML estimator gives estimates for individual components of the parameter which behave very much like values selected at random from the posterior distribution of the component.

Although this result was derived only for multivariate Normal problems (or, equivalently, a number of unrelated scalar Normal problems considered together), and for a Uniform prior, we will show later a much more general version of the result.

The exact properties of the SMML code discussed here depend on the very simple nature of the multivariate Normal problem. The form of the model $f(x|\theta)$ and the assumption of a Uniform prior lead to use of a Θ^* set which is very regular, with constant spacing of estimates throughout the model space. Most multi-parameter problems do not have this property, and so their optimal Θ^* sets will not be regular lattices. However, the broad thrust of the conclusions reached here do carry over to more general multivariate and multi-parameter problems. In particular, the surprisingly good performance of random Θ^* sets does not depend on the simplicity of the present problem. A random selection of points in a model space has no preferred orientation, no regular structure, and indeed only one property: the mean density of points as a function of position in the space. Provided this mean density is chosen correctly, a random Θ^* will achieve an expected message length very close to that of a truly optimal SMML code, with efficiency increasing with increasing model space dimension D and with $(I_1 - I_0)$ close to the $\frac{1}{2} \log(D\pi)$ behaviour of the Normal model.

3.3.5 Summary of Multivariate Mean Estimator

The SMML estimator for the D-dimensional mean of a multivariate Normal with known covariance has an excess expected message length $I_1 - I_0$ bounded by

$$\frac{1}{2} \log(D\pi) - \gamma + O(1/D) > (I_1 - I_0) > \frac{1}{2} \log(D\pi) - 1 + O(1/D)$$

assuming a Uniform prior density for the mean. Thus, $(I_1 - I_0)$ grows only logarithmically with D. If the data and mean coordinate systems are such that, or are transformed so that, the covariance matrix is diagonal, the estimator problem is equivalent to D unrelated component problems, each concerned to estimate the mean of a univariate Normal. Nonetheless, the SMML estimate for each component mean depends on data for all components. For large D, if the data for components other than k are treated as random, the estimate for component mean k behaves roughly as if randomly selected from its Bayes posterior distribution given the data component k.

Further, if the D component problems are treated separately, the overall expected message length excess

$$(I_1 - I_0) = D \log \sqrt{\pi e / 6}$$

grows linearly with D.

These results apply approximately to many multivariate problems and sets of univariate problems where the log likelihood has an approximately quadratic behaviour near its maximum and the prior is slowly varying.

3.3.6 Mean of a Uniform Distribution of Known Range

The foregoing examples have all dealt with model distributions having a minimal sufficient statistic. For such models, it is not surprising that an estimator can be found whose estimates capture most of the relevant information in the data. We now treat a problem without a minimal sufficient statistic.

The data $x = \{y_n : n = 1, \ldots, N\}$ are an ordered set of N values drawn from a Uniform density of known range (which for simplicity we take to be one) and unknown mean θ. The data are given to a precision $\delta \ll 1/N$.

Define $a = \min\{y_n : n = 1, \ldots, N\}$; $b = \max\{y_n : n = 1, \ldots, N\}$.

We assume that the rounding of data to precision δ is such that $(b-a) \leq 1$. Then

$$f(x|\theta) = \begin{cases} \delta^N & \text{if } (\theta - \frac{1}{2}) \leq a \leq b \leq (\theta + \frac{1}{2}) \\ 0 & \text{otherwise} \end{cases}$$

Clearly, the pair (a, b) is a sufficient statistic, but is not minimal as it has two components. No minimal sufficient statistic exists. However, rather than using (a, b), we will use the pair $(c = (a+b)/2, w = (b-a))$, which is also a sufficient statistic as it is a one-to-one function of (a, b). Then

$$f(x|\theta) = \begin{cases} \delta^N & \text{if } c - \frac{1}{2}(1-w) \leq \theta \leq c + \frac{1}{2}(1-w) \\ 0 & \text{otherwise} \end{cases}$$

We assume the prior density $h(\theta)$ of the mean is uniform in some range $L \gg 1$, and ignore effects arising near the end of this range. Hence, the marginal density of the sample centre m is also substantially uniform in the same range.

The marginal distribution of the data is

$$\begin{aligned} r(x) &= \int_L h(\theta) f(x|\theta) \, d\theta \\ &= \int_{c-(1-w)/2}^{c+(1-w)/2} (1/L)\delta^N \, d\theta \\ &= (\delta^N/L)(1-w) \end{aligned}$$

and the length of the shortest non-explanation message encoding x is

$$I_0(x) = \log L - N \log \delta - \log(1 - w)$$

To obtain I_0, the average value of $I_0(x)$ over all x, note that for given θ, the marginal distribution of the sample range w is easily shown to be

$$g_w(w|\theta) = \delta N(N-1)(1-w)w^{N-2}$$

Since this is independent of θ, the prior marginal distribution of w has the same form:

$$r_w(w) = \delta N(N-1)(1-w)w^{N-2}$$

Averaging over this distribution (using an integral to approximate the sum)

$$\begin{aligned} I_0 &= \log L - N \log \delta - N(N-1) \int_0^1 (1-w)w^{N-2} \log(1-w) \, dw \\ &= \log L - N \log \delta - W \text{ (say)} \end{aligned}$$

The integral

$$W = N(N-1) \int_0^1 (1-w)w^{N-2} \log(1-w)\, dw = \sum_{k=2}^N 1/k$$

is rational, but is well approximated by

$$W \approx \log(N + \frac{1}{2}) - 1 + \gamma \approx \log(N + \frac{1}{2}) - 0.42278\ldots$$

where γ is Euler's constant.

The approximation is within 0.006 for $N \geq 2$, so we may take

$$I_0 = \log L - N \log \delta + \log(N + \frac{1}{2}) - 0.423$$

Now consider the construction of an explanation code. If the explanation asserts an estimate $\hat{\theta}$, the detail will encode each value y_n $(n = 1, \ldots, N)$ independently, to precision δ in the range $\hat{\theta} \pm \frac{1}{2}$. The length of the detail, $-\log f(x|\hat{\theta})$, is just $(-N \log \delta)$, provided every y_n value lies in the range $\hat{\theta} \pm 1/2$. Values outside this range cannot be encoded in the detail. Thus, given data x with sample range w and sample centre c, any estimate $\hat{\theta}$ in the range $c \pm \frac{1}{2}(1-w)$ will give the same detail length, but estimates outside this range are unusable. This estimation problem is unusual in that the range of estimate values with which a given data value might be encoded can be extremely small. Indeed, as w approaches 1, the range of usable estimates approaches zero. As this may happen for any c, the set Θ^* must have a very close spacing of estimate values everywhere. The spacing necessary to ensure that a usable estimate can be found for every possible x is $\delta/2$. It might seem, therefore, that as $\delta \rightarrow 0$, the expected assertion length needed to specify one of such a closely spaced estimate set would grow without bound. However, this does not happen.

The best known SMML code Θ^* comprises a hierarchy of estimate values $\{\hat{\theta}\}$ with differing coding probabilities and different sizes of data groups $\{t(\hat{\theta})\}$. At the top of the hierarchy, with the largest coding probabilities $\{q(\hat{\theta})\}$, is a set of estimate values spaced at unit intervals in Θ. We may take these as having integer values of $\hat{\theta}$ (although their actual values would depend on location of the prior range L). Each of these top-level (level 0) estimates $\hat{\theta}$ has a data group $t(\hat{\theta})$ comprising all members x of X for which $f(x|\hat{\theta}) > 0$.

For all lower levels of the hierarchy (levels $1, 2, \ldots, K, \ldots$), estimates at level K are odd multiples of 2^{-K} and each $\hat{\theta}$ at level K has a data group including all x for which $f(x|\hat{\theta}) > 0$ and which are not included in some data group at a higher level (less than K). This code is not known to be optimal, but obeys all three optimizing relations R1, R2, R3. Clearly, this construction of Θ^* results in all estimates at the same level having equal coding probabilities $q()$, and in the $q()$ at level $(K+1)$ being less than the $q()$ at level K. The lowest level of the hierarchy has $2^{-K} \leq \delta$, as at that point Θ^* contains estimates at intervals at most $\delta/2$ everywhere. In a prior range $L \gg 1$, there are approximately L level 0 estimates, L level 1 estimates, $2L$ level 2, $4L$ level 3, and so on.

The effect of this choice of Θ^* and the resulting partition of the sufficient statistics (c, w) plane is that a data vector x results in an estimate $\hat{\theta}$ which, if expressed as a binary number, has just sufficient digits following the binary point to ensure that $f(x|\hat{\theta}) > 0$. For a sample size N, the most probable value of w is about $1 - 2/(N+1)$, so most of the marginal probability of data is concentrated in data with similar values of w. Since the estimate can differ from the sample centre c by at most $\pm(1-w)/2$, the number of digits after the binary point in $\hat{\theta}$ (which equals the hierarchic level K) is typically about $\log_2(N)$. Note however that the data group of an estimate at a high level (small K) still includes some data vectors with values of w close to one. Estimates at levels near $\log_2(N)$ have collectively most of the coding probability, but since there are many of them for large N, their individual coding probabilities may be small.

Calculation of the message lengths for data requires a series of integrations to evaluate the $q()$ values of the estimates at each level. The difference $I_1 - I_0$ depends weakly on N. A few values are shown in Table 3.5. These results were obtained for the limit of small δ, treating the variate values as continuous.

Table 3.5. Message lengths for Uniform distribution.

Sample Size	$I_1 - I_0$
2	0.463
3	0.528
5	0.579
10	0.617
30	0.642
100	0.651
$N \to \infty$	0.655

The Uniform problem is an example where an efficient SMML code must provide a code for assertions which can specify an estimate with a range of precisions. In this case, the estimated mean $\hat{\theta}$ is stated to just enough binary digits to ensure that $f(x|\hat{\theta}) > 0$. (Note, however, that the assertion string used to encode an estimate has a length (in bits) different from the length of this binary representation, as the probability than an estimate will require k fractional binary digits is not proportional to 2^{-k}.)

It will be seen that, although the SMML estimate is a single number, it usually manages to capture much of the information in both components of the sufficient statistic (c, w). The resulting $I_1 - I_0$ difference is worse (i.e., greater) than is the difference for the more regular Normal problem, but is still less than $\log 2$. Thus, the estimates produced have (loosely) posterior probabilities greater than 0.5 in logarithmic average.

3.4 Some General Properties of SMML Estimators

The SMML estimator $m(x)$ minimizes the expected explanation length

$$I_1 = E\ I_1(x) = -\sum_j q_j \log q_j - \sum_j \sum_{x \in t_j} r(x) \log f(x|\hat{\theta}_j)$$

where $r(x) = \int h(\theta)f(x|\theta)\ d\theta$,

$$\Theta^* = \{\hat{\theta}_j : j = 1, 2, \ldots;\ \exists x : m(x) = \hat{\theta}_j\}$$

$$t_j = \{x : m(x) = \hat{\theta}_j\}; \qquad q_j = \sum_{x \in t_j} r(x)$$

or, if x is treated as continuous, the SMML estimator minimizes the difference between the expected explanation length of the data and the length I_0 of the optimal non-explanatory coding, that is,

$$
\begin{aligned}
I_1 - I_0 \ &= \ E\ (I_1(x) - I_0(x)) \\
&= \ -\sum_j q_j \log q_j - \sum_j \int_{x \in t_j} r(x) \log \left(f(x|\hat{\theta}_j)\ /\ r(x) \right) dx
\end{aligned}
$$

where now $f(x|\theta)$ and $r(x)$ are densities rather than probabilities, and

$$I_0 = -\int_X r(x) \log r(x)\ dx; \qquad q_j = \int_{x \in t_j} r(x)\ dx$$

3.4.1 Property 1: Data Representation Invariance

I_1 and the estimator $m()$ are invariant under one-to-one changes in the representation of the data. That is, if, instead of being represented by the value x, the data is represented by a value $y = u(x)$ where u is an invertible function, and the SMML process applied to y yields an estimator $m_y(y)$, then for all x

$$m_y(u(x)) = m(x)$$

and $I_1 - I_0$ is unchanged.

For discrete data, i.e., when the set X of possible values of x is countable, this property is obvious. The transformation $y = u(x)$ amounts simply to a relabelling of the discrete values, and for any θ, the model probabilities of an x value and its corresponding y value are equal. Hence, the marginal probabilities $r(x)$ and $r_y(y = u(x))$ are also equal. The model and marginal probabilities are the only ways in which the data enter the definition of I_1 and the construction of the estimator.

When the data are treated as continuous, so that $f(x|\theta)$ and $r(x)$ are densities on X, data representation invariance (data invariance for short) still holds, providing the model and marginal densities of $y = u(x)$ are appropriately transformed:

$$f_y(y|\theta) = \frac{dx}{dy} f(x|\theta); \qquad r_y(y) = \frac{dx}{dy} r(x)$$

For continuous data, I_1 is strictly speaking undefined, but the difference $(I_1 - I_0)$ is defined. The invariance property is, of course, shared by all estimation processes in which the data enters the process only via the model distribution $f(x|\theta)$.

3.4.2 Property 2: Model Representation Invariance

I_1 and the estimator $m()$ are invariant under one-to-one changes in the representation of the models. If the representation of models is changed from θ to $\phi = g(\theta)$ where $g()$ is an invertible function, and an SMML estimator $m_\phi()$ constructed to estimate ϕ, then for all x, $m_\phi(x) = g(m(x))$, and I_1 is unchanged. Note that the prior density of ϕ, $h_\phi(\phi)$, must be the appropriately transformed density:

$$h_\phi(\phi) = \frac{d\theta}{d\phi} h(\theta)$$

This property follows by observing that (a) the model distribution $f(x|\theta)$ is unchanged:

$$f_\phi(x|g(\theta)) = f(x|\theta)$$

and (b) the prior density $h()$ enters the definition of I_1 only via the integral

$$r(x) = \int_\Theta h(\theta) f(x|\theta) \, d\theta = \int_\Phi h_\phi(\phi) f_\phi(x|\phi) \, d\phi$$

The marginal distribution of data $r(x)$ depends on our prior expectations about the possible models of the data, and how the data might behave under each of these models, but is not affected by how we choose to describe or parameterize the models.

Model representation invariance (model invariance for short) is lacking in some commonly used estimation methods. For instance, choosing an estimator to have zero bias is not model-invariant: the bias $E(\theta - \hat{\theta})$ of an estimator is defined with respect to a particular parameterization of the model. For θ a scalar parameter, an estimator which had zero bias for θ would in general have non-zero bias for an alternative parameter $\phi = g(\theta)$. Similarly, choosing an estimator to have small variance $E((\theta - \hat{\theta})^2)$ is tied to the particular parameterization θ.

Among Bayesian methods, minimum-cost estimation, which seeks to minimize the expected value of a loss function $L(\theta, \hat{\theta})$ given the data, is model invariant, but choosing as the estimate the mean or mode of the posterior density $h(\theta)f(x|\theta)/r(x)$ is not.

Unless there are very good grounds for preferring one parameterization of the models over all others, induction and estimation methods which are not model-invariant appear undesirable. In essence, an induction or estimation method asks the data a question, and obtains an inference or estimate as answer. Changing the model representation amounts to changing the language used to frame the question and answer. If the method is not model-invariant, changing the language will change not merely the representation but the meaning of the answer.

3.4.3 Property 3: Generality

The SMML method can be used for a wide variety of problems. Many other methods are restricted in the classes of models to which they can be applied. For instance, the classical method of seeking a "Minimum-Variance Unbiased" estimator succeeds only for a small family of model distributions, and only for particular parameterizations of these models.

The Maximum Likelihood method requires the set Θ of possible models to be either countable or a continuum of fixed dimension. That is, it cannot directly be used to choose among models with different numbers of parameters.

By contrast, the SMML method requires only that (a) the data can be represented by a finite binary string: (b) there exists a language for describing models of the data which is agreed to be efficient, i.e., there exists a prior density $h(\theta)$: (c) the integrals $r(x)$ exist for all possible data values, and satisfy $r(x) \geq 0$, $\sum_X r(x) = 1$.

3.4.4 Property 4: Dependence on Sufficient Statistics

If $y = y(x)$ is a sufficient statistic, then the SMML estimate is a function of y. That is, if y is known to contain all the data information relevant to θ, the SMML estimate depends on no aspect of the data other than y. This property is shared by all methods in which the data enters only via the model distribution $f(x|\theta)$.

3.4.5 Property 5: Efficiency

In inductive and statistical inference, we attempt to infer from a body of data a hypothesis or estimate which we hope will hold for all similar data from the same source. Typically, perhaps always, the inference asserts that the data conforms to some pattern or model and is to this extent non-random.

However, we rarely expect the pattern to entail all details of the given data. There will remain fluctuations in the data, ascribable to sampling, measurement error and the like, for which we offer no detailed account, and which we can only regard as random, unpredictable noise.

The given data contains information about both the pattern to which it conforms and the "noise". The task of an inductive or statistical inference procedure may be regarded as the separation of the data information into "pattern" and "noise" components. The former, suitably represented, becomes the inference or estimate.

The SMML method attempts the separation of pattern and noise information explicitly. The data is recoded into a two-part message, in which the first part (assertion) concerns pattern, and the second part (detail) encodes whatever details of the data cannot be deduced from the assertion. We now argue that, by choosing the shortest possible explanation, we effect the best possible separation between pattern and noise.

The essential theorems for the argument are the classic results of Information Theory: (a) the expected length of a message encoding data from some source is minimized when the probability model assumed in the encoding agrees with the probability distribution characterizing the source; and (b) when data from some source is encoded optimally, the encoded string has the statistical properties of a purely random sequence.

The separation of pattern and noise information into the assertion and detail parts of an explanation can fall short in basically two ways. The assertion may contain noise as well as pattern information, and the assertion may fail to contain all the pattern information.

Suppose the latter, i.e., that the assertion does not contain all the pattern information which is present in the data. The explanation encodes the data in its entirety, and so any pattern information not present in the assertion must appear in the detail. But, if the detail exhibits any pattern, it cannot have the statistical properties of a purely random sequence, and so cannot be the result of optimal coding. Hence, the explanation as a whole cannot be optimally coded, and so cannot be the shortest possible. Hence, the shortest explanation cannot fail to include in its assertion all available pattern information.

Suppose the former, i.e., that the assertion contains some noise information. By "noise", we here mean some information not relevant to knowledge of the true value of θ. Since the SMML estimate $\hat{\theta}$ is a deterministic function of the data x, the content of this noise information cannot be extraneous. That is, it cannot concern some random choice made by the statistician or some random event affecting the computation of the estimate. Rather, the noise content must concern some proposition $M(x)$ which happens to be true of x but which is uninformative about θ. Now, the detail encodes x using a code optimal if $\theta = \hat{\theta}$. But if the assertion, as well as estimating θ, contains the proposition $M(x)$, then an optimal code for the detail would be a code optimal if $\theta = \hat{\theta}$ and $M(x)$. Since the code actually used to encode the detail does not assume $M(x)$, but this proposition is true of x, the evidence of the truth of $M(x)$ is recoverable from the detail. Thus, the information about $M(x)$ is repeated in the explanation: it occurs both in the assertion and in the detail. It follows that the explanation cannot be the shortest possible, since it repeats some information. Since an SMML explanation is by definition and construction the shortest possible, the assumption that its assertion contains noise information cannot be true.

The above arguments, that the SMML assertion contains all and only the information relevant to θ which can be recovered from the data, are inexact. Ideally optimal coding of information into a binary string is in general impossible, since it would in general require a non-integral number of binary digits. The arguments thus only imply that the assertion lacks at most one bit of the relevant information, and contains at most one bit of noise.

An illustration may clarify the argument. Suppose the data comprise 12 values independently sampled from a Normal distribution with S.D. known to be 1.0, and unknown mean μ, and that the prior density of μ is uniform over some large range, say, ± 100. Let the observed data have sample mean $m = 13.394\ldots$. If an explanation were coded so that the estimated mean was rounded to the nearest 10, the shortest available explanation would assert $\hat{\mu} = 10$. The 12 data values would then be encoded in the detail as if drawn from $N(10, 1)$. The receiver of the explanation would be able to decode the explanation without difficulty, but might then notice that the sample mean m differed from $\hat{\mu}$ by 3.394, although the RMS difference to be expected with 12 values and S.D. $= 1$ is only about 0.3. The receiver might well consider this good evidence that the true mean differed significantly from 10, and that more information was available about μ than had been given in the assertion.

On the other hand, suppose the code was chosen to state $\hat{\mu}$ rounded to the nearest 0.1. The explanation would state $\hat{\mu} = 13.4$ so as soon as the receiver read the assertion, she could conclude that the sample mean lay in the range $13.35 < m < 13.45$. The detail, encoding the 12 values as if drawn from $N(13.4, 1)$, could encode with almost equal brevity any set of 12 values whose sample mean lay within 13.4 ± 0.3. Thus, as far as the coding of the detail is concerned, some of the information in the detail is used in conveying the fact that m lies within the narrower range 13.4 ± 0.05, thereby repeating information already contained in the assertion. The repeated information, which is essentially the ".4" in the estimate $\hat{\mu} = 13.4$, is almost uninformative about μ, since under the conditions assumed, we can only expect to estimate μ to roughly the nearest integer.

In fact, the SMML explanation will state $\hat{\mu}$ to the nearest integer ($\sigma\sqrt{12/N} = 1$), i.e., $\hat{\mu} = 13$. Its estimate is accurate enough to capture almost all the information about μ, and when the receiver discovers that $m = 13.394$, the difference gives no grounds for rejecting the SMML estimate. Further, in the SMML explanation, there is little repetition of information. The detail does in fact use some of its information to repeat the fact that

12.5 < m < 13.5, which has been already implied by the assertion, but the amount of information involved is small. Given that $N = 12$, $\sigma = 1$ and $\mu = 13$ (the assumptions used in coding the detail), the probability of the proposition "12.5 < m < 13.5" is so high (0.92) that little information is wasted in repeating it (0.13 bit).

If it were exactly the case that the SMML assertion extracted from the data all and only the information relevant to θ, it would be a universally sufficient estimate, in the sense of Section 3.2.6. Since some distributions do not admit minimal sufficient statistics, this is in general impossible. However, as the example of Section 3.3.6 on the Uniform distribution shows, the SMML assertion can convey some information about the sufficient statistics even when they are more numerous than the number of parameters estimated. In that example, the assertion states only one parameter estimate, the estimated mean, yet the manner of its encoding conveys information about both sufficient statistics, the sample midrange and the sample range.

3.4.6 Discrimination

In the SMML explanation code, assertions are encoded in a strictly optimal fashion. The length of the assertion of $\hat{\theta}$ is $-\log q(\hat{\theta})$, where $q(\hat{\theta}) = \sum_{x:m(x)=\hat{\theta}} r(x)$ is the probability that $\hat{\theta}$ will be asserted. One result is that $q(\hat{\theta})$ bears no direct relation to the prior probability density $h(\theta)$ at or in the vicinity of $\theta = \hat{\theta}$. The SMML construction "sees" $h(\theta)$ only through its convolution with $f(x|\theta)$, i.e., through the data marginal distribution $r(x)$. In general, $r(x)$ does not preserve special features of $h()$, as these get blurred by the convolution. An embarrassing consequence is that, while SMML estimators can be constructed for model sets Θ which are the nested union of continua of different dimensionality, the SMML estimator may not behave as we might wish. By a "nested union" we mean a model set comprising two or more continua of increasing dimensionality where a low-dimensioned continuum is a zero-measure subset of a higher-dimensioned member of the model set, for instance when the low-dimensioned member is equivalent to a higher-dimensioned member with one or more parameters of the latter fixed at special values. An example is the model set comprising all first-, second- and third-degree polynomials in some variable z, where the space of second-degree polynomials is equivalent to the subset of the space of third-degree polynomials in which the coefficient of z^3 is zero, and the space of first-degree polynomials is equivalent to the subset of the space of second-degree polynomials which has the coefficient of z^2 (and z^3) zero. Although these subsets have zero measure in the space of third-degree polynomials, we may have a reasonable prior belief that the "true" model for the data in hand has a finite probability of lying in one of these subsets. For instance, we may believe the probabilities that the true degree is 1, 2 or 3 are respectively 0.5, 0.3 and 0.2.

Normally, if Θ contains several nested model classes of different dimension, i.e., different numbers of parameters, we would like an estimator to tell us which of these classes (most likely) contains the model of the data source. In general, the SMML estimator will not do so.

3.4.7 Example: Discrimination of a Mean

Suppose that the data are a set of N values drawn from a Normal population of known Standard Deviation σ and unknown mean μ. Our prior belief might be that there is probability $1/2$ that $\mu = a$, a known value, and that if $\mu \neq a$, then μ is equally likely to lie anywhere in some range $b \leq \mu \leq c$, with $(c - b) \gg \sigma/\sqrt{N}$ Then Θ is the union of two sets: a singleton set, dimension zero, containing the value a, and a one-dimensional continuum containing all values from b to c. The two sets have equal prior probability $1/2$. In the first set it is a mass on a, in the second it is spread uniformly over the interval b to c. We suppose $b < a < c$.

There is no great difficulty in constructing an SMML code. The receiver is assumed to know the distinguished value a already. The marginal density $r(\bar{y})$ of the sample mean can be easily computed. It has a smooth peak of width roughly $\sigma/\sqrt{N}$ centred around $\bar{y} = a$, falling off smoothly to a uniform plateau over most of the interval b to c, with smooth shoulders near the ends. Either by using the "boundary rule of thumb" followed by relaxation using R1–R3, or otherwise, an optimal estimate set Θ^* can be found and the $\bar{y}$ line divided into the corresponding intervals $\{t(\hat{\mu}) : \hat{\mu} \in \Theta^*\}$.

In general, no member of Θ^* will have $\hat{\mu} = a$. There will, of course, be some member of Θ^* close to a, with a high q value, and the two members of Θ^* immediately above and below a may be unusually far apart, but a itself will not in general be a possible assertion. Thus, the SMML estimator will, whatever the data, select a model from the continuum, and never choose the distinguished model $\hat{\mu} = a$. We show this effect in the following example, where for simplicity we have taken $\sigma/\sqrt{N} = 1$, so the sample mean has a distribution Normal$(\mu, 1)$ for true mean μ. We have taken $h(\mu)$ to be

$$h(\mu) = \frac{1}{2}\text{Normal}(0, 100) + \frac{1}{2}\delta(\mu - 4.0)$$

where $\delta()$ is the Dirac delta function. Thus, we have a prior belief that μ is equally likely to be 4.0 or to have a value sampled from a Normal density with zero mean, S.D. = 10. Table 3.6 shows part of the SMML estimator, giving the data groups t_j, estimates $\hat{\mu}_j$, coding probabilities q_j, and the width of each data group.

The estimate 3.991 is close to the distinguished value 4, and has a large q. Any data mean in the range $1.33 < \bar{y} < 6.73$ gives this estimate, in effect accepting the hypothesis $\mu = 4$, but the estimate is not actually the distinguished value.

Table 3.6. SMML Estimator for Normal Mean, with Prior Normal plus Delta.

t_j	$\hat{\mu}_j$	q_j	Width of t_j
$-12.52 \ldots -9.04$	-10.67	0.039	3.48
$-9.04 \ldots -5.56$	-7.22	0.053	3.48
$-5.56 \ldots -2.08$	-3.78	0.064	3.48
$-2.08 \ldots 1.33$	-0.33	0.069	3.41
$1.33 \ldots 6.73$	3.991	0.59	5.39
$6.73 \ldots 10.14$	8.31	0.049	3.41
$10.14 \ldots 13.62$	11.76	0.034	3.48
$13.62 \ldots 17.11$	15.21	0.022	3.49
$17.11 \ldots 20.60$	18.67	0.012	3.49

The same unwillingness to choose simple distinguished models or model classes, even of high prior probability, will be found in all SMML estimators. The trouble arises because of the very generality and model invariance properties which are attractive features of the SMML method. The SMML construction, so to speak, cares so little for how models are represented that it does not notice a condensation of prior probability on some distinguished subset of Θ, and treats a model near the distinguished subset as just as acceptable an assertion as one in the subset. The marginal data distribution on which the SMML code is constructed has almost the same form as for a prior density $h()$ which, instead of placing condensed probability mass in the subset, merely has a high, narrow, but finite peak around the subset.

The cure for this lack of discrimination in SMML estimators is simple. If we wish the assertion to discriminate among two or more subsets of Θ which we regard as conceptually distinct or representing different kinds of hypothesis about the data, we can constrain the encoding of the assertion so that it makes a definite choice among the subsets.

The assertion is now encoded in two parts. The first part names one of the subsets of Θ, using a string of length the negative log of the prior probability of the subset. The second part of the assertion, specifying a member of the subset, and the detail encoding the data, are constructed by the standard SMML method using the prior density, model data distribution and data marginal distribution appropriate to the named subset. To take the simple example above, the explanation will take the following form:

The first part of assertion names *either* the distinguished subset $\theta = a$ or the continuum $[b, c]$. Since both subsets have equal prior probability $1/2$, the length of this part is $\log 2$ in either case.

If the continuum was named, the second part of the assertion, naming some $\hat{\mu}$ in $[b, c]$, and the detail giving the N data values, use an SMML code constructed for the estimation of a Normal mean μ given known σ, a sample of size N, and a prior density $h(\mu)$ uniform in $[b, c]$. The sample mean marginal distribution $r(\bar{x})$ will be uniform in $[b, c]$ except near the ends, and the SMML code construction is entirely unaffected by the distinguished value a, since this has been ruled out by the first part of the assertion.

On the other hand, if the distinguished subset $\mu = a$ was named, the second part of the assertion is null, since the data model is now fully specified, and the detail encodes the data using the model $N(a, \sigma^2)$.

In effect, we have modified the code so that the explanation first names the estimated structural or model class, then uses an SMML code which assumes the data indeed came from some member of the named class.

Note that the modification has introduced a fundamental shift in the way assertions are coded. In the unmodified SMML method, the coding of estimates or models is based on a partition of the data set X into subsets $\{t_j : \hat{\theta}_j \in \Theta^*\}$, and the coding probability $q(\hat{\theta})$ of an estimate, and hence the length of its assertion, is based on the total marginal probability of its associated subset. In the modified code, however, the coding of an assertion is in part based on a partition of the model space Θ into subsets, and the coding probability of a subset is taken simply as the total prior model probability within the subset. Thus, the coding of the first part of an assertion, which names a subset of Θ, is much more directly and intuitively tied to the prior density $h()$. The resulting explanation code is slightly less efficient than the unmodified SMML code, for two reasons. First, the assertion of a model class is based on the prior probability of that class, which may not exactly equal the total marginal probability of all those data sets in X which will lead the estimator to assert that class. Second, the modified SMML code may provide for estimates which in fact will never be used. In our simple example, the SMML code for the model class $b \leq \mu \leq c$, $h(\mu) = 1/(c - b)$ may well have a Θ^* containing some estimate $\hat{\mu}$ very close to a. This estimate will never be asserted, as any sample mean close enough to $\hat{\mu}$ to fall within its $t(\hat{\mu})$ data group will instead be encoded using the $\mu = a$ model class. The inefficiencies arising from both causes are small, and decrease with increasing sample size. They are too small to outweigh the greater clarity and simplicity of the modified code.

Henceforth, in any problem requiring discrimination among nested model classes, we will assume the modified code, which begins by asserting an inferred class, is used.

3.5 Summary

The Strict Minimum Message Length (SMML) estimator exactly minimizes the expected length of an explanation within a Bayesian framework (except for a small inefficiency when the explanation begins by asserting a model class within a nested set of model classes). The given set Θ of possible models is replaced by a discrete subset Θ^* of models which may be used in explanations. The coding probability, analogous to a prior probability, given to each model in Θ^* is just its probability of being asserted in an explanation. The SMML estimator partitions the set X of possible data into subsets, one for each model in Θ^*. All data strings in a subset result in estimating that model.

The SMML estimator is very general, being defined even when Θ is the union of several (possibly nested) continua, and is invariant under one-to-one measure-preserving transformations of Θ. Its estimates capture almost all information in the data which is relevant to the choice of model. Unlike most Bayesian estimators, it neither requires use of a "cost" or "utility" function, nor depends on how Θ is parameterized.

Its difficulties are great computational difficulty, discontinuous behaviour of the estimator function, and an inability (unless modified) to make an unequivocal choice among nested continua of possible models. A further apparent disadvantage is that when several scalar parameters are to be estimated, the SMML estimate of one of them will depend somewhat on aspects of the data relevant only to other parameters. However, this dependence is small, and does not result in the estimate of a parameter being inconsistent with the data relevant to that parameter.

4. Approximations to SMML

This chapter develops a couple of approximations to the Strict Minimum Message Length (SMML) method of estimation and model selection. The SMML method developed in the previous chapter, while having many desirable properties, is computationally very difficult and yields parameter estimates which are discontinuous functions of the data, even when the data are treated as continuously variable real-valued quantities. Similarly, the explanation lengths given by an SMML estimator are rather bumpy functions of the data. A data value which happens to fall close to the centre of an SMML data group will tend to have a smaller, perhaps even negative, value of "excess" message length $(I_1(x|\hat{\theta}) - I_0(x))$ than a data value near the edge of the data group.

For practical use, we would prefer to use methods which, while retaining as much as possible of SMML's virtues, result in smoother estimator functions and smoother dependence of message length on the data, even if this means accepting some approximation error in the calculation of message length and of the parameter estimates which minimize the length. Here, two approximate methods are presented. The first leads to parameter estimates which are continuous functions of continuous data, but does not directly lead to an approximation to the length of the explanation based on these estimates. The second method gives a smooth approximation to the explanation length, but does not directly lead to estimates of the parameters. Both approximate methods avoid the need, present in SMML, to construct a partition of the set X of possible data into data groups.

Sections 4.1 to 4.9 deal with the approximate estimator, its application to a somewhat difficult problem, and the performance of a couple of other estimators on this problem. Section 4.10 deals with the message length approximation.

4.1 The "Ideal Group" (IG) Estimator

This method of constructing an estimator was suggested by David Dowe. In a genuine SMML code, the three minimizing relations R1, R2 and R3 of Section 3.2.1 apply to all data groups in the partition. The approximation considered here constructs an "ideal" group with estimate $\hat{\theta} = \theta$ which obeys only the "Boundary" rule of Section 3.3.2. It ignores the fact that in a real SMML code, the various data groups must together form a partition of the set X of possible data vectors, that is, that the groups must fit together without overlap and without omitting any member of X. This constraint on a real SMML code, together with relations R1, R2 and R3, results in the selection of a set of data groups and their associated set of estimates Θ^* which are sensitive to the behaviour of the prior $h(\theta)$ over the entire hypothesis space Θ, including regions of Θ which contain no credible estimate for the current observed data vector.

4.1.1 SMML-like codes

As has been seen in the previous chapter, there are for most estimation problems many near-optimal partitions of X which, although having different choices of Θ^*, give expected message lengths only insignificantly worse than the strictly optimum code, and indeed message lengths $I_1(x)$ for every $x \in X$ which differ insignificantly from the lengths given by the true SMML code. I shall call such near-optimal codes "SMML-like". In some problems, some of these near-optimal codes even obey the three relations. The features of a code which are really important in determining the expected message length I_1 are the sizes and general shapes of the data groups. The approximation aims to abstract these important determinants of message length while avoiding the need to construct a complete partition of X.

4.1.2 Ideal Data Groups

An "ideal" data group which obeys the Boundary Rule optimizes, for some given estimate value $\hat{\theta}$, the expected code length of data vectors within the group, or more precisely, the expected difference between the code length $I_1(x)$ and the length $I_0(x) = -\log(r(x))$ of the best non-estimating code for x. Such a data group may be expected to resemble in size and general shape the data group which would be found in an SMML-like code if the given $\hat{\theta}$ happened to occur in the estimate set Θ^* of the SMML-like code. In particular, the total marginal probability of the data vectors within the ideal group may well approximate the "coding probability" $q(\hat{\theta})$ which $\hat{\theta}$ would have in an SMML-like code. Note that the estimate $\hat{\theta}$ for an ideal group need not in general satisfy relation R3. It does not necessarily maximize the average log-likelihood for the data in the group.

The Boundary Rule says that a data group t should ideally be such that

$$x \in t \text{ iff } I_1(x|\hat{\theta}) - I_0(x) < Av_{y \in t}[I_1(y|\hat{\theta}) - I_0(y)] + 1$$

where

$$I_0(x) = -\log r(x)$$

$$r(x) = \int_{\theta \in \Theta} h(\theta) f(x|\theta)$$

$$I_1(x|\hat{\theta}) = -\log q(\hat{\theta}) - \log f(x|\hat{\theta})$$

$$q(\hat{\theta}) = \sum_{y \in t} r(y)$$

The Rule may be re-written as

$$x \in t \text{ iff } g(x|\hat{\theta}) < Av_{y \in t}[g(y|\hat{\theta})] + 1 \text{ where } g(y|\theta) = \log(r(y)/f(y|\theta))$$

Thus, for any given $\theta \in \Theta$, the Boundary Rule allows us to construct an ideal data group having θ as its estimate, and hence to compute a "prior probability" $q(\theta)$ which approximates the coding probability θ would have were it to occur in the estimate set of some near-optimal SMML-like code.

In the form originally suggested by Dowe, the function $g(y|\theta)$ was taken simply as $-\log f(y|\theta)$, omitting the marginal data probability $r(y)$. This form is incorrect, for instance because it yields different results depending on whether y is taken as the full data set or a sufficient statistic derived from the full data. The inclusion of the dependence on $r(y)$ removes this defect.

4.1.3 The Estimator

The Ideal Group estimator, given data x, chooses as its estimate that value $\hat{\theta} \in \Theta$ which minimizes

$$DI_1(x|\hat{\theta}) = -\log(q(\hat{\theta})) - \log(f(x|\hat{\theta}))$$

where $q(\hat{\theta})$ is computed via the ideal data group having $\hat{\theta}$ as its estimate. It should be obvious that the coding probability thus computed for any $\theta \in \Theta$ is not altered by non-linear transformations of the hypothesis space, and that hence the Ideal Group estimator preserves the invariance property of the SMML estimator. It is also clear that the Ideal Group estimate is a function of the sufficient statistics only.

Like the SMML estimate, it has the inconvenient property that when the hypothesis space is a nested set of continua with different dimensionality, it is unlikely that the estimate will ever lie in a low-dimension subspace, because, like SMML, the approximation sees the structure of the hypothesis space and its prior density only via the smeared-out form of the marginal data distribution $r(x)$. It therefore does not automatically answer the question of what order of model should be inferred from the data. As with SMML, this difficulty may be overcome by requiring the imaginary explanation to begin by naming the model order or dimension with a message fragment of length the negative log of the prior probability given to the stated model subspace. To find the best explanation, one finds the best estimate (or its Ideal Group approximation) for each model order, considering only hypotheses within that order in computing $r(x)$ for data vectors. The message length achieved by the best estimate for each model order is increased by the length of the fragment needed to name the order, and the order giving the shortest total length chosen. However, as noted below, the value of $DI_1(x|\hat{\theta})$ defined above cannot be used directly as the message length for a given model order with best estimate $\hat{\theta}$, but requires a correction term.

As yet the general properties of the Ideal Group estimator have not been explored. When applied to specific estimation problems, it appears to give good results. An example of some interest is given in the next section. The message length defined as $DI_1(x|\hat{\theta})$ will generally be less than the length $I_1(x)$ given by a true SMML code. $DI_1()$ is optimistic, in that it assumes a code is used which just happens to have available in its estimate set Θ^* an estimate ideally tailored to encode the given data. A true SMML code has only a limited estimate set, so the true SMML message for the data will in general have to use an estimate differing from the ideal value found by the Ideal Group. For estimation problems with sufficiently regular likelihood functions, the correction appears to be an increase in length of 0.5 nit for each real-valued parameter of the model. More generally, one might use as an additive correction to $DI_1()$ the expression

$$\text{Average}_{y \in t}(-\log(f(y|\hat{\theta})/r(y))) + \log(f(x|\hat{\theta})/r(x))$$

where t is the ideal data group using the estimate $\hat{\theta}$, x is the given data, data vector y ranges over t, and y's contribution to the average is weighted by $r(y)$.

4.2 The Neyman-Scott Problem

The Neyman-Scott problem is quite simple, yet it presents difficulties for some normally satisfactory principles for estimation such as Maximum Likelihood (ML). The given data comprises N instances. Each instance comprises J data values drawn from a Normal density peculiar to that instance. It is believed that the Normal densities for the different instances all have the same Standard Deviation σ, but different means. The different instance means are not believed to have any relation to one another, nor to be clustered. The mean of the density for instance n ($n = 1, \ldots, N$) will be denoted by μ_n and the N-vector of these means simply by μ. The $(N+1)$ parameters of the model are therefore σ and the N components of μ. All $(N+1)$ parameters are unknown, and are to be estimated from the data.

Let the data values found in instance n be $x_{n,j}$, ($j = 1, \ldots, J$). Define

$$m_n = (1/J) \sum_j x_{n,j}$$

$$s^2 = (1/NJ) \sum_n (x_{n,j} - m_n)^2$$

The N-vector of instance data means $(m_n : n = 1, \ldots, N)$ will be written as m. Then s and m are together sufficient statistics for the problem. It will be convenient to treat the given data as comprising just these statistics, and to define the data marginal distribution and likelihood function in terms of these rather than the raw $x_{n,j}$ values.

We assume the vector of means μ to have a Uniform prior density over some large region of N-space, and assume σ to have the usual non-informative prior density $(1/\sigma)$ over some large range. The priors are of course unnormalized. With these priors, it is easily shown that the marginal distribution of the sample mean vector m is Uniform in N-space, and, independently, the marginal density of s is proportional to $(1/s)$. Hence, the marginal data density (unnormalized) can be written as

$$r(m, s) = 1/s$$

For given parameters μ and σ, the sample means vector m has an N-dimensional Gaussian density with mean μ and covariance matrix $Diag(\sigma^2/J)$. Independently of m, the average within-instance variance s^2 is distributed so that NJs^2/σ^2 has a Chi-Squared distribution with $N(J-1)$ degrees of freedom. Hence, the negative log probability density of m, s can be written (neglecting constant terms) as:

$$
\begin{aligned}
&- \log(f(m, s | \mu, \sigma)) \\
&= (N/2) \log(\sigma^2/J) + (J/(2\sigma^2)) \sum_n (m_n - \mu_n)^2 - \log(2s) \\
&\quad + NJs^2/(2\sigma^2) - ((NJ - N)/2 - 1) \log(NJs^2/\sigma^2) + 2 \log \sigma \\
&= NJ \log \sigma + (J/(2\sigma^2))(\sum_n (m_n - \mu_n)^2 + Ns^2) \\
&\quad - (NJ - N - 1) \log(s) - (N/2) \log(J) \\
&\quad - \frac{1}{2}(NJ - N - 2) \log(NJ) - \log 2
\end{aligned}
$$

The final (constant) three terms in the latter form will be dropped.

4.3 The Ideal Group Estimator for Neyman-Scott

Given the Uniform prior on μ and the scale-free $1/\sigma$ prior on σ, we do not need to explore the details of an ideal group with estimate (μ, σ). It is sufficient to realise that the only quantity which can give scale to the dimensions of the group in (m, s)-space is the Standard Deviation σ. All $(N + 1)$ dimensions of the data space, viz., $((m_n : n = 1, \ldots, N), s)$ are commensurate with σ.

Hence, for some N and J, the shape of the ideal group is independent of μ and σ, and its volume is independent of μ but varies with σ as σ^{N+1}. Since the marginal data density $r(m, s)$ varies as $1/s$, the coding probability $q(\mu, \sigma)$, which is the integral of $r(m, s)$ over the group, must vary as σ^N. The Ideal Group estimate for data (m, s) obviously has $\hat{\mu} = m$, and the estimate of σ is found by maximizing $q(\mu, \sigma) f(m, s | m, \sigma)$ as

$$\hat{\sigma}_{IG}^2 = J s^2 / (J - 1)$$

This estimate is consistent, since for true σ, the expected value of s^2, is $(J - 1)\sigma^2/J$. This result is not critically dependent on the prior assumed for σ. For instance, if a Uniform (and hence scale-free) prior is assumed, the above argument shows that $q(\mu, \sigma)$ then varies as σ^{N+1}, giving

$$\hat{\sigma}_{IG}^2 = N J s^2 / (N(J - 1) - 1)$$

Although biased towards high values, this estimate is still consistent, as its expected value still converges towards the true σ^2 as the number of instances increases (but it no longer exists for $N = 1$, $J = 2$).

A numerical computation of the ideal group and estimate gives some further details of the estimator's behaviour. For the $1/\sigma$ prior and $J = 2$, we find that the group includes data with widely varying values of s^2/σ^2. With just one instance $(N = 1)$, the ratio ranges from 0.002 to 3.59 with a mean of 0.5 and Standard Deviation of 0.7. With more data $(N = 100)$, the range contracts slightly to $[0.07, 1.6]$, again with mean 0.5 but now with a lower SD of 0.07. This shows that the mean squared difference of s^2/σ^2 from its expected value (0.5 for $J = 2$) varies as $N^{-1/2}$. That is, the IG estimate, if it occurred as an estimate in an SMML code, would be used for any of a group of data vectors whose within-instance variance have a spread of the same order as would be expected were the data obtained from a source whose σ equalled the given IG estimate.

Similar behaviour is found with larger J, but of course as J increases, the range of s within an ideal group becomes smaller.

The numerical solutions showed that the formula in Section 4.1.3 for an additive correction to $DI_1()$ indeed gave values within 0.1 nit of half the number of parameters estimated for either prior. The difference dropped to less than 0.002 nit when $N = 100$.

4.4 Other Estimators for Neyman-Scott

The Neyman-Scott problem is of interest because it is a simple example of an estimation problem in which the number of unknown parameters increases with increasing data sample size. Each additional instance gives J more data values, but introduces an additional parameter, the instance mean μ_n. The approximate MML methods of the next chapter have been applied to the problem by Dowe and Wallace [13] and shown to be consistent, but problems with this character prove difficult for some otherwise respectable estimation principles.

4.5 Maximum Likelihood for Neyman-Scott

The Maximum Likelihood (ML) estimate for the Neyman-Scott problem chooses estimates which maximize $f(m, s|\mu, \sigma)$. This gives $\hat{\mu} = m$, which is fine, but $\hat{\sigma}^2 = s^2$ for any J and N. The latter estimate is inconsistent, and for fixed J will not converge towards the true σ^2 no matter how large N may be.

Maximum Likelihood estimators are generally prone to under-estimation of parameters of scale, such as σ. In most problems, this tendency results in no more than a modest bias which decreases with sample size and does not lead to inconsistency. However, when new data brings with it more unknown parameters, the bias may never be overcome with enough data, as in the Neyman-Scott problem. Other, more realistic, estimation problems with this character appear in later chapters.

4.5.1 Marginal Maximum Likelihood

In problems where each new data instance brings another unknown parameter, such parameters are sometimes termed "nuisance" parameters. If Maximum Likelihood attempts to estimate the nuisance parameters (the instance means μ_n in Neyman-Scott) simultaneously with the "global" parameters (the common SD σ in Neyman-Scott), inconsistent estimation of the global parameters can result. However, it is sometimes possible to find a statistic $z(x)$ whose probability distribution depends only on the global parameters, and which is a sufficient statistic for the global parameters. If so, maximizing the likelihood of the global parameters given $z(x)$ and ignoring all other aspects of the data may give a consistent estimate. Such an estimate is called a *Marginal Maximum Likelihood* estimate. In the Neyman-Scott case, the total within-instance variance NJs is such a statistic. The probability density of NJs is

$$(1/\sigma^2)\, \chi^2_{NJ-N}(NJs/\sigma^2)$$

i.e., (NJs/σ^2) has a Chi-Squared distribution with $(NJ - N)$ degrees of freedom independently of the nuisance parameters. If this probability density is maximized for given s, the resulting marginal maximum likelihood estimate of σ is $Js/(J - 1)$, which is consistent.

The success of Marginal Maximum Likelihood depends on the existence of such a statistic, which may not always be available. Further, if estimates of the nuisance parameters are required, there is no general reason to suppose

that good estimates of them will be got by maximizing the total data probability assuming that the global parameters have their marginal ML estimate
values, although this practice is sometimes used. In the Neyman-Scott case
it succeeds.

If no statistic such as NJs can be found, a partial Bayesian approach
may be used. If γ is the set of global parameters and η the vector of nuisance
parameters, and there is a reasonable Bayesian prior density $h(\eta|\gamma)$, the nui-
sance parameters can be integrated out to give a marginal likelihood for the
global parameters, namely

$$\Pr(x|\gamma) = \int f(x|\gamma, \eta)\, h(\eta|\gamma)\, d\eta$$

Maximization of this with respect to γ may then give a good estimate of $\hat{\gamma}$,
but it remains the case that there is then no valid reason to estimate the
nuisance parameters by maximizing $f(x|\hat{\gamma}, \eta)$ with respect to η.

4.6 Kullback-Leibler Distance

This is a convenient place to introduce a new function which has many uses.
It will lead to a new estimator which will be compared with the Ideal Group
estimator.

The Kullback-Leibler distance (KLD) is a measure of the difference of one
probability distribution from another distribution for the same variable. Let
$a(x)$ be a probability distribution of a discrete random variable x and $b(x)$
be another. Then the Kullback-Leibler distance of $b()$ from $a()$ is defined as

$$KLD(a, b) = \sum_x a(x) \log(a(x)/b(x))$$

If x is real-valued or a vector of real-valued components, then $a()$, $b()$ are
probability densities rather than discrete distributions, and

$$KLD(a, b) = \int a(x) \log(a(x)/b(x))\, dx$$

Note that in the latter case, $KLD(a, b)$ is not altered by a 1-to-1 non-linear
transformation of the random variable, as such a transformation does not
change the ratio of the densities.

In either case, $KLD(a, b)$ is zero if and only if $a()$ and $b()$ are identical,
and is otherwise positive. It can become infinite if there is no value of x for
which both $a(x)$ and $b(x)$ are non-zero, that is if $a(x)b(x) = 0$ everywhere.
The distance is not symmetrical: in general, $KLD(a, b) \neq KLD(b, a)$.

KLD is a useful measure of difference, in part because of its invariance
under transformations of the variable, but importantly it is a measure of the

bad effects of assuming x is distributed as $b(x)$ when in fact it is distributed as $a(x)$. $KLD(a, b)$ tells us how surprising observed values of x will appear if we think they are being generated according to $b()$ when in fact they come from $a()$. KLD commends itself as a general-purpose "cost function" giving a cost of mistaking $a()$ for $b()$, since it does not depend on how the distributions are parameterized, unlike for instance cost functions which depend on the squared difference between estimated and true parameter values. In a message-length context, $KLD(a, b)$ shows how much longer, on average, an encoding of x will be if we use a code optimized for $b()$ rather than the true distribution $a()$. Whatever the real cost of mistaking $a()$ for $b()$, which may be unknown when we try to estimate the distribution of x, it is likely to increase with increasing distance of the estimated distribution from the true one.

For these reasons, it is attractive to consider estimating the distribution of x by trying to find a model distribution $g(x)$ with a small KLD from the unknown true distribution $f(x)$. Assuming that $f(x)$ is believed to be some member $f(x|\theta)$ of some set Θ of models with unknown parameter θ and prior probability $h(\theta)$, and that we wish to select $g(x)$ from the same set, so $g(x) = f(x|\hat{\theta})$, Bayesian Decision Theory (Section 1.14) suggests we choose the estimate $\hat{\theta}$ which minimizes the expected KLD "cost" of the estimate from the unknown "true" value. Since we do not know the true value, we can only minimize the expectation with respect to the posterior distribution or density of the true value.

4.7 Minimum Expected K-L Distance (MEKL)

The expected KLD of estimate θ given data x is

$$
\begin{aligned}
EKLD(\theta|x) &= \int_{\phi \in \Theta} p(\phi|x)d\phi \int_X f(y|\phi) \log(f(y|\phi)/f(y|\theta))dy \\
&= \int_{\phi \in \Theta} p(\phi|x)d\phi \int_X f(y|\phi) \log f(y|\phi)dy \\
&\quad - \int_{\phi \in \Theta} p(\phi|x)d\phi \int_X f(y|\phi) \log f(y|\theta)dy
\end{aligned}
$$

where X is the range of x, $p(\phi|x) = h(\phi)f(x|\phi)/r(x)$ is the posterior density of ϕ given x, and $r(x)$ is, as usual, the marginal data distribution. Dropping terms not involving θ, it suffices (subject to regularity) to minimize the expression

$$
\int_X R(y|x) \log(y|\theta) \, dy
$$

where

$$
R(y|x) = \int_{\phi \in \Theta} p(\phi|x) \, f(y|\phi) \, d\phi
$$

Thus, $R(y|x)$ is the probability of obtaining new data y averaged over the posterior of the unknown parameter. It is the probability (or probability density) of new data, given the known data x. In the Bayesian literature, it is called the *predictive distribution*. The predictive distribution can be regarded as the probability distribution of a data value y generated by a two-step random sampling process: first sample a parameter value θ from the posterior of θ given x, then sample y from the conditional distribution $f(y|\theta)$.

The expression to be maximized in getting the MEKL estimate is therefore the expected log likelihood of new data, the expectation being taken over the predictive distribution of new data. The minimizing estimate is a maximum *expected* log-likelihood estimate, based not on the data we have, but on what data we might expect to get from the same source. N.B. The predictive distribution in general is not of the same mathematical form as the conditional data distribution $f(x|\theta)$. That is, there is in general no $\theta \in \Theta$ such that for given data x, $f(y|\theta) = R(y|x)$ for all $y \in X$. The estimate so found is called the "Minimum Expected K-L Distance" estimate, or MEKL. Strictly speaking, it is a minimum-expected-cost value deduced from the data rather than an inductively justified guess at the true model, but it can certainly be used as an estimate and has some virtues, e.g., invariance under re-parameterization, reliance on sufficient statistics, and the fact that in a useful sense it minimizes the difference between the estimated and true data sources.

4.8 Minimum Expected K-L Distance for Neyman-Scott

The MEKL estimator for the Neyman-Scott problem will now be obtained. For N data instances each of J data values, (using the same assumptions as in Section 4.3) the data may be replaced by the sufficient statistics $\{m_n = (1/J)\sum_j x_{n,j}\}$ and $S = \sum_n \sum_j (x_{n,j} - \mu_n)^2$. Obviously, the MEKL estimate will have $\hat{\mu} = m$. Hence, the MEKL estimate of σ will maximize the expected likelihood of new data $\{y_{n,j}\}$ by choosing

$$\hat{\sigma}^2 = (1/NJ)E\left(\sum_n \sum_j (y_{n,j} - m_n)^2\right)$$

To evaluate the expectation above, first note that for arbitrary (μ, σ),

$$E\left(\sum_j (y_{n,j} - m_n)^2\right) = J\,E\left((\mu_n - m_n)^2\right) + J\sigma^2$$

Now, the posterior for σ^2 is independent of m and μ, since S is a sufficient statistic for σ. Thus, to sample (μ, σ) from their joint posterior, one may first sample σ from its posterior, then sample μ from the posterior for μ given m and σ. Suppose a value σ has been randomly chosen from its posterior. Then the posterior for μ_n is Normal$(m_n, \sigma^2/J)$, so $J E\left((\mu_n - m_n)^2\right) = \sigma^2$ for all n. Hence, summing over the N instances,

$$E\left(\sum_n \sum_j (y_{n,j} - m_n)^2\right) = N(J+1)\sigma^2$$

It remains to find the expected value of σ^2 when σ^2 is sampled from its posterior.

For given σ, S/σ^2 has a Chi-Squared density with $N(J-1)$ degrees of freedom. Writing $D = (N(J-1)/2)$:

$$Dens(S|\sigma^2) = (1/(2^D \sigma^2\, \Gamma(D)))(S/\sigma^2)^{(D-1)} \exp\left(-S/2\sigma^2\right)$$

Writing $C = 1/\sigma^2$, and using the uninformative prior $h(\sigma) = 1/\sigma$ which implies the prior on C is $1/C$, the posterior density of C given data S can be found as

$$Dens(C|S) = (S/(2^D\, \Gamma(D)))\, (SC)^{(D-1)} \exp\left(-SC/2\right)$$

Hence, the expected value of $\sigma^2 = 1/C$ is:

$$
\begin{aligned}
E(\sigma^2|S) &= (S/(2^D\, \Gamma(D))) \int (1/C)\, (SC)^{(D-1)} \exp\left(-SC/2\right)\, dC \\
&= (S/(2^D\, \Gamma(D)))\, 2^{(D-1)}\, \Gamma(D-1) \\
&= S\, /\, (2(D-1)) \\
&= S\, /\, (NJ - N - 2)
\end{aligned}
$$

Substituting this expected value of σ^2 gives

$$E\left(\sum_n \sum_j (y_{n,j} - m_n)^2\right) = (NJ + N)\, S\, /\, (NJ - N - 2)$$

and hence

$$\hat{\sigma}^2 = (1/NJ)\, (N(J+1))\, S\, /\, (N(J-1) - 2)$$

Now, if the data were obtained from a distribution with true $\sigma = \sigma_T$, the expected value of S is $N(J-1)\sigma_T$, giving an expected MEKL estimate

$$
\begin{aligned}
E\, \hat{\sigma}^2 &= \sigma_T^2 \frac{N(J+1)\, N(J-1)}{NJ\, (N(J-1) - 2)} \\
&= \sigma_T^2 \frac{J+1}{J(1 - 2/(NJ - N))}
\end{aligned}
$$

For fixed J, the estimate approaches $(1+1/J)\sigma_T^2$ from above as N increases, so the MEKL estimate of σ^2 is inconsistent, but, unlike the inconsistent estimate given by Maximum Likelihood, over-estimates rather than under-estimates the true value. If $N(J - 1) < 3$, the expected value of the MEKL estimate does not exist.

MEKL estimation appears to have a general tendency to over-estimate parameters of scale such as σ. In most situations, the tendency produces only a bias towards large values which reduces with increasing sample size, but in the Neyman-Scott and similar situations, the bias is never overcome. This result does not imply that MEKL is failing its task. MEKL aims to find, within the set Θ of available models, the model which minimizes the expected value of a cost function, namely the code length required by new data. Equivalently, it minimizes the surprise expected to be occasioned by new data, i.e., maximizes the expected log-likelihood of new data. As has been pointed out before, such minimum-expected-cost techniques really belong to Bayesian Decision Theory rather than to Statistical Induction. Their job is to minimize an expected cost, not to make a good guess about the source of the present data.

Note that MEKL is, in effect, trying to find that model $\theta \in \Theta$ which best approximates the predictive distribution of new data given present data. In general, the predictive distribution will not have the same form as any model in Θ.

4.9 Blurred Images

A comparison of the operations of Maximum Likelihood (ML), Strict Minimum Message Length (SMML) or its IG approximation, and Minimum Expected Kullback-Leibler Distance (MEKL) can give some insight into why often ML underestimates and MEKL overestimates parameters of scale, but SMML appears not to. The following discussion is informal and proves nothing, but at least suggests a reason.

Imagine a very simple estimation problem, with parameter $\theta \in \Theta$, prior $h(\theta)$ varying only slowly in the region of interest, and data $x \in X$. Suppose x is a minimal sufficient statistic for θ commensurate with θ and, really to simplify the situation, suppose that for all θ, $E(x) \approx \theta$. Let the model form be $f(x|\theta)$.

The real world has some true, unknown, θ_T. When we do an experiment or observation, the Universe gives us some x which is a realization of $f(x|\theta_T)$. That is, it randomly (as far we know) picks x from a blurred image of θ_T, the density $f(x|\theta_T)$. The task of an estimator is to attempt to recover θ_T as far as possible. The Bayesian argument, given x, gives us a blurred idea of θ_T, the posterior density $P(\theta|x) = h(\theta)f(x|\theta)/r(x)$. This image, by in a sense mimicking the blurring of θ_T done by the Universe, gives us all the information about θ_T which is available, but does not satisfy a pragmatic need for a single good guess. The various estimator methods ML, SMML and MEKL make such a guess.

ML simply picks the value of θ which maximizes $f(x|\theta)$, ignoring the width of the blurring done by the Universe (and also our prior beliefs). Blurring is simply not addressed.

SMML tries to mimic, and thereby allow for, the Universe's blurring. It replaces Θ by a discrete subset of models Θ^* and partitions X into subsets, each being a blurred image of some model in Θ^*. The blurring is not quite the same as done by the Universe, as each "image" in X is trimmed to a finite size and made to fit with its neighbours, but the width of each SMML image matches closely the RMS width of the blurring produced by $f(x|\theta)$. The IG approximation is similar, but allows all $\theta \in \Theta$ to have a blurred image in X.

MEKL first makes a blurred image of x in Θ, i.e., the posterior, then makes a blurred image of the posterior back in X, i.e., the predictive data distribution given x. Finally, it picks the value of θ which maximizes the average log-likelihood of data values in the doubly-blurred predictive distribution.

By not allowing for blurring, ML ignores the fact that fitting a location parameter to the data such as a mean may reduce the apparent spread of data values about their mean location, and hence lead to an underestimation of the scale of the spread.

MEKL, by performing two blurring operations, in fact may double the apparent spread around the mean, and so over-estimate the scale.

SMML does just one blurring, that which produces a blurred image in X of a model θ, which at least roughly imitates what the Universe does in producing x from θ_T. In so doing, it well compensates for the Universe's blurring and recovers a fair estimate of its scale.

I do not claim for this argument any more than an arm-waving plausibility. At most it gives some reason to prefer estimation methods which involve just one blurring operation by $f(x|\theta)$.

4.10 Dowe's Approximation I1D to the Message Length

This section presents a fairly simple method of calculating an approximation to the message length $I_1(x)$ which would be got when data x is encoded in an explanation using an SMML code. It is similar in spirit to the IG approximate estimator, which does not directly yield a message length, and the method presented here does not directly yield an estimate. The approximation was suggested by David Dowe and has been used in a number of applications of MML, such as the inference of a univariate polynomial approximation to noisy data points [16]. It emerges from the following scenario:

Consider an estimation problem with, in our usual notation, a prior $h(\theta)$ over a set or space Θ of possible models parameterized by θ, a set X of possible values for a data vector x, and a conditional probability model $f(x|\theta)$. Imagine that a data value x has been observed, and is to be encoded in a two-part message: an "assertion" stating an estimate $\hat{\theta}$ followed by a "detail" encoding x with length $-\log f(x|\hat{\theta})$. The value x is given to a statistician who is not expert in binary coding. She can work out the detail length for any assumed estimate, but does not want to be involved in actually encoding an estimate. She therefore, knowing x, Θ, $h(\theta)$ and $f(x|\theta)$, determines some region A_x in Θ which she considers to contain good estimates. She describes this region to a coding expert, and asks him to construct an assertion which specifies a value of θ just precisely enough to ensure that the specified value lies in A_x and then to use this specified value ($\hat{\theta}$ say) in constructing a detail which encodes x. How should she choose the region A_x? If she chooses a large region, the coder will be able easily to specify some θ within it using only a short assertion, but the value he chooses may be a poor fit to the data. The statistician's best choice will depend on how she expects the assertion length to vary with the size of A_x.

4.10.1 Random Coding of Estimates

There is a crude way the coder may use to find an assertion $\hat{\theta}$ which lies in A_x and to encode it in a form decodable by the receiver. It does not require him to use, or even know, the form of the probability model function $f(x|\theta)$. He need only use the prior $h(\theta)$, which the receiver is also assumed to know. The coder and receiver agree before the data x is known on a coding scheme for assertions. They agree on a good pseudo-random number algorithm and a seed value for starting it. The algorithm could be one such as is often used in computer simulation programs, producing a sequence of real numbers which appear to be randomly drawn from a Uniform distribution between 0 and 1. Using standard techniques, they further construct and agree on a derived algorithm which will generate a sequence of values

$$\theta_1, \ \theta_2, \ \theta_3, \ldots, \theta_n, \ldots$$

which appear to be drawn randomly from the prior density $h(\theta)$.

Now, given a region $A_x \in \Theta$, the coder simply uses the generator to generate the sequence of θ values until he finds one which lies in A_x. Suppose this is θ_n, the nth in the sequence. He then chooses $\hat{\theta}$ to be θ_n and uses it to encode x with a detail length of $\log f(x|\hat{\theta}_n)$. To encode the assertion, he simply encodes the integer n. If they are lazy, he and the receiver may agree on a Universal code for the integers, in which case the length of the encoded assertion is about $\log^*(n) \approx \log(n) + \log\log(n)$. If they care to use their knowledge of the form $f(x|\theta)$ and this form is sufficiently regular, they can work out a code with length about $\log n$, which is a little shorter. The receiver can decode n from the assertion, then use the agreed algorithm to generate θ_n. Now knowing this estimate, the receiver can decode the detail to recover the data x.

4.10.2 Choosing a Region in Θ

If the statistician chooses a region A_x and the coder uses the coding scheme above, the expected value of the sequence number n is easily shown to be $E(n) = 1/q(A_x)$ where $q()$ is now defined as

$$q(A_x) = \int_{\theta \in A_x} h(\theta)\, d\theta$$

Knowing how this method of choosing and coding assertions works, the statistician expects that choosing a region A_x will result in an assertion length of about $- \log q(A_x)$. She also knows that any θ within A_x may be employed to code x in the detail, with values of θ having high prior density being more likely to be chosen than those of low prior density. She therefore expects the length of the detail to be

$$(1/q(A_x)) \int_{\theta \in A} h(\theta)\, (- \log f(x|\theta)) d\theta$$

Hence, she should choose the region A_x to minimize

$$- \log q(A_x) - (1/q(A_x)) \int_{\theta \in A_x} h(\theta)\, \log f(x|\theta) d\theta$$

An argument parallel to that used in 3.3.2 to derive the Boundary Rule for SMML data groups shows that the minimum is achieved when

$$\theta \in A_x \text{ iff } \log f(x|\theta) > (1/q(A_x)) \int_{\phi \in A_x} h(\phi) \log f(x|\phi)\, d\phi - 1$$

That is, the region A_x should contain all models whose log-likelihood given x is not less than the average log-likelihood within the region minus one. The minimum value thus found is Dowe's approximation to the explanation length of the data x and will be written as $I1D(x)$.

$I1D(x)$ is quite a good approximation to the message length for data x which would occur in an SMML code. In effect, the region A_x is constructed to contain all those estimate values which might be used for x in an SMML-like code, and uses the average detail length of these estimates as an approximation to the SMML detail length. The method departs from SMML in that the "coding probability" whose negative log determines the assertion length is based on the prior model probability within a region of the parameter space Θ rather than on the marginal data probability within a region or subset of the data space X. However, as should be evident from the scenario for the construction of an explanation of x, the resulting message is quite efficiently encoded despite the curious use of pseudo-random sequences. The expected message length for x given by this construction is little more than the length of the strictly optimal SMML explanation. The main source of inefficiency in the coding envisaged lies in the encoding of the estimate's sequence number n. Use of a Universal code for n will never be optimal, and gives an extra length of order $-\log(-\log A_x)$ (which I have ignored in describing the optimum region). In principle, if the coder used knowledge of the function $f(x|\theta)$ as well as knowledge of the prior, he could work out the probability distribution over n implied by these, and use a code for n optimized for this distribution. This would not be easy, as the probability distribution over n itself depends on the code used to encode n. In the approximation for the message length described in this section, the slightly optimistic assumption has been made that the code for n will have length only $\log n$.

This method for approximating the explanation length $I_1(x)$ is rather simpler than the IG approximate estimator, as it does not require calculation of the marginal data distribution, and requires only one construction of a region. The IG estimator, unless the problem permits an analytic solution, requires an iterative search for the desired estimate, and the construction via the Boundary Rule of a data group for each estimate considered in the search.

The construction leading to $I1D(x)$ however, does not yield an estimate. It only yields a region in parameter space, not a unique point. It would be nice if a general, invariant method could be devised for deriving from the region a point estimate which fairly represented its "centre", but no such method has been found. However, the indecision about the best-guess model reflected in the size of the region is not without excuse. At least roughly, the region represents the set of estimates which might arise in SMML-like explanations of x. Given x and the region A_x, it appears that for almost any $\theta \in A_x$, there will be some SMML-like code in which θ appears as the estimate for x.

The implication of the above argument is that, to the extent that it is valid and Minimum Message Length a good principle, any point in A_x can be regarded as an acceptable estimate of θ. If this is the case, the failure of the method described here to yield a "region centre" as a point estimate is of no great consequence. If a unique point estimate is desired, this argument suggests that one may as well pick a random point in the region, e.g., by sampling from the prior density distribution in the region. (This choice would have the same effect as the method of finding and coding an estimate used by the "coder" in our imaginary scenario.) Further, since the log-likelihood of the worst-fitting models at the edge of the region have log-likelihoods only one nit short of the average, no model in the region could be reasonably rejected in the light of the data.

To sharpen this last point a little, suppose that the log-likelihood given x has quadratic behaviour about its maximum, and that the prior on θ is slowly varying. Then the region A_x will be a P-dimensional ellipsoid, where P is the dimension of the parameter space, i.e., the number of scalar parameters. It is then easily shown that the average log-likelihood of models in A_x is less that the maximum log-likelihood by $(P/2)$ nits, so the log-likelihood of every model in the region falls short of the maximum by at most $P/2 + 1$. A standard significance test for the fit of a model θ to given data x is based on the "log-likelihood ratio" λ defined as

$$\lambda = \log(f(x|\theta_{ML})/f(x|\theta)) = \log f(x|\theta_{ML}) - f(x|\theta)$$

where θ_{ML} is the Maximum Likelihood estimate from x. Under the conditions assumed here, it is known that 2λ is distributed as a Chi-Squared variate with P degrees of freedom. Hence, the likelihood-ratio test rejects the model θ if 2λ has a value exceeding what is probable for a χ_P^2 variate. For $\theta \in A_x$, 2λ never exceeds $P + 2$. But for no number P of free parameters is $P + 2$ an improbably large value for a χ_P^2 variate. Hence, no model in A_x would be rejected by a likelihood-ratio test.

4.11 Partitions of the Hypothesis Space

The Strict Minimum Message Length construction for an explanation code is based on a partition of the set X of possible data into groups, and assigns a single codeable estimate for each data group. However, when the set Θ of possible models is the union of several distinct subspaces, e.g., the union of several model classes with different structural forms or numbers of parameters, we are prepared to modify the SMML code construction in order to force the resulting estimator to assert an unambiguous choice of model subspace (Sections 3.4.6 and 3.4.7). We force the explanation to begin by asserting a subspace, using a code length which is the negative log of the prior probability of all models within the subspace, and then proceed to encode the data using an SMML code constructed as if the subspace were the entire set of possible models. The first part of the explanation, asserting the subspace, is thus based on the partition of Θ into regions (the subspaces) and uses a coding probability given by the total prior probability within the asserted region.

In early work on the development of MML, I attempted to develop an approximation to SMML based entirely on a partition of Θ rather than on a partition of X. I hoped that the code construction would be simpler than SMML and have a more obvious relation to prior beliefs. In this coding scheme, the set of possible models Θ is partitioned into a countable set of regions $\{R_j : j = 1, \ldots\}$. The set Θ^* of codeable estimates has one member $\hat{\theta}_j$ for each region R_j, and the coding probability $q_j = q(\hat{\theta}_j)$ assumed for $\hat{\theta}_j$ is equated to the prior probability within the corresponding region R_j. The regions and codeable estimates are then chosen to minimize an approximation to the expected message length. In this approximation, the simplifying assumption is made that any data generated from a source whose true model is in region R_j will be encoded in an explanation which asserts estimate $\hat{\theta}_j$.

Hence, the partition of Θ into regions and the choice of codeable estimates is made to minimize

$$
\begin{aligned}
I_2 &= \sum_j \int_{\theta \in R_j} h(\theta) \left[-\log q_j - \sum_{x \in X} f(x|\theta) \log f(x|\hat{\theta}_j) \right] d\theta \\
&= -\sum_j q_j \log q_j - \sum_j \int_{\theta \in R_j} h(\theta) \left[\sum_{x \in X} f(x|\theta) \log f(x|\hat{\theta}_j) \right] d\theta
\end{aligned}
$$

where $q_j = \int_{\theta \in R_j} h(\theta) \, d\theta$.

The coding scheme is intuitively attractive in that each codeable estimate $\hat{\theta}_j \in \Theta^*$ can be interpreted as representing an "uncertainty region" in hypothesis space whose extent represents the likely estimation error, and is coded with a coding probability which is the prior probability that the true model in fact lies within this region. However, I no longer consider this code construction to be generally advantageous.

For some estimation problems, the estimate set Θ^* obtained by minimization of I_2 closely resembles in spacing and coding probabilities those given by SMML. The quantity I_2 is a pessimistic approximation to the expected message length achieved by the coding scheme, since it assumes that data drawn from a model in R_j will always be encoded using estimate $\hat{\theta}_j$, even if the data is such that some other member of Θ^* would give a shorter explanation. In fact, for simple model classes such as the Binomial and Normal distributions, the scheme achieves an average explanation length within a small fraction of a nit of an SMML code. Further, the notion of an "uncertainty region" has proved useful in approximating the explanation lengths for many model classes, even when the actual construction of an optimal partition of the hypothesis space is infeasible. It also led to Dowe's approximation I1D. Unfortunately, the code construction based on a partition of Θ into "uncertainty regions" has flaws which vitiate the approach.

– A code for an estimation problem must be agreed before the data is known. If it uses a partition of Θ, it is in effect deciding on the precision with which the estimate is to be stated in a manner which may depend on the estimate (different codeable estimates may have regions of different sizes) but may not otherwise depend on the data. This restriction is tolerable for models in the exponential family, for which sufficient statistics exist having the same dimensionality as the parameter θ. The optimal precision for stating an estimate must depend on the data only via the sufficient statistics, and the estimate giving the shortest explanation for such models approximately encodes the sufficient statistics. Hence, the best estimate for some data determines its own precision. However, for model classes having no minimal sufficient statistic, there may be no partition of Θ which provides an appropriately precise estimate for all possible data. The size of

the "uncertainty regions" of two different data sets, as found for instance
by Dowe's I1D construction, may be quite different even if both data sets
can be well explained by the same estimate.

- For any model class, the construction of an SMML code need only be concerned with the set of possible sufficient statistic vectors. When a minimal
sufficient statistic exists, this set has the same dimensionality as the parameter space, so a construction which partitions the parameter space is not
essentially simpler than the SMML construction. When a minimal sufficient statistic does not exist, as noted above no generally efficient partition
of Θ may exist.

- In extreme cases, no feasible partition of Θ may exist. The partition construction via minimization of I_2 requires that for each region R_j a single
model $\hat{\theta}_j$ be found which can be used to encode any data sourced from any
model in R_j. If the model class is the Uniform distribution with unknown
mean μ and known range 1.0, the data which might be sourced from any
model within a finite range of μ will have highest and lowest values differing by more than 1.0, so no model within the model class can encode all
such data.

- The estimate $\hat{\theta}_j$ for some region R_j minimizes I_2 when it is chosen to
give the highest possible average log likelihood over all data which could
come from models in R_j. If θ includes both a parameter of location and a
parameter of scale, the variation within R_j of model locations may make
the spread of all data sourced from models within R_j greater than the
spread expected from any one model in R_j. By fitting $\hat{\theta}_j$ to the collection
of all data sourced within the region, the construction may overestimate the
scale parameter. In fact, it can be shown that this construction, if applied
to the Neyman-Scott problem, results in an inconsistent overestimation of
the scale parameter σ similar to that shown by the MEKL estimator of
Section 4.8.

4.12 The Meaning of Uncertainty Regions

Although the idea of constructing an MML code via a partition of the hypothesis space proved to be of very limited use, the associated idea that the
explanation of some given data can in some sense express an "uncertainty
region" in hypothesis space remains fruitful, and has been used in the MML
analysis of many complex models. Its principal application has been to approximate the length of the first part of an explanation by the negative log
of the total prior probability of all the hypotheses or models which lie within
the uncertainty region. This is a simpler approach than the strict one of using the negative log of the total marginal probability of all the data vectors
which would result in the asserted estimate, as is done in SMML. It also
gives a more direct and transparent connection between prior beliefs about

the model and the length of the code needed to describe it. The idea of an uncertainty region is, however, open to misinterpretation in its connection to the construction and length of an explanation. The matter has been touched upon in Section 3.1.7 but is worth another look. In thinking about how best to encode some data (perhaps as yet unknown) in an explanation message, an uncertainty region can be conceived in at least two ways.

4.12.1 Uncertainty via Limited Precision

Suppose the hypothesis space Θ is a one-dimensional continuum. The first part of an explanation will encode an estimated parameter value $\hat{\theta}$ specifying some model in Θ which fits the data well, but the need to keep the first part short means that the estimate can be stated only to some limited precision. Analysis of the effects of limiting the precision of $\hat{\theta}$ may reveal that the explanation length will be (approximately) minimized if the estimated value of θ is "rounded off" to the nearest value within a range of $\pm\Delta/2$. If the analysis falls short of actually constructing an SMML code with its set Θ^* of assertable estimates, the exact first-part estimate value $\hat{\theta}$ cannot be known, but it is reasonable to suppose that, whatever it turned out to be, it could be encoded fairly efficiently by giving it a code length equal to the negative log of the total prior probability lying in an interval of size Δ and centred on the unrounded value we would ideally like to assert. This coding probability is of course the prior probability that the true parameter value lies in the interval, so should be a reasonable approximation to the probability (prior to observing the data) that the unrounded estimate will lie in the interval, and hence result in the assertion of the rounded-off value. This kind of approach is developed in some detail in Chapter 5.

4.12.2 Uncertainty via Dowe's I1D Construction

In Section 4.10, Dowe's I1D approximation to the explanation length of some given data x finds a region in Θ in which every model has a log-likelihood $\log f(x|\theta)$ no more than one nit worse than the (prior-weighted) average log-likelihood in the region. The region so found may be considered an "uncertainty region" and the I1D approximation supposes that the explanation of x will assert some model in effect randomly chosen from the prior within the region. Although the line of argument leading to the choice of region is somewhat different from that leading to a "precision" Δ, in model classes where the "precision" approach is usable, the two types of region are numerically very similar or identical.

4.12.3 What Uncertainty Is Described by a Region?

In both types of uncertainty region arising from the MML inference from given data, the actual specification of the region does not form part of the explanation message. The assertion in the explanation specifies a single model.

It does not specify a range or set of models and should not be understood to do so. The detail (second part) of the explanation encodes the data using a code which would minimize the expected length of the detail were the data generated randomly by the asserted model source. The coding of the detail depends only on the single asserted model and not at all on the statistician's uncertainty about what the true model might be.

The uncertainty represented by an uncertainty region is the uncertainty about what exact model might be asserted in an explanation of the data using an SMML-like code. It should be seen as specifying a range or set of models any one of which might be used in constructing an explanation of the data with near-optimal brevity. Of course, were a genuine SMML code constructed *a priori* and used in the explanation, there would be no uncertainty: the asserted estimate would be the best available in the SMML Θ^* set of codeable estimates. Hence, the notion of an "uncertainty region" does not appear in the SMML construction and there is no association of regions of hypothesis space with the members of Θ^*. The uncertainty region notion arises only with some SMML-like code constructions which aim to achieve explanation lengths almost as short as a genuine SMML code by simpler means.

As shown in Section 4.10.2, the I1D construction for well-behaved model classes leads to an uncertainty region such that no model within the region would be rejected in favour of the maximum-likelihood model by a conventional likelihood-ratio statistical test. In fact, the region contains almost all models which would be acceptable by such a test. The region thus corresponds quite closely to the region in hypothesis space which conventional statistical analysis would regard with fair "confidence" as containing the true model. A similar result can be shown for uncertainty regions derived from the "precision" argument. Hence, although in their derivation uncertainty regions do not in principle indicate the statistician's uncertainty about the true model, for many model classes these regions do correspond closely to regions of uncertainty about the true model, at least if it is assumed that the model class contains a model of the true source. Given this correspondence, it may seem odd, even perverse, to insist that in an explanation message the assertion should name a single model in the region, and that the detail be encoded as if the data were drawn from just that exact model. If, after examining the data, the statistician concludes with some confidence that the data came from some model in the region, but cannot further pin the model down, should we not allow the inductive inference stated in the assertion to assert a confidence (or uncertainty) region of models and then encode the data using a code which is optimized for the distribution of data expected to arise as some sort of average over the distributions of the various models in the region? Such a message construction is entirely possible. However, as shown in Section 3.1.7, minimization of the length of a message of this form does not lead to a useful inference. The minimum is reached only when the uncertainty region expands to contain the entire hypothesis space. Of course, a more sensible choice of region could be made by standard statistical techniques of constructing "confidence" regions, and would lead to a message encoding the data briefly. However, the message would not be the shortest possible message which asserted a region, so minimization of explanation length could not be used as an inference principle.

It appears that minimization of explanation length, as an inference principle, works only if the explanation asserts a single model and uses only that model in encoding the data in the detail. In later chapters, some models will be discussed in which "nuisance" parameters occur whose number increases with the data sample size. In these models, it is possible to remove the nuisance parameters from the likelihood function by summing or integrating the likelihood of the data over the posterior distribution of these parameters and then to estimate the remaining parameters using the modified likelihood. Experience with these models has shown that MML estimates based on the modified likelihood are somewhat less reliable than MML estimates based on the original likelihood function in which all parameters appear and are estimated. Although this experience is based on a small number of model classes, it supports the principle that an MML assertion should specify a single, fully detailed hypothesis, and suggests that the principle should be observed even when the model class can apparently be simplified by the elimination of some parameters of the original hypothesis space. However, the generality of this idea has not been established.

4.13 Summary

This chapter has described two methods of approximating SMML estimation which avoid the need to construct a complete partition of the set X of possible data, and give results which are smoother functions of the given data. Both methods retain the invariance property of SMML and much of its generality.

The Ideal Group (IG) estimator produces an estimated model by considering "ideal" data groups which do not form a partition. The construction allows a finite coding probability to be associated with any model, not just those models in the SMML set Θ^*, and then finds as estimate the model with highest "posterior probability" calculated by taking the coding probability of a model as its (finite) prior probability. The estimate it finds for the well-known tricky Neyman-Scott problem is consistent. The estimates given by two other methods, Maximum Likelihood and Minimum Expected Kullback-Leibler Distance, are shown to be inconsistent in this problem. The inconsistency of Maximum Likelihood can sometimes be corrected by marginalization at the expense of losing estimates of nuisance parameters.

The IG estimator does not directly yield a good approximation to the total message length. A length calculated as $-\log(q(\hat{\theta})f(x|\hat{\theta}))$ where $\hat{\theta}$ is the IG estimate and $q(\hat{\theta})$ its coding probability is optimistic, and must be increased by an additive correction, which in regular cases is close to half a nit per scalar parameter estimated.

The second method, due to David Dowe, gives an approximation $I1D(x)$ for the message length for data x by considering a region of parameter space containing models of high likelihood. It then takes the assertion length as the negative log of the total prior probability within the region, and the detail length as the negative log probability of the data averaged over all models within the region. The method does not yield an estimate from the data, but suggests that any model within the region can provide an explanation of the data with a length close to that which would be given by a true SMML code.

The notion of an "uncertainty region" of hypothesis space such as is derived in the Dowe I1D approximation is discussed and contrasted with the classical notion of a "confidence" interval or region. The assertion made in the first part of an MML explanation is a single, exactly specified model, and should not be read as asserting that the true model lies within an uncertainty region. It is suggested that this principle may imply that nuisance parameters should not be eliminated from the model space even when this seems possible, but should be retained and estimated values specified in the assertion.

5. MML: Quadratic Approximations to SMML

The previous two chapters have described how to construct estimate sets Θ^* and estimators which allow data to be encoded in explanations of short expected length I_1. The SMML method accurately minimizes I_1, but requires an intuitively obscure assignment of coding probabilities to assertions: the coding probability of an assertion is taken as the probability that it will be asserted. Further, it requires the generation of a code for the complete set X of possible data sets before any estimation can be made for the data in hand, and results in an estimator and an explanation length which are both bumpy functions of the data. Chapter 4 offered approximations which avoid the bumpiness and the need to generate a complete code book by concentrating on just the on-average properties of the SMML estimate set Θ^*, but are still computationally difficult and fail to reflect properties of SMML which arise from the partitioning of X. Both SMML and the approximate methods are data- and model-invariant, and the estimators depend on the data only via sufficient statistics. The SMML estimator captures almost all and only the information in the data which is relevant to the model.

None of these methods is suitable for ordinary statistical or inductive inference, being mathematically difficult except in the simplest problems. In short, the whole coding-based approach, as described in these two chapters, seems to result in a great deal of pain for doubtful gain.

In this chapter, I attempt to develop approximate methods which avoid the mathematical difficulties, but retain much of the generality, invariance, and efficiency of SMML. It must be confessed that the attempt has only limited success. No mathematically simple formula has been found for obtaining estimators which are continuous functions of continuous data, model-invariant, and applicable to any form of model distribution. This goal, if it can be achieved, must be reached by others. However, the methods presented here are useable in sufficiently regular problems, are model-invariant or nearly so, and have led to estimators which are competitive with and in some cases superior to those obtained by more conventional methods.

Note that throughout this chapter, I assume that the hypothesis or model space is a continuum of some fixed dimension and model probability function $f(x|\theta)$. Choice among model spaces of differing dimension or functional form must be made by finding the space which contains the model giving the shortest explanation, where the model estimation within each space ignores the existence of the others, as described in Section 3.4.6 on Discrimination.

5.1 The MML Coding Scheme

Two simplifications to the SMML approach form the basis of this chapter. The first changes the basis for assigning coding probabilities to assertable estimates to a scheme based on the prior probability density $h(\theta)$ rather than the marginal distribution of data $r(x)$. The second is the use of a quadratic approximation to the log-likelihood function $\log f(x|\theta)$ in the neighbourhood of θ.

The SMML coding scheme is based on partitioning the set X of possible data into "data groups", each being served by a single estimate in the set Θ^* of assertable estimates of θ. The construction of this partition is difficult, and requires the computation of the marginal distribution $r(x)$ over X, which itself may be quite hard. The basic idea of the approximations in this chapter is to avoid this construction. But if the data groups for assertable estimates are not constructed, even in the "ideal" form of Section 4.1, we cannot equate the coding probability $q(\hat{\theta})$ of an estimate to the total marginal probability of its data group. Instead, we will attempt to relate the coding probability of an estimate more directly to the prior probability density $h(\theta)$ in its neighbourhood.

Initially, we will suppose that the model form $f(x|\theta)$ is such that an SMML-like code would give a Θ^* in which neighbouring estimates have similar coding probabilities and a fairly even spacing. Such is the case with, for instance, the simple Normal and binomial examples of Sections 3.3.1, 3.3.3 and 3.2.3. We defer for now problems such as the Uniform problem of Section 3.3.6 where neighbouring estimates can have widely varying values of $q()$.

In the fairly regular cases assumed for the time being, we have seen that the efficiency of an SMML-like code depends mostly on the sizes of data groups rather than on the precise location of the boundaries between them. In general, the larger the size of data groups, the fewer and more widely spaced are the assertable estimates, so we may say that the efficiency of the code is primarily dependent on the spacing, or density, of estimates in Θ^*. We will therefore attempt to abstract from the code construction the effect of varying the estimate spacing, without assuming anything about the precise location of the estimates.

First, consider the case where θ is a single scalar parameter, so Θ is (some interval of) the real line. Then Θ^* is some set of points arranged along the θ line. Different SMML-like codes could have different selections of points without much affecting their efficiencies, but given some arbitrary point θ on the line, the separation between the Θ^* members immediately above and immediately below θ will be about the same in all SMML-like codes. In other words, the spacing between assertable estimates in the vicinity of θ will be roughly the same in all near-optimum codes. Thus, we can consider the behaviour of SMML-like codes for the given problem to be summarized by a function $w(\theta)$ defined for all $\theta \in \Theta$ such that in any SMML-like code, the separation between the assertable estimates bracketing θ is approximately $w(\theta)$. This "spacing" function abstracts the notion of estimate spacing (or equivalently data group size) which is common to all SMML-like codes for the given problem, but the function contains no hint as to where precisely the assertable estimates of a code might lie.

It is then natural to set the coding probability $q(\hat{\theta})$ of an assertable estimate $\hat{\theta}$ in some code equal to the total prior probability in an interval of θ centred on $\hat{\theta}$ and of width $w(\hat{\theta})$. If the prior density $h(\theta)$ varies little over an interval of this width, we can approximate the coding probability of an estimate $\hat{\theta}$ by:

$$q(\hat{\theta}) = h(\hat{\theta})\,w(\hat{\theta})$$

With this assignment of coding probabilities to estimates, there is no guarantee that the coding probabilities of all estimates in Θ^* will sum exactly to one. However, the discrepancy should be small if the prior density and spacing function vary only slowly from one estimate to the next. In any case, the discrepancy will be ignored, as worse approximations are to come!

Imagine a statistician who wants to set up a coding scheme for two-part explanation messages encoding data to be collected in a situation specified by our usual X, Θ, $h(\theta)$ and $f(x|\theta)$, but (knowing little of binary coding techniques) is not prepared to embark on the construction of a true SMML-like code. Fortunately, she is assisted by a coder who, while knowing little about statistical estimation, knows standard techniques for constructing efficient codes for any specified probability distribution. It is believed that the model family $f(x|\theta)$ is one for which SMML-like codes may be summarized by a spacing function $w(\theta)$, but since the statistician cannot construct SMML-like codes, the appropriate spacing function is unknown. The two agree to divide the task of encoding the (as yet unknown) data between them. The statistician will devise some spacing function $w(\theta)$ and give it and the prior density function $h(\theta)$ to the coder. He will use some standard algorithm for selecting a set $\Theta^* = \{\hat{\theta}_j : j = 1, 2, \ldots\}$ from the Θ line with spacing conforming to $w()$, and will then construct an optimal code for the assertion based on the probability distribution

$$\Pr(\hat{\theta}_j) = q(\hat{\theta}_j) = h(\hat{\theta}_j)w(\hat{\theta}_j)$$

Then, when the statistician receives the data x, she will pick on some "target" estimate θ' and tell it and the data x to the coder. The coder will then pick $\hat{\theta}$ as the assertion, where $\hat{\theta}$ is the member of Θ^* closest to θ', and encode x using a standard-algorithm code for the distribution $f(x|\hat{\theta})$.

This cooperative procedure will result in an explanation length of

$$I_1(x) = -\log(h(\hat\theta_j)w(\hat\theta_j)) - \log(f(x|\hat\theta))$$

and does not require the statistician to be concerned with the exact placing of Θ^* estimates in Θ. It should be obvious that the resulting message is decodable by a receiver who knows X, Θ, $h()$ and $f(|)$, the method used by the statistician to choose the function $w(\theta)$ and the standard methods used by the coder to select the members of Θ^* given $w()$ and to construct efficient binary codings of the assertions and detail from the distributions $q(\hat\theta_j)$ and $f(x|\theta)$.

Consider how the statistician should best choose her "spacing function" $w(\theta)$ and (once the data is known) choose her "target" estimate θ'. (She must of course choose $w()$ without reference to the data.)

She would like to minimize $I_1(x)$ but this is a function of $\hat\theta$ which she does not know. What she does know is that $\hat\theta$ is the member of Θ^* closest to her target θ', so the difference or "roundoff error" $\varepsilon = \hat\theta - \theta'$ lies in a range roughly $\pm w(\theta')/2$. So she may get an approximate value for $I_1(x)$ by expanding the expression for $I_1(x)$ as a power series in ε around θ'. For the present, we will assume $h(\theta)$ and $w(\theta)$ to vary so slowly that, to second order in ε, their variation may be neglected. Then

$$
\begin{aligned}
I_1(x) &= -\log(w(\hat\theta)h(\hat\theta)) - \log f(x|\hat\theta) \\
&\approx -\log(w(\theta')h(\theta')) - \log f(x|\theta') - \varepsilon\frac{\partial}{\partial\theta'}\log f(x|\theta') \\
&\quad -\frac{1}{2}\varepsilon^2\frac{\partial^2}{(\partial\theta')^2}\log f(x|\theta') + O(\varepsilon^3)
\end{aligned}
$$

The statistician may expect the rounding-off to be unbiased, and ε to be equally likely to take any value in the range. Thus, she expects

$$E_c\varepsilon = 0; \qquad E_c\varepsilon^2 = w(\theta')^2/12$$

where E_c denotes the expected result of replacing θ' by the closest codeable value $\hat\theta$. Then, the statistician expects

$$
\begin{aligned}
E_c I_1(x) &\approx -\log(w(\theta')h(\theta')) - \log f(x|\theta') - E_c(\varepsilon)\frac{\partial}{\partial\theta'}\log f(x|\theta') \\
&\quad -\frac{1}{2}(E_c\varepsilon^2)\frac{\partial^2}{(\partial\theta')^2}\log f(x|\theta') \\
&\approx -\log(w(\theta')h(\theta')) - \log f(x|\theta') \\
&\quad -\frac{1}{24}w(\theta')^2\frac{\partial^2}{(\partial\theta')^2}\log f(x|\theta')
\end{aligned}
$$

This expression is minimized with respect to $w(\theta')$ by setting

$$w(\theta')^2 = -12 / \left[\frac{\partial^2}{(\partial\theta')^2}\log f(x|\theta')\right]$$

However, as mentioned above, the explanation code must be decided *before* the data x is available, or, equivalently, be based only on information already available to the receiver of the message. Thus, $w(\theta)$ must be chosen without using knowledge of x, and hence without using the value of $-\frac{\partial^2}{(\partial\theta')^2}\log f(x|\theta')$.

While the exact value of this second differential is unavailable, the statistician can compute its expectation

$$F(\theta') = -E\frac{\partial^2}{(\partial\theta')^2}\log f(y|\theta')$$

$$= -\sum_{y\in X} f(y|\theta')\frac{\partial^2}{(\partial\theta')^2}\log f(y|\theta')$$

The function $F(\theta)$ is well known as the "Fisher Information". It is a function of θ whose form depends only on the form of the model probability function $f(y|\theta)$. Being the expected value of minus the second differential of the log likelihood, it indicates how sharply peaked we expect the negative log likelihood to be, as a function of θ. If F is large for some θ, we expect to find, in analysing data drawn from a source described by θ, that the negative log likelihood will have a sharp and narrow peak. Thus, in framing an explanation of such data, we expect to have to quote our estimate of θ quite precisely in order to achieve a high value of $f(x|\hat{\theta})$, and hence a short detail. If F is small, even quite large "rounding off errors" in coding assertion will not be expected to increase the length of the detail much.

The use of the word "information" in the conventional name for $F(\theta)$ is entrenched but slightly misleading. The quantity is not a measure of information in its modern, Shannon, sense although it is clearly related to how informative we expect the data to be about θ.

By analogy with this usage, the actual value of the second derivative of the negative log likelihood is sometimes called the "observed" or "empirical" Fisher Information, or just observed or empirical information. It is of course a function of the data x as well as of the parameter θ, allowing the overloaded but unambiguous notation

$$F(\theta, x) \stackrel{\text{def}}{=} -\frac{\partial^2}{(\partial\theta)^2}\log f(x|\theta)$$

Using a "spacing function" $w(\theta) = \sqrt{12/F(\theta)}$, we have, in expectation over roundoff effects,

$$E_c I_1(x) = -\log(w(\theta')h(\theta')) - \log f(x|\theta') - \frac{1}{2}\left[\frac{F(x,\theta')}{F(\theta')}\right]$$

$$I_1(x) \approx -\log h(\theta') - \log f(x|\theta') + \frac{1}{2}\log F(\theta') - \frac{1}{2}\log 12$$

$$+ \frac{1}{2}\left[\frac{F(\theta',x)}{F(\theta')}\right]$$

$$\approx \left[-\log \frac{h(\theta')}{\sqrt{F(\theta')/12}} \right] + [-\log f(x|\theta')] + \left[\frac{1}{2} \frac{F(\theta', x)}{F(\theta')} \right]$$

(**Formula I1A**)

Formula I1A approximates the explanation length when data x is encoded using asserted estimate $\hat{\theta}$ closest to θ' in a set Θ^* of assertable estimates selected with spacing function $w(\theta) = \sqrt{12/F(\theta)}$. The three square-bracketed terms in the formula give respectively the length of the assertion, the ideal length of the detail, and a "roundoff" term showing by how much the actual detail length is expected to exceed the ideal length as a result of replacing the target estimate θ' by the nearest available assertable estimate $\hat{\theta}$.

Given data x, the statistician should choose her target θ' to minimize I1A, and this θ' may be taken as the MML estimate. Note that for continuous data, θ' will be a continuous function of x, unlike the SMML estimate.

A further simplification is possible, convenient, and *usually* innocuous. For model classes which have a minimal sufficient statistic s, we expect the estimate θ' to be an invertible function of s. That is, we do not expect two different values of s to result in the same estimate, since the two values of s say different things about θ and θ' will be a smooth function of s. If indeed θ' is an invertible value of s, then s is a function of θ'. Therefore, $F(\theta', x)$ may be expressed as a function of θ' alone, since it depends on x only via s. And of course the Fisher Information is a function of θ' alone. Hence, we have both the expected and empirical informations as functions of s alone. Barring a very strong influence from the prior $h(\theta)$, we can reasonably assume the two informations will be almost equal. Then we can approximate the empirical information in the third term of I1A by the Fisher Information, giving the simpler approximation for message length below.

$$I_1(x) \approx \left[-\log \frac{h(\theta')}{\sqrt{F(\theta')/12}} \right] + [-\log f(x|\theta')] + \frac{1}{2} \quad (\textbf{Formula I1B})$$

The estimate θ' which minimizes I1B has sometimes been referred to as the "MML87" estimate since this formula was first published in 1987 [55]. Although its derivation involves some crude approximations, it has been used with considerable success, as will be shown in some examples in later chapters. However, it is important to understand the assumptions and approximations made. In any intended application, one should check that the nature of the problem does not seriously invalidate these simplifications.

5.1.1 Assumptions of the Quadratic MML Scheme

The assumptions underlying the derivation of formula I1A are:
(a) The form of the model class $f(x|\theta)$ is such that, were an SMML code constructed, its Θ^* could be described by a spacing function. Essentially, this

means that if two data sets x_1, x_2 should result in the same target estimate θ', the same precision of specification for the estimate is acceptable in both cases.

(b) For all $x \in X$, the likelihood function $f(x|\theta)$ has approximately quadratic dependence on θ near its maximum.

(c) The space Θ is such that it has a locally Euclidian metric, permitting the meaningful use of terms like "distance" and "nearest".

(d) The Fisher Information is defined everywhere in Θ except perhaps at boundaries of the range of θ.

(e) The prior density $h(\theta)$ and the Fisher Information $F(\theta)$ vary little over distances in Θ of the order of $1/\sqrt{F(\theta)}$.

The simpler formula I1B requires all of the above plus the further assumption:

(f) For all $x \in X$, if $\hat{\theta}$ is a reasonable estimate of θ given x, then $F(\hat{\theta}, x) \approx F(\theta)$.

5.1.2 A Trap for the Unwary

Consider a data set of N independent cases, where the data for each case x_n comprises a binary value b_n and some other, possibly vector-valued data y_n. The probability model is that the binary values have independently some probability α of being 1, otherwise 0, and that y_n is selected from a distribution of known form $f(y|\theta_b)$ where the parameter θ_b is θ_0 if $b_n = 0$ but θ_1 if $b_n = 1$. The model parameters α, θ_0 and θ_1 are all unknown, with some priors which need not concern us. Suppose the parameter θ_b has K scalar components.

The expected Fisher determinant factorizes into three factors. The first is just the Binomial-distribution form for the parameter α, viz., $N/(\alpha(1 - \alpha))$. The second relates to θ_0, so since this parameter has K components, the second factor will contain the factor $((1 - \alpha)N)^K$ since the expected number of cases relevant to θ_0 is $(1 - \alpha)N$. Similarly, the third factor, from θ_1, will contain the factor $(\alpha N)^K$. Overall, the Fisher Information will depend on α via the factor $(\alpha(1 - \alpha))^{K-1}$. If K is large, minimization of Formula I1B, which contains half the log of the Fisher Information, will be biased by this factor towards extreme values of α. If the sample contains n_0 cases with $b = 0$ and n_1 with $b = 1$, and either of these counts is not much larger than K, minimization of I1B may give a silly estimate of α and of at least one of θ_0, θ_1.

This deficiency in I1B was noted by P. Grunwald [18] who attempted to circumvent it by assuming an unrealistic prior over α. The fault was not his, since the article [55] from which he worked failed to make clear that I1B is a sometimes dangerous simplification of the more accurate Formula I1A. The problem of course arises because the sensitivity of the "detail" length to roundoff of the θ_b parameters depends on the actual counts n_0, n_1, not on their expected values $(1 - \alpha)N$, αN. Use of I1A in some form solves the problem.

A simpler approach also solves the problem without recourse to I1A. In the situation considered, the parameters θ_0, θ_1 are irrelevant to the model, and hence the best coding, of the $\{b_n : n = 1, \ldots, N\}$ data. We can therefore simply change the order of items in the explanation message. Instead of insisting that the assertion of all estimated parameters precede any of the data, we first encode the estimate of α as the probability parameter of a Binomial distribution with sample size N, then encode the b_n values using the asserted α. Having received this much, the receiver knows n_0 and n_1, so we can now encode the estimates of θ_0 and θ_1 to the precision the receiver will expect knowing the actual relevant sub-sample sizes. Finally, the y_n values are encoded using the appropriate θ estimate for each.

In general, there is no reason in framing an explanation message why the assertion of a parameter should not be deferred until it is needed by the receiver. The details of such of the data whose distribution does not depend on the parameter may well precede the assertion of the parameter, and may assist the receiver to know how precisely the parameter will be asserted. A rearrangement of this sort is exploited in a later example (Section 7.4).

5.2 Properties of the MML Estimator

The discussion in this section mainly relates to the estimates found by minimizing formula I1B and the resulting explanation lengths. Where formula I1A is discussed, it will be explicitly mentioned.

5.2.1 An Alternative Expression for Fisher Information

It is a well-known result that the Fisher Information $F(\theta)$ can also be expressed as the expectation of the square of the first derivative of the log likelihood. By our definition,

$$F(\theta) = -\sum_X f(x|\theta) \frac{\partial^2}{\partial \theta^2} \log f(x|\theta)$$

But $\quad \dfrac{\partial}{\partial \theta} \log f(x|\theta) = \left(\dfrac{\partial}{\partial \theta} f(x|\theta) \right) / f(x|\theta) = f_\theta(x|\theta)/f(x|\theta) \quad$ say

$$\frac{\partial^2}{\partial \theta^2} \log f(x|\theta) = \left(\frac{\partial}{\partial \theta} f_\theta(x|\theta) \right) / f(x|\theta) - (f_\theta(x|\theta)/f(x|\theta))^2$$

$$= \left(\frac{\partial}{\partial \theta} f_\theta(x|\theta) \right) / f(x|\theta) - \left(\frac{\partial}{\partial \theta} \log f(x|\theta) \right)^2$$

Hence,

$$F(\theta) = -\sum_X \frac{\partial}{\partial\theta} f_\theta(x|\theta) + \sum_X f(x|\theta) \left(\frac{\partial}{\partial\theta} \log f(x|\theta)\right)^2$$

$$= -\frac{\partial}{\partial\theta} \sum_X f_\theta(x|\theta) + \mathrm{E}\left(\frac{\partial}{\partial\theta} \log f(x|\theta)\right)^2$$

$$= -\frac{\partial^2}{\partial\theta^2} \sum_X f(x|\theta) + \mathrm{E}\left(\frac{\partial}{\partial\theta} \log f(x|\theta)\right)^2$$

$$= -\frac{\partial^2}{\partial\theta^2}(1) + \mathrm{E}\left(\frac{\partial}{\partial\theta} \log f(x|\theta)\right)^2$$

$$= \mathrm{E}\left(\frac{\partial}{\partial\theta} \log f(x|\theta)\right)^2$$

Obviously, $F(\theta) \geq 0$ for all θ.

Note also that if the data are treated as continuous,

$$F(\theta) = -\int_{x \in X} dx\, f(x|\theta) \frac{\partial^2}{\partial\theta^2} \log f(x|\theta)$$

$$= \int_{x \in X} dx\, f(x|\theta) \left(\frac{\partial}{\partial\theta} \log f(x|\theta)\right)^2$$

5.2.2 Data Invariance and Sufficiency

This is obvious, since the data enter the estimation only via the model probability function $f(x|\theta)$.

5.2.3 Model Invariance

Formula I1B and the MML87 estimator which minimizes it are invariant under one-to-one changes in the parameterization of the model space.

This follows because the function

$$\frac{h(\theta)}{\sqrt{F(\theta)}} f(x|\theta)$$

is unchanged by such a change, as we will now show.

Let $\phi = g(\theta)$ be a new parameterization of the model class, where $g()$ is an invertible function. Then in terms of the new parameter,

(a) The model probability function is $f_1(x|\phi)$ where $f_1(x|g(\theta)) = f(x|\theta)$ for all x, θ.
(b) The prior probability density is $h_1(\phi)$ where

$$h_1(g(\theta)) = \frac{d\theta}{d\phi} h(\theta) = h(\theta) \Big/ \frac{d}{d\theta} g(\theta)$$

(c)   The Fisher Information is $F_1(\phi)$, where

$$F_1(\phi) = \sum_X f_1(x|\phi) \left(\frac{\partial}{\partial \phi} \log f_1(x|\phi) \right)^2$$

Putting $\phi = g(\theta)$,

$$
\begin{aligned}
F_1(\phi) &= \sum_X f_1(x|g(\theta)) \left(\frac{\partial}{\partial \phi} \log f_1(x|g(\theta)) \right)^2 \\
&= \sum_X f(x|\theta) \left(\frac{d\theta}{d\phi} \frac{\partial}{\partial \theta} \log f(x|\theta) \right)^2 \\
&= \sum_X f(x|\theta) \left(\frac{\partial}{\partial \theta} \log f(x|\theta) \right)^2 \left(\frac{d\theta}{d\phi} \right)^2 \\
&= F(\theta) / \left(\frac{d}{d\theta} g(\theta) \right)^2
\end{aligned}
$$

Hence, for any x,

$$
\begin{aligned}
\frac{h_1(\phi)}{\sqrt{F_1(\phi)}} f_1(x|\phi) &= \frac{h(\theta) / \left(\frac{d}{d\theta} g(\theta) \right)}{\sqrt{F(\theta) / \left(\frac{d}{d\theta} g(\theta) \right)^2}} f(x|\theta) \\
&= \frac{h(\theta)}{\sqrt{F(\theta)}} f(x|\theta)
\end{aligned}
$$

So, if θ' is the MML estimate maximizing the right hand side, $\phi' = g(\theta')$ maximizes the left hand side, and is the MML estimate of ϕ.
    Formula I1A is not model-invariant.

5.2.4 Efficiency

Here, we ask how well explanations based on MML estimates approach the minimum possible message length. No very general result is available, but we can get an approximation which is useful for a number of common model classes.
    For some data value $x \in X$, the most efficient non-explanation code will give a message of length $I_0(x) = -\log r(x)$ where $r(x) = \int_{\theta \in \Theta} h(\theta) f(x|\theta) d\theta$
    We will attempt to obtain an approximation to $r(x)$ in terms which facilitate comparison with the MML explanation. Suppose that for given x, the log likelihood $\log f(x|\theta)$ has an approximately quadratic behaviour about its maximum, and that this maximum occurs at $\theta = \theta_0$ and has value $v = \log f(x|\theta_0)$. Then, for values of θ close to θ_0, we can approximate the log likelihood by

$$\log f(x|\theta) \approx v + \frac{1}{2}(\theta - \theta_0)^2 \frac{\partial^2}{\partial\theta^2} \log f(x|\theta)$$

$$\approx v - \frac{1}{2}a(\theta - \theta_0)^2 \quad \text{say}$$

$$\text{where } a = \left(\frac{\partial^2}{\partial\theta^2} \log f(x|\theta)\right)_{\theta=\theta_0}$$

Hence, approximately, $f(x|\theta) = \exp\left(v - a(\theta - \theta_0)^2/2\right)$, and assuming $h(\theta)$ to be slowly varying near θ_0, and that the main contribution to $r(x)$ comes from the vicinity of θ_0,

$$\int_{\theta \in \Theta} d\theta \, h(\theta) f(x|\theta) \approx h(\theta_0) \int_{\theta \in \Theta} \exp\left(v - a(\theta - \theta_0)^2/2\right) d\theta$$

$$\approx h(\theta_0) \exp(v) \sqrt{2\pi/a}$$

whence $r(x) \approx h(\theta_0) f(x|\theta_0) \sqrt{2\pi/a}$ and

$$I_0(x) \approx -\log h(\theta_0) - \log f(x|\theta_0) + \frac{1}{2}\log a - \frac{1}{2}\log 2\pi$$

By comparison, an explanation of x using the MML estimate θ' has expected length

$$E_c I_1(x) = -\log h(\theta') - \log f(x|\theta') + \frac{1}{2}\log(F(\theta'/12) + \frac{1}{2}$$

The MML estimate θ' and the maximum-likelihood value θ_0 are not in general the same, but are usually quite close. In any case, since θ' is chosen to minimize $E_c I_1(x)$, we have

$$E_c I_1(x) \leq -\log(h(\theta_0) f(x|\theta_0) + \frac{1}{2}\log F(\theta_0/12) + \frac{1}{2}$$

$$E_c I_1(x) - I_0(x) \leq \frac{1}{2}[\log F(\theta_0) - \log a + \log(2\pi e/12)]$$

Recalling that $-F$ is the expected second derivative of the log likelihood for the parameter value θ_0, and that $-a$ is the actual second derivative of the log likelihood for the given data x, we can conclude that, on average over all x, the difference between the expected and actual second derivatives should be very small. Hence, defining I_1 as the expectation over all x of $E_c I_1(x)$, we have approximately

$$I_1 - I_0 \approx \frac{1}{2}\log(\pi e/6) \approx 0.1765$$

This is the same difference as was encountered in several SMML single-parameter estimators.

Thus, to within the approximation possible under our assumptions, we are unable to distinguish between SMML and MML explanation lengths on average.

5.2.5 Multiple Parameters

The generalization of the MML approximation to model classes with several parameters is straightforward. If θ is a vector with D components $(\theta_1, \theta_2, \ldots, \theta_k, \ldots, \theta_D)$ we define the Fisher matrix as the matrix of expected second partial derivatives of the negative log likelihood.

$$\mathbf{F}(\theta) = [f_{kl}(\theta)] \quad \text{where}$$

$$f_{kl}(\theta) = -\sum_{x \in X} f(x|\theta) \frac{\partial^2}{\partial \theta_k \partial \theta_l} \log f(x|\theta)$$

or, for continuous data

$$f_{kl}(\theta) = -\int_{x \in X} f(x|\theta) \frac{\partial^2}{\partial \theta_k \partial \theta_l} \log f(x|\theta) dx$$

It is easily shown that $\mathbf{F}(\theta)$ is symmetric and positive definite. Thus, for any θ there exists a local linear non-singular transformation of the parameter space

$$\phi = \mathbf{A}^{-1}\theta$$

such that the Fisher matrix for ϕ

$$\mathbf{F}_1(\phi) = \mathbf{A}^T \mathbf{F}(\theta) \mathbf{A}$$

is a multiple of the D-dimension identity matrix $\mathbf{I}_D$. We choose the transformation $\mathbf{A}$ to have unit Jacobian, i.e., unit determinant, so the determinants of $\mathbf{F}_1(\phi)$ and $\mathbf{F}(\theta)$ are equal. Then

$$\mathbf{F}_1(\phi) = \lambda \mathbf{I}_D \quad \text{where} \quad \lambda^D = |\mathbf{F}(\theta)|$$

and the prior density of $\phi h_1(\phi) = h(\mathbf{A}\phi) = h(\theta)$.

Paralleling the SMML solution for the Multivariate Normal problem discussed in Section 3.3.4, we find that the statistician must instruct the coder to construct the estimate set Θ^* so that, when transformed into ϕ-space, it appears as a set Φ^* which is an optimum quantizing lattice with Voronoi regions of some volume $w(\phi)$. When the coder is given the statistician's estimate θ' based on data x, he is instructed to pick from Φ^* the member $\hat{\phi}$ closest to $\phi' = \mathbf{A}^{-1}\theta'$, and to use the corresponding member $\hat{\theta} = \mathbf{A}\hat{\phi}$ of Θ^* in framing the explanation. Then the length of the detail formed by the coder will be, to second order in $(\hat{\theta} - \theta')$,

$$-\log f(x|\theta') + \frac{1}{2}(\hat{\theta} - \theta')^T \mathbf{F}(\theta')(\hat{\theta} - \theta')$$

where the first order term is omitted as it will have zero expectation.

In ϕ-space, the detail length is

$$-\log f_1(x|\phi') + \frac{1}{2}(\hat{\phi} - \phi')^T \mathbf{F}_1(\phi')(\hat{\phi} - \phi') = -\log f(x|\theta') + \frac{1}{2}\lambda(\hat{\phi} - \phi')^2$$

Hence, the increase in detail length due to the "rounding-off" of θ' to the closest codeable estimate $\hat{\theta}$ is $\frac{1}{2}\lambda(\hat{\phi} - \phi')^2$.

The statistician knows only that ϕ' is a point in the Voronoi region of $\hat{\phi}$, and that this region, of volume w, is a Voronoi region of an optimum quantizing lattice in D dimensions. Hence, her expectation of the roundoff term is (following Section 3.3.4)

$$E_c[\frac{1}{2}\lambda(\hat{\phi} - \phi')^2] = \frac{1}{2}\lambda E_c(\hat{\phi} - \phi')^2 = \frac{1}{2}\lambda D\kappa_D w^{2/D}$$

where κ_D depends on the lattice geometry.

The coding probability associated with $\hat{\theta}$ is that associated with $\hat{\phi}$:

$$q(\hat{\theta}) = q_1(\hat{\phi}) = wh_1(\hat{\phi}) = wh(\hat{\theta})$$

so the statistician expects the length of the assertion to be about $-\log(wh(\theta'))$.

The terms $-\log(wh(\theta'))$ and $\frac{1}{2}\lambda D\kappa_D w^{2/D}$ are the only terms through which the choice of w affects the explanation length. Hence, the statistician chooses w to minimize their sum, giving

$$-1/w + \lambda\kappa_D w^{2/D-1} = 0$$

$$w = (\lambda\kappa_D)^{-D/2}$$

$$E_c[\frac{1}{2}\lambda(\hat{\phi} - \phi')^2] = \frac{1}{2}\lambda D\kappa_D w^{2/D} = D/2$$

Recalling that $\lambda^D = |\mathbf{F}(\theta)|$, we have $w = (\kappa_D)^{-D/2}/\sqrt{|\mathbf{F}(\theta)|}$

$$q(\hat{\theta}) \approx wh(\theta') = (\kappa_D)^{-D/2}h(\theta')/\sqrt{|\mathbf{F}(\theta')|}$$

Hence, the statistician, given data x, will choose the MML estimate θ' to maximize

$$\frac{h(\theta')}{\sqrt{|\mathbf{F}(\theta')|}}f(x|\theta')$$

and the expected explanation length will be

$$E_c I_1(x) = -\log h(\theta') + \frac{1}{2}\log|\mathbf{F}(\theta')| - \log f(x|\theta') + (D/2)\log\kappa_D + D/2$$

Note that the determinant $|\mathbf{F}(\theta)|$ is the Fisher Information for multiple parameters and will henceforth be denoted by $F(\theta)$, as for a single parameter.

5.2.6 MML Multi-Parameter Properties

The single-parameter arguments generalize directly to show that in the multi-parameter case, the MML estimate using formula I1B is a function of sufficient statistics, and is data- and model-invariant.

To consider the coding efficiency of multi-parameter explanations, we can generalize the argument of Section 5.2.4 to obtain an approximation to the marginal probability $r(x)$ of the given data. If the log likelihood has approximately quadratic behaviour about a maximum value v at θ_0, we may write for θ close to θ_0

$$\log f(x|\theta) \approx v - \frac{1}{2}(\theta - \theta_0)^T \mathbf{B}(\theta - \theta_0)$$

where $\mathbf{B}$ is the matrix of partial second derivatives of $\log f(x|\theta)$ at θ_0. This approximation leads to

$$
\begin{aligned}
r(x) &= \int_\Theta h(\theta) f(x|\theta)\, d\theta \\
&\approx h(\theta_0) \int_\Theta \exp(v - (\theta - \theta_0)^T \mathbf{B}(\theta - \theta_0)/2)\, d\theta \\
&\approx h(\theta_0) e^v (2\pi)^{D/2} / \sqrt{|\mathbf{B}|}
\end{aligned}
$$

whence $I_0(x) \approx -\log h(\theta_0) - \log f(x|\theta_0) + \frac{1}{2}\log|\mathbf{B}| - (D/2)\log 2\pi$.

Making the same assumptions as in Section 5.2.4, most importantly that on average $\mathbf{B}$ is close to its expectation $\mathbf{F}(\theta)$ at $\theta = \theta'$, we find that, on average over all $x \in X$,

$$Av(E_c I_1(x) - I_0(x)) \approx (D/2)\log \kappa_D + D/2 - (D/2)\log 2\pi$$

$$I_1 - I_0 \approx (D/2)(1 + \log(2\pi\kappa_D))$$

As shown in Section 3.3.4, the known bounds on κ_D then lead to

$$\frac{1}{2}\log(D\pi) - \gamma > (I_1 - I_0) > \frac{1}{2}\log(D\pi) - 1$$

(Note that in the above we have tacitly re-defined the generic symbol I_1 to stand, not for the average message length achieved by some particular explanation code, but rather the average message length which we expect to be achieved by an explanation code following the MML spacing function $w(\theta)$ but otherwise unspecified. We will use I_1 with this meaning in any context concerned with the MML approach rather than a fully specified code. Similarly, in an MML context, we may use $I_1(x)$ to stand for the explanation length for data x which we expect to achieve using a code following the MML spacing function. That is, we will use $I_1(x)$ to stand for $E_c I_1(x)$.)

We again find that, to within the approximation of this section, the expected coding efficiency of the MML approach is indistinguishable from that of the strictly optimal SMML code.

It is worth remarking that the argument of this section suggests that the coding efficiency of the MML approach for some specific data x differs from its average value $(I_1 - I_0)$ mainly to the extent that the determinant $|\mathbf{B}| = F(\theta', x)$ of the actual log-likelihood differentials differs from the determinant $|\mathbf{F}(\theta')| = F(\theta')$ of expected differentials at the estimated parameter value. For well-behaved model classes, the sampling distribution of $|\mathbf{B}|$ for some fixed θ should become concentrated around $|\mathbf{F}(\theta)|$ as the sample size increases, so for large samples we expect that the coding-efficiency achieved on the actual data will approach the average. That is, we expect

$$I_1(x) - I_0(x) \to I_1 - I_0 \quad \text{for large samples}$$

5.2.7 The MML Message Length Formulae

As mentioned above, in use of the MML approach, we will in future drop the "E_c" symbol denoting an averaging over all codes with the same spacing function, and use simply $I_1(x)$ for the E_c expected length of an MML explanation of data x, and I_1 for the expected length of an explanation averaging over all MML codes and all data.

5.2.8 Standard Formulae

The formula usually used in constructing MML estimators for D scalar parameters will be (from Section 5.2.5):

$$
\begin{aligned}
I_1(x) &= -\log h(\theta') + \frac{1}{2}\log F(\theta') - \log f(x|\theta') + (D/2)\log \kappa_D + D/2 \\
&= -\log \frac{h(\theta')}{\sqrt{F(\theta')\kappa_D^D}} - \log f(x|\theta') + D/2
\end{aligned}
$$

Here, the first term gives the length of the assertion; the second and third give the length of the detail including an expected correction $(D/2)$ for the effect of rounding off θ' to $\hat{\theta}$.

The first term is based on approximating the coding probability of the assertion by the local prior density $h(\theta')$ times a volume $w(\theta') = \left(1/\sqrt{F(\theta')\kappa_D^D}\right)$ of Θ surrounding θ'. This approximation may be poor under certain conditions, and some possible improvements are suggested below.

5.2.9 Small-Sample Message Length

If there is little data, the volume $w(\theta')$ as given above may exceed the volume of a local peak in the prior density $h(\theta')$, or even exceed the volume of the parameter space Θ. In such case, $h(\theta')w(\theta')$ is a poor approximation to the total coding probability which should be accorded to θ' and may even exceed one. A crude but useful improvement is then to approximate the coding probability by

$$q(\theta') = h(\theta')w(\theta')/\sqrt{1 + (h(\theta')w(\theta'))^2}$$

which at least cannot exceed one. This gives

$$I_1(x) \approx \frac{1}{2}\log\left(1 + \frac{F(\theta')\kappa_D^D}{(h(\theta'))^2}\right) - \log f(x|\theta') + D/2$$

Note that this revised expression is still invariant under nonlinear transformations of the parameter.

5.2.10 Curved-Prior Message Length

If the prior density $h(\theta)$ varies substantially over the volume $w(\theta)$, its curvature can affect the optimum choice of $w(\theta)$. For instance, if $\log h(\theta)$ has a large negative second derivative, the effect of a rounding-off error $\varepsilon = \hat{\theta} - \theta'$ will not only be to increase the detail length, but also, on average, to decrease $\log h(\hat{\theta})$ and hence increase the assertion length. If so, the expected message length can be reduced by making the volume $w(\theta')$ (and hence expected roundoff effects) slightly smaller.

A rough correction for curvature in the prior can be made by replacing the Fisher Information $F(\theta)$ by a corrected expression

$$G(\theta) = |\mathbf{G}(\theta)|$$

where the elements of $\mathbf{G}$ are given by

$$g_{k,l}(\theta) = -\sum_{x \in X} f(x|\theta)\frac{\partial^2}{\partial\theta_k \partial\theta_l}\log f(x|\theta) - \frac{\partial^2}{\partial\theta_k \partial\theta_l}\log h(\theta)$$

That is, the second derivative of $\log h()$ is added to the expected second derivative of the log likelihood.

A valid objection to this revision is that the resulting expression for $I_1(x)$ is no longer invariant under transformations of the parameters. However, the revision is capable of giving an improved approximation to $I_1(x)$ and an improved estimate of θ in parameterizations where $F(\theta)$ varies only slowly.

In the special case that the prior has the invariant conjugate form (Section 1.15.6)

$$h(\theta) = Ch_0(\theta)f(z|\theta)$$

where $h_0()$ is an uninformative prior (Section 1.15.4), z is some real or imagined "prior data" and C is a normalization constant, the prior data z contributes to the explanation length in exactly the same way as the given data x. Thus, the calculation of $F(\theta)$ can properly be based on an effective sample size which includes both the size of x and the "sample size" of z. The $F(\theta)$ so calculated then accounts for the *expected* curvature of $\log h(\theta)$, and does so in an invariant manner. Of course, the expected second derivative is a function of θ and the sample size of the prior data, not of z, and so may differ from the actual second derivative of $-\log h(\theta)$.

5.2.11 Singularities in the Prior

If the prior density $h(\theta)$ becomes infinite at some point in Θ, say, θ_0, the approximation to the coding probability $q \approx h(\theta)w(\theta)$ will of course break down at θ_0 (unless $F(\theta)\to\infty$ at θ_0 in such a way as to compensate for the singularity in $h(\theta)$). A singularity in $h(\theta)$ may be removed by a non-linear change in parameter to $\phi = g(\theta)$, giving a non-singular prior density for ϕ, but such a transformation will not remove the problem, since $h(\theta)/\sqrt{F(\theta)}$ is invariant under non-linear transformations. It follows that a similar difficulty arises if for some $\theta_1 \in \Theta$, $F(\theta_1)\to 0$ and $h(\theta_1) > 0$.

We have found no general remedy for this kind of problem within the MML framework. Recourse to the more robust approximations of Chapter 4, which do not use the Fisher Information, should give satisfactory results but only with a fairly complex calculation.

5.2.12 Large-D Message Length

When the number of scalar parameters (D) is four or more, the approximation to the lattice constant κ_D derived in Section 3.3.4 can be used to simplify the expression for $I_1(x)$ based on formula I1B, giving

$$I_1(x) \approx -\log \frac{h(\theta')}{\sqrt{F(\theta')}} - \log f(x|\theta') - (D/2)\log 2\pi + \frac{1}{2}\log(\pi D) - 1$$

with an error of less than 0.1 nit. Use of this approximation avoids the need to consult tables of lattice constants.

5.2.13 Approximation Based on I_0

For the multi-variate Normal distribution with flat prior, it was shown in Section 5.2.6 that

$$\frac{1}{2}\log(D\pi) - \gamma > (I_1 - I_0) > \frac{1}{2}\log(D\pi) - 1$$

Although the derivation was based on assumptions of large D, Normality and a flat prior, the inequality has been found to hold approximately for many forms of likelihood function and for non-uniform priors. Even for $D = 1$, it is in error by less than one nit when applied to a Binomial distribution. In calculating message lengths for complex models, we often need to calculate the length for some simple distribution which forms part of the model, but have no great interest in the estimates of its parameters. For some simple distributions, it can be easier or more convenient to calculate I_0 rather than I_1. It is then useful to approximate the explanation length as

$$I_1 \approx I_0 + \frac{1}{2}\log(D\pi) - 0.4$$

Here, the constant 0.4 has been chosen to make the approximation correct within 0.1 nit for $D = 1$ and Normal or Binomial distributions. The approximation becomes slightly pessimistic (over-estimating I_1 by perhaps 0.5 nit) for large D and any likelihood function whose log has roughly quadratic behaviour around its peak.

5.2.14 Precision of Estimate Spacing

The optimum spacing of MML estimates in a D-parameter model is approximated in by choosing the spacing function $w(\theta)$ to minimize the sum of two terms in the expected message length $E_c I_1(x)$, viz.,

$$- \log w + (C/2)w^{2/D}$$

where we have dropped the dependence on θ for brevity. The first term, arising from the coding probability of the estimate $\hat{\theta}$, favours large w, i.e., wide spacing. In the second term, C depends on the Fisher Information, and the term represents the expected increase in detail length resulting from rounding-off the MML estimate θ' to the nearest available estimate $\hat{\theta}$. It favours small w, i.e., fine spacing. These are the only terms in $E_c I_1(x)$ which depend on $w()$.

Minimization by choice of w gives $w_0 = (D/C)^D/2$, which results in the second term's having the value $D/2$. Now consider the consequences of a non-optimal choice of w, say, $w = w_0 \beta^{D/2}$. Then the sum of the $w()$-dependent terms becomes

$$- \log(w_0 \beta^{D/2}) + (D/2)\beta$$

an increase over the minimum of

$$-(D/2)\log\beta + (D/2)(\beta - 1) = (D/2)(\beta - \log\beta - 1)$$

The message length, and hence the posterior probability of the explanation, is not significantly degraded by an increase of, say, 0.5 nit, so a variation of w causing an increase up to 0.5 would usually be tolerable. For a tolerable increase of δ, the value of β must lie between the two solutions of

$$\beta - \log\beta - 1 = 2\delta/D$$

If these solutions are $\beta_1 < 1$, $\beta_2 > 1$, then the value of w may range over a factor $(\beta_2/\beta_1)^{D/2}$ without exceeding an increase of δ nit. The linear dimensions of a coding region, which roughly speaking determine the roundoff precision for individual parameters, vary as $w^{1/D}$, and hence can range over a factor $(\beta_2/\beta_1)^{1/2}$. For $\delta/D \ll 1$,

$$\beta_1 \approx 1 - 2\sqrt{\delta/D}, \quad \beta_2 \approx 1 + 2\sqrt{\delta/D}, \quad \beta_2/\beta_1 \approx 1 + 4\sqrt{\delta/D}$$

Hence, the volume w may range over a factor of about

$$w_2/w_1 \approx (1 + 4\sqrt{\delta/D})^{D/2} \approx (2\sqrt{\delta D})$$

and the linear dimensions over a factor of about

$$r_2/r_1 = \sqrt{\beta_2/\beta_1} \approx 1 + 2\sqrt{\delta/D}$$

A few values r_2/r_1 and w_2/w_1 are shown in Table 5.1 for $\delta = 0.5$.

Number of Parameters D	r_2/r_1	w_2/w_1
1	4.45	4.45
2	2.80	7.81
10	1.57	89.7
30	1.30	2346
100	1.15	1397218

Table 5.1. Ranges of precision volume for < 0.5 nit change

Even for a single scalar parameter ($D = 1$) the coding region size can vary over a 4:1 range with little effect on the message length, and for 100 parameters, the range exceeds one million to one. However, for increasing D, the linear size of a coding region, i.e., the roundoff precision of a single parameter, must be tightly controlled, as shown by the decreasing values of r_2/r_1 at high D.

The above calculation assumes that the roundoff precisions of all parameters are varied by the same ratio. However, if the roundoff precisions of the individual parameters differ randomly and independently from their optimum values within a range of say, r_0/α to $r_0\alpha$ where r_0 is the optimum precision for the parameter, then the ratio of the resulting region volume w to its optimum w_0 is expected to fall roughly in the range $(1/a)^{\sqrt{D}}$ to $a^{\sqrt{D}}$. In this case, to achieve an expected increase in message length of less than δ we need only require

$$\alpha^{\sqrt{D}} < \exp(\sqrt{\delta D}); \quad \alpha < \exp(\sqrt{\delta})$$

Thus, for $\delta = 0.5$, a *random* error in choosing the precision of each parameter of order $\alpha = \exp(\sqrt{0.5}) \approx 2$ is probably tolerable, i.e., each parameter precision may vary over a 4:1 range.

The conclusion of this analysis is that we need not be too fussy about choosing the precision with which parameters are stated in an explanation, getting each parameter precision within a factor of two of its optimum value will usually be good enough. By corollary, we may accept fairly rough approximations in evaluating the Fisher Information in MML analyses, especially for large D. For $D = 100$, an error of 1000:1 in calculating $F(\theta)$ will scarcely matter. However, a *systematic* error in choosing the precisions of a large number of parameters can have a serious effect on the explanation length, and may badly corrupt an MML comparison of competing models with widely differing numbers of parameters.

5.3 Empirical Fisher Information

The MML approximation uses the expected second derivative of the negative log-likelihood, i.e., the Fisher Information $F(\theta)$, to determine the precision with which an assertion should specify real-valued parameters. It is used because the parameter estimates must be asserted *before* the data has been stated in the detail, and therefore the code used in the assertion cannot depend on the data. In fact, the extent to which the length of the explanation detail is affected by rounding-off estimates depends (approximately) on the actual second derivative of the negative log-likelihood, $F(\theta, x) = -\frac{\partial^2}{\partial \theta^2} \log f(x|\theta)$, which is, of course, a function of the data x. For well-behaved likelihood functions, $F(\theta)$ is an adequate approximation to $F(\theta, x)$ for values of θ which are reasonable estimates given x. However, there are circumstances where it may be preferable to use the actual or "empirical" information $F(\theta, x)$, or some other approximation to it, rather than $F(\theta)$.

5.3.1 Formula I1A for Many Parameters

Even when the spacing function is based on the expected information $F(\theta)$, the increase in detail length expected because of coding roundoff is better estimated by the roundoff term of formula I1A of Section 5.1

$$\frac{1}{2} \frac{F(\theta', x)}{F(\theta')}$$

rather than the simpler term $\frac{1}{2}$ of formula I1B. (Both terms are for a single scalar parameter.)

To generalize to the multi-parameter case with D scalar parameters (i.e., with Θ a D-dimensional space) appears to involve a complication. In D dimensions, both the expected and empirical Fisher matrices are D-by-D symmetric matrices. In dealing with the multi-parameter form for formula I1B (Section 5.2.5), we used a local linear transformation of Θ to transform the expected matrix $\mathbf{F}(\theta)$ to a multiple of the identity matrix, so that in the transformed space we did not need to worry about the direction of the roundoff vector $\varepsilon = \hat{\phi} - \phi'$, only its length. However, there is no guarantee that the transform which makes the expected matrix have this form will make the empirical matrix also have this form. Thus, it seems we have to try to estimate the roundoff term in the transformed space by a term like

$$\delta^T \mathbf{F}_\phi(\phi, x) \, \delta$$

where ϕ is the transformed parameter, δ is a roundoff vector which, in the transformed space, has a uniformly distributed direction, and $\mathbf{F}_\phi(\phi, x)$ is the empirical Fisher matrix in the transformed space, but is not necessarily a multiple of the identity matrix.

This complication may be avoided if we are prepared to accept a code in which, rather than making Θ^* as nearly as possible an optimal quantizing lattice having the Voronoi region volume for each $\hat{\theta} \in \Theta^*$ equal to $w(\hat{\theta})$, we have as Θ^* a *random* selection of points in Θ where the estimates are randomly selected from a density given by $1/w(\theta)$. As shown in Section 5.2.5, such a selection does nearly as well as an optimum lattice, especially when D is large. Suppose such a Θ^* is used. To estimate the roundoff term for some target estimate θ', transform Θ by a locally linear unit-Jacobian transformation to a space Φ in which the *empirical* information matrix $\mathbf{F}_\phi(\phi', x)$ is a multiple of the identity matrix, where ϕ' is the transform of θ'. In this new space, the random set Θ^* will appear as a set Φ^* which is randomly distributed in Φ with local density near ϕ' equal to $1/w(\theta')$, the same density as in the original Θ space, because the unit-Jacobian transformation does not change densities. Now we need only estimate the expected squared length of the roundoff vector $\hat{\phi} - \phi'$, where $\hat{\phi}$ is the member of Φ^* closest to ϕ' in terms of ordinary Euclidian distance. This follows because $\mathbf{F}_\phi(\phi', x)$ is diagonal with equal diagonal elements

$$(F_\phi(\phi', x))^{1/D} \;=\; (F(\theta', x)^{1/D}$$

It then easily follows that, with optimal choice of $w(\theta')$, based on the expected Fisher Information $F(\theta')$, the roundoff term is expected to be the simple expression

$$\text{Roundoff term} = \frac{D}{2} \left(\frac{F(\theta', x)}{F(\theta')} \right)^{(1/D)}$$

The resulting multi-parameter version of **Formula I1A** is then

$$\mathrm{E}_c I_1(x) \;\approx\; \left[-\log \frac{h(\theta')}{\sqrt{F(\theta')/12}} + \frac{D}{2} \log \kappa_D \right] + \left[-\log f(x|\theta') \right]$$
$$+ \left[\frac{D}{2} \left(\frac{F(\theta', x)}{F(\theta')} \right)^{(1/D)} \right]$$

where κ_D is the constant appropriate for a random Θ^* in D dimensions.

5.3.2 Irregular Likelihood Functions

First, for some likelihood functions, $F(\theta, x)$ may differ greatly from $F(\theta)$ even when θ is close to the "true" parameter value. In this case, a code whose Θ^* set of assertable estimates is based on $F(\theta)$ may lead to assertions whose estimates are much more precise, or much less precise, than is warranted by the given data. As the analysis of Section 5.2.14 has shown, the explanation length is little affected by changes in estimate precision by factors of two or more, but for particularly difficult model classes, $F(\theta, x)$ and $F(\theta)$ may well differ by orders of magnitude. If so, the length of the explanation may be much longer if the estimate precision is based on $F(\theta)$ than it would be were the precision tailored to $F(\theta, x)$. A possible remedy is then to adopt a three-part explanation structure:

Part 0 specifies a precision quantum w.

Part 1 states an estimate $\hat{\theta}$ to precision w, e.g., using a Θ^* code where the Voronoi region of each estimate has volume about w, and hence the estimate has a coding probability about $h(\hat{\theta})w$, giving an assertion length about $-\log(h(\hat{\theta})w)$.

Part 2 encodes the data as usual using a code optimal if $\theta = \hat{\theta}$, and has length $-\log f(x|\hat{\theta})$.

In practice, rather than having Part 0 specify the region volume w directly, it would often be better to have it specify one of a family of precision functions, $w_m(\theta), m = 1, 2, \ldots$. For instance, if $F(\theta)$ exists for the model family $f(.|\theta)$ it may suffice to specify one of the family of spacing functions

$$w_m(\theta) = 4^m / \sqrt{F(\theta)}, \quad m = \ldots, -2, -1, 0, 1, 2, \ldots$$

This family provides a choice of spacing functions with spacing in Θ near θ based on $F(\theta)$, but varying by powers of 4. It would then always be possible to choose a spacing within a factor of two of the spacing optimal for the actual information $F(\theta, x)$.

The family of spacing functions should, if possible, be chosen so that the marginal distribution of the optimum index m is independent of θ. If such a choice can be made, Part 0 conveys no information about θ and so the coding of Part 1, which assumes a prior $h(\theta)$ independent of m, remains efficient. The index m, (or rather, that function of the data which leads the coder to choose a value for m) plays a similar role to Fisher's concept of an "ancillary statistic": it conveys some information about the likelihood function given x (in this case, about the sharpness of its peak) but is not directly informative about θ.

If a three-part explanation is used, a code for Part 0 must be determined, based on the marginal distribution of the index m. That is, to produce an optimum version of this three-part code, it is necessary to calculate, for random θ sampled from $h(\theta)$ and random x sampled from $f(x|\theta)$, how often we would expect to use the spacing function $w_m(\theta)$. The Part 0 code for m would then be based on this probability distribution over m. However, in many cases a rough guess at the proper coding of the index m might be adequate, since the length of Part 0 will typically be small compared to the other components of the explanation.

5.3.3 Transformation of Empirical Fisher Information

There is a problem with expressions for explanation lengths which involve use of the empirical Fisher Information, such as formula I1A and the three-part message form above. The problem will carry over into estimator functions which seek to minimize these expressions.

Formula I1B, which uses only the expected Fisher Information, is model-invariant. Its value is not changed by a regular non-linear transformation of the model space Θ with parameter θ to a different parameterization space Φ with parameter ϕ, as was shown in Section 5.2.3. The invariance arises because the Fisher Information $F(\theta)$ transforms to $F_\phi(\phi)$ in the same way as the square of a density. That is, in the one-dimensional case,

$$F_\phi(\phi) = F(\theta) \left(\frac{d\theta}{d\phi} \right)^2$$

so ratios such as $h(\theta)/\sqrt{F(\theta)}$ are unaffected by the transformation.

The *empirical* information does not transform as the square of a density. In the one-dimensional case, if $L = -\log f(x|\theta) = -\log f_\phi(x|\phi)$ then

$$
\begin{aligned}
F_\phi(\phi, x) = \frac{\partial^2 L}{\partial \phi^2} \quad &= \quad \left(\frac{d\theta}{d\phi} \right)^2 \frac{\partial^2 L}{\partial \theta^2} + \frac{d^2\theta}{d\phi^2} \frac{\partial L}{\partial \theta} \\
&= \quad \left(\frac{d\theta}{d\phi} \right)^2 F(\theta, x) + \frac{d^2\theta}{d\phi^2} \frac{\partial L}{\partial \theta}
\end{aligned}
$$

While the first term of the final right hand side shows variation as the square of a density, the second term does not. It follows that expressions like formula I1a and the purely empirical approximation

$$I_1(x) \approx -\log \left(\frac{h(\theta)}{\sqrt{F(\theta, x)}} f(x|\theta) \right) + \text{constants} \quad \textbf{(Formula I1C)}$$

are not model-invariant. However, the troublesome second term in the expression for $F_\phi(\phi)$ is proportional to the first derivative of the log-likelihood with respect to θ. For θ a reasonable estimate given the data x, we would usually expect θ to be close to the Maximum Likelihood estimate, and hence expect $\partial L/\partial \theta$ to be small. Moreover, if x is drawn from the distribution $f(x|\theta_T)$ then

$$\mathrm{E}\left[\frac{\partial L}{\partial \theta}\right]_{\theta=\theta_T} = \sum_X \left\{ f(x|\theta_T) \left[\frac{\partial}{\partial \theta}(-\log f(x|\theta))\right]_{\theta=\theta_T} \right\}$$

$$= \sum_X \left\{ f(x|\theta_T) \left[\frac{\partial}{\partial \theta}(-f(x|\theta))\right]_{\theta_T} / f(x|\theta_T) \right\}$$

$$= -\sum_X \left[\frac{\partial}{\partial \theta} f(x|\theta)\right]_{\theta_T}$$

and subject to regularity

$$= \left[\frac{\partial}{\partial \theta}\sum_X f(x|\theta)\right]_{\theta_T} = \left[\frac{\partial}{\partial \theta}(1)\right]_{\theta_T} = 0$$

so the first-derivative term is expected to be small for $\theta \approx \theta_T$, which we may hope to be true of our estimate.

It appears that using the empirical Fisher in place of the expected Fisher will result in a loss of model-invariance, but the loss may well be small enough to be tolerable. Note that the loss can be expected to be small only when the parameter values or estimates involved are close to their "true" values. It could be dangerous to estimate θ by minimizing an expression like formula I1C unless the search for a minimum is restricted to "reasonable" values. For values of θ far from any "reasonable" estimate based on x, the empirical information $F(x, \theta)$ may be grossly different from $F(\theta)$, or even negative.

5.3.4 A Safer? Empirical Approximation to Fisher Information

Suppose the data x is a sequence of N independent observations or cases $x = (y_1, y_2, \ldots, y_n, \ldots, y_N)$. Then $f(x|\theta) = \prod_{n=1}^{N} g(y_n|\theta)$, and, using the alternative form for the Fisher Information

$$F(\theta) = \mathrm{E}\left(\frac{\partial}{\partial \theta}\log f(x|\theta)\right)^2 = \mathrm{E}\left(\sum_n \frac{\partial}{\partial \theta}\log g(y_n|\theta)\right)^2$$

Writing s_n for $\frac{\partial}{\partial \theta}\log(y_n|\theta)$ for some fixed θ,

$$F(\theta) = \mathrm{E}\left(\sum_n s_n\right)^2 = \mathrm{E}\sum_n s_n^2 + 2\sum_n \sum_{m \neq n} \mathrm{E}(s_n s_m)$$

But, since the data $\{y_n : n = 1, \ldots, N\}$ are independent, $\{s_n : n = 1, \ldots, N\}$ are uncorrelated. Hence, $\mathrm{E}(s_n s_m) = (\mathrm{E}s_n)(\mathrm{E}s_m)$ and we have seen that for any distribution (subject to regularity) $\mathrm{E}[\frac{\partial}{\partial \theta}\log g(y|\theta)] = 0$. Hence,

$$F(\theta) = \mathrm{E}\sum_n (\frac{\partial}{\partial \theta}\log g(y_n|\theta))^2$$

Assuming the observations $\{y_n\}$ are a typical sample from the distribution $g(y|\theta)$, the expected sum above may be approximated by the empirical sum over the actual data, giving the approximation

$$F(\theta) \approx F_e(\theta, x) = \sum_n \left(\frac{\partial}{\partial \theta} \log g(y_n|\theta) \right)^2$$

where x is included in the definition $F_e(\theta, x)$ as a reminder that the expression, while perhaps approximating the expected Fisher Information, is actually an empirical function of the data.

The approximation can only be expected to be close if θ is close to the true value or model which produced the data, but it has the virtue of being guaranteed non-negative. Further, under regular non-linear transformations of the model space, $F_e(\theta, x)$ transforms as the square of a density, so the expression

$$\frac{h(\theta)}{\sqrt{F_e(\theta, x)}} f(x|\theta)$$

is invariant under such parameter transformations, and leads to a model-invariant means of approximating explanation lengths and finding estimates which approximately minimize explanation lengths. We have found in a number of cases that estimates and message lengths obtained using $F_e(\theta, x)$ in place of $F(\theta)$ in the MML method are acceptable approximations to the correct MML values. Also, use of $F_e(\theta, x)$ in place of $F(\theta, x)$ in the roundoff term of Formula I1A makes the formula model-invariant.

There is a problem with the F_e form. While the empirical information $F(\theta, x)$ depends on the data x only via sufficient statistics, $F_e(\theta, x)$ in general depends on aspects of the data other than its sufficient statistics. That is, $F_e(\theta, x)$ cannot in general be expressed as a function of θ and the sufficient statistics alone. Hence, estimates of θ based on the minimization of explanation length approximations involving $F_e(\theta, x)$ are in general affected by properties of the data which convey no information about θ. Nevertheless, the F_e form has been used without trouble in many analyses of identically and independently distributed (i.i.d.) data sets.

$F_e(\theta, x)$ may be grossly different from $F(\theta, x)$ in extreme cases. For instance, if the data $\{y_n : n = 1, \ldots, N\}$ are values taken from a Normal distribution of unknown mean μ and known Standard Deviation 1, it can conceivably happen that all y values are the same, i.e., that $y_n = m$ for all n. In that case, m is the obvious estimate for μ, but $F_e(\mu, x) = 0$ when $\mu = m$, whereas $F(\mu, x) = 2N$. The failure of the approximation in this case occurs because the y values do not resemble at all a typical or representative selection of values from Normal$(m, 1)$. When the supposedly i.i.d. data are found to depart so strongly from what might be expected from any model $g(y|\theta) : \theta \in \Theta$, the statistician, rather than abandoning the $F_e()$ form, might reasonably begin to doubt whether the true source of the data lies within the model space she is supposed to assume.

5.4 A Binomial Example

For the binomial problem with N trials, s successes and unknown success probability p, s is a minimal sufficient statistic and we will take it as the data value. Then the negative log likelihood is

$$L = -\log f(s|p) = -s\log p - (N-s)\log(1-p) - \log\binom{N}{s}$$

and

$$\frac{dL}{dp} = -s/p + (N-s)/(1-p)$$

$$F(p,s) = \frac{d^2L}{dp^2} = \frac{s}{p^2} + \frac{N-s}{(1-p)^2}$$

$$F(p) = EF(p,s) = \frac{Np}{p^2} + \frac{N(1-p)}{(1-p)^2} = \frac{N}{p(1-p)}$$

With the flat prior $h(p) = 1$ $(0 \le p \le 1)$ the explanation length is (using the simple MML formula I1B)

$$
\begin{aligned}
I_1(s) &= -\log h(p) + (1/2)\log F(p) - \log f(s|p) + (D/2)\log \kappa_D \\
&\quad + (D/2) - \log\binom{N}{s} \\
&= \frac{1}{2}\log\frac{N}{p(1-p)} - s\log p - (N-s)\log(1-p) \\
&\quad + \frac{1}{2}\log\frac{1}{12} + \frac{1}{2} - \log\binom{N}{s} \\
&= \frac{1}{2}\log N - (s+\frac{1}{2})\log p - (N-s+\frac{1}{2})\log(1-p) \\
&\quad - \log\binom{N}{s} - 0.7425\ldots
\end{aligned}
$$

The MML estimate is $p' = (s+\frac{1}{2})/(N+1)$.

If some prior information about p is known, it may be possible to represent it within the form of the conjugate prior for the Binomial distribution, namely the Beta density

$$h(p) = (1/B(\alpha,\beta))p^{\alpha-1}(1-p)^{\beta-1} \quad (\alpha,\beta > 0)$$

where $B(\alpha,\beta)$ is the Beta function

$$B(\alpha,\beta) = \frac{\Gamma(\alpha)\,\Gamma(\beta)}{\Gamma(\alpha+\beta)} = \frac{(\alpha-1)!\,(\beta-1)!}{(\alpha+\beta-1)!}$$

This form is mathematically convenient, and with suitable choice of α and β can express a wide range of prior expectations about p. For instance, $(\alpha =$

$1, \beta = 1$) gives a uniform prior density in $(0, 1)$, $(\alpha = 10, \beta = 20)$ expresses a strong expectation that p will be about $1/3$, $(\alpha = \frac{1}{2}, \beta = \frac{1}{2})$ expresses an expectation that p is more likely to have an extreme value close to 0 or 1 than to be close to $1/2$, and so on.

With a $B(\alpha, \beta)$ prior,

$$
\begin{aligned}
I_1(x) \ &= \ L - \log h(p) + \frac{1}{2} \log F(p) + \frac{1}{2}(1 + \log \kappa_1) \\
&= \ -s \log p - (N - s) \log(1 - p) - (\alpha - 1) \log p \\
&\quad - (\beta - 1) \log(1 - p) + \log B(\alpha, \beta) + \frac{1}{2} \log \left(\frac{N}{p(1 - p)} \right) \\
&\quad + \frac{1}{2}(1 + \log \kappa_1) \\
&= \ -(s + \alpha - \frac{1}{2}) \log p - (N - s + \beta - \frac{1}{2}) \log p + C
\end{aligned}
$$

where the constant C is independent of p:

$$
C = \log B(\alpha, \beta) - 0.7425 \ldots
$$

The MML estimate is then

$$
p' = \frac{s + \alpha - \frac{1}{2}}{N + \alpha + \beta - 1}
$$

5.4.1 The Multinomial Distribution

These results generalize directly to a Multinomial distribution with M possible outcomes for each of N trials. If s_m is the number of outcomes of type m, and p_m is the unknown probability of outcome m,

$$
L = -\sum_{m=1}^{M} s_m \log p_m - \log \binom{N}{s_1, s_2, \ldots, s_M}
$$

$$
F(p_1, p_2, \ldots, p_M) = \frac{N^{M-1}}{\prod_m p_m}
$$

$$
p'_m = (s_m + (1/2))/(N + M/2) \quad \text{for a uniform prior}
$$

For the conjugate prior (a generalized Beta function)

$$
h(p_1, p_2, \ldots, p_M) = \frac{\Gamma(A)}{\prod_m \Gamma(\alpha_m)} \prod_m p_m^{\alpha_m - 1} \quad (\alpha_m > 0 \text{ for all } m)
$$

where $A = \sum_m \alpha_m$. Then the MML estimates are

$$
p'_m = \frac{s_m + \alpha_m - \frac{1}{2}}{N + A - M/2} \quad \text{(all } m\text{)}
$$

5.4.2 Irregularities in the Binomial and Multinomial Distributions

The Binomial and Multinomial distributions illustrate that the two forms for the Fisher Information, one based on the expected second derivative of L, the other on the expected squared first derivative, can differ if regularity conditions are not met. In the binomial case, for $p = 1$, the expected and indeed only possible success count is $s = N$. Then both actually and in expectation

$$\left[\frac{d^2L}{dp^2}\right]_{p=1} = N; \quad \left(\left[\frac{dL}{dp}\right]_{p=1}\right)^2 = N^2$$

Neither of these expressions equals the limit of the normal expression $F(p) = N/(p(1-p))$ as $p \to 1$, which is infinite. The discrepancies occur because at $p = 1$ (and at $p = 0$) the regularity required for $\mathrm{E}\, d/dp\,(-\log f(s|p)) = 0$ is violated.

The sample-based approximation $F_e(p, s)$ for s successes in N trials

$$F_e(p, s) = s(1/p^2) + (N - s)(1/(1 - p)^2) = s/p^2 + (N - s)/(1 - p)^2$$

becomes equal to N when $p = 1$ and, necessarily, $s = N$.

This breakdown of regularity at the extremes of the possible range of a parameter causes no problems for SMML. In the binomial and multinomial cases, no real problem arises for MML either, as the MML estimates of the outcome probabilities avoid the extreme values of 0 and 1 with well-behaved priors. However, in the multinomial case, especially for large M with some small s_m counts, numerical calculations have shown that the "standard" explanation length formula

$$I_1(x) \approx \log \frac{\prod_m \Gamma(\alpha_m)}{\Gamma(A)} - \sum_m (s_m + \alpha_m - \frac{1}{2}) \log p'_m + \frac{1}{2}(M - 1)(1 + \log(N\kappa_k))$$

tends to under-estimate the message length. It is safer, while accepting the MML probability estimates, to calculate the explanation length from the length of the optimal non-explanation code, as in Section 5.2.13:

$$\begin{aligned}
I_1(x) &\approx I_0(x) + \frac{1}{2}\log(D\pi) - 0.4 \\
&\approx \log \frac{\Gamma(N + A)}{\Gamma(A)} + \sum_m \log \frac{\Gamma(\alpha_m)}{\Gamma(s_m + \alpha_m)} \\
&\quad + \frac{1}{2}\log((M - 1)\pi) - 0.4
\end{aligned}$$

The form for $I_0(x)$ comes from the length of a message which encodes the result of the nth outcome with a probability distribution over the M possible values given by

$$p_{nm} = \frac{s_{nm} + \alpha_m}{n - 1 + A}$$

where s_{nm} is the number of results of type m in the first $n - 1$ outcomes.

5.5 Limitations

The MML approach can be used only for model classes where the Fisher Information exists for all $\theta \in \Theta$. Thus, it cannot be used for the estimation of distributions like the Uniform, where the log likelihood is a discontinuous function of the parameters. (Actually, although the quadratic expansion of the log likelihood cannot be used for Uniform distributions, the other MML simplification of equating the coding probability of an estimate to the total prior probability in some "coding region" of parameter space, will work and leads to sensible message lengths and estimates.) Further, the coding efficiency of explanations using MML estimates depends on a fairly close correspondence between the actual and expected second differentials of the log likelihood. For small samples, or model classes with extreme mathematical properties, the correspondence may be poor. The MML approach relies on the expected second differentials to determine an appropriate spacing of estimates in Θ^*, i.e., to determine the precision with which estimates should be asserted. We saw in Section 5.2.14 that the latitude in choosing the estimate spacing or precision is quite broad: a factor of two change makes little difference to the lengths of explanations. However, in extreme cases, the MML spacing function $w(\theta)$ derived from the expected differentials could lead to a seriously sub-optimal estimate precision and perhaps to a poor estimate.

The MML approach as developed here assumes that the prior density $h(\theta)$ and the optimum estimate spacing function $w(\theta)$ vary negligibly over a region of Θ of size $w(\theta)$. In problems where either function is rapidly varying, the MML approximation to the message length may be poor, and the MML estimate biased and/or inefficient. For instance, in the examples of Section 3.3.3 on the estimation of the mean of a Normal with Normal prior, we saw that if the prior is strongly peaked, the optimum SMML estimate spacings increase somewhat for estimates far from the peak of the prior. This phenomenon is not captured by the MML approximation, which gives an estimate spacing function independent of the behaviour of the prior. At least for the Normal distribution, the practical consequences of this defect of MML seem to be small: the estimates remain sensible and the coding efficiency good despite the sub-optimal estimate spacing function.

Finally, the MML approximation assumes the log likelihood to have quadratic behaviour in the model parameters. While approximately quadratic behaviour will be found in most model classes, it must be realized that the MML approximation assumes that the quadratic behaviour obtains over the whole of a region of size $w(\theta)$. That is, we have assumed quadratic behaviour over a scale comparable to the spacing between the estimates in an SMML code. This spacing is relatively large, and is not small compared with the expected width of the likelihood function peak. Third and higher order terms in the log likelihood expansion about its peak may become significant at this scale, in which case the MML approximation may need revision. However, the quadratic approximation used above, and leading us to choose the estimate which maximizes $h(\theta')f(x|\theta')/\sqrt{F(\theta')}$, is adequate for a number of common model classes.

5.6 The Normal Distribution

The data are a set of N independent values drawn from a Normal density of unknown mean μ and unknown Standard Deviation σ. Let the data be

$$x = (y_1, y_2, \ldots, y_N)$$

$$f(x|\mu, \sigma) = \prod_n \frac{\delta}{\sqrt{2\pi}\sigma} \exp\left(-\sum_n (y_n - \mu)^2/2\sigma^2\right)$$

where we have assumed each datum y to have been measured to precision $\pm\delta/2$, and $\delta \ll \sigma$. Define

$$
\begin{aligned}
L &= -\log f(x|\mu, \sigma) \\
&= (N/2)\log(2\pi) + N\log\sigma - N\log\delta + \sum_n (y_n - \mu)^2/(2\sigma^2)
\end{aligned}
$$

$$\frac{\partial L}{\partial \mu} = -\sum_n (y_n - \mu)/\sigma^2$$

$$\frac{\partial^2 L}{\partial \mu^2} = N/\sigma^2$$

$$\frac{\partial L}{\partial \sigma} = N/\sigma - (1/\sigma^3)\sum_n (y_n - \mu)^2$$

$$\frac{\partial^2 L}{\partial \sigma^2} = -N/\sigma^2 + (3/\sigma^4)\sum_n (y_n - \mu)^2$$

$$\frac{\partial^2 L}{\partial \mu \partial \sigma} = (1/\sigma^2)\sum_n (y_n - \mu)$$

For the Normal distribution, $E(y_n - \mu) = 0$ and $E(y_n - \mu)^2 = \sigma^2$ Hence,

$$E\frac{\partial^2 L}{\partial \sigma^2} = -N/\sigma^2 + (3/\sigma^4)N\sigma^2 = 2N/\sigma^2$$

$$E\frac{\partial^2 L}{\partial \mu \partial \sigma} = 0$$

$$F(\mu, \sigma) = \begin{pmatrix} N/\sigma^2 & 0 \\ 0 & 2N/\sigma^2 \end{pmatrix}$$

Hence, Fisher $F(\mu, \sigma) = 2N^2/\sigma^4$.

The MML estimates μ', σ' are found by minimizing

$$I_1(x) = L - \log h(\mu, \sigma) + \frac{1}{2} \log F(\mu, \sigma) + (1 + \log \kappa_2)$$

$$\text{from formula I1B}$$

$$= (N/2) \log(2\pi) - N \log \delta + N \log \sigma + \sum_n (y_n - \mu)^2/(2\sigma^2) +$$

$$\frac{1}{2} \log(2N^2/\sigma^4) - \log h(\mu, \sigma) + 1 + \log \kappa_2$$

The prior $h(\mu, \sigma)$ should be chosen to express any prior expectations aris-ing from the particular estimation problem. Where there is little prior knowl-edge, it is common to assume that μ and σ have independent priors, that μ has a Uniform prior density (i.e., there is no preference for any particular location for the distribution) and that $\log \sigma$ has a Uniform density (i.e., there is no preference for any particular scale.) A uniform density of $\log \sigma$ implies a density of σ proportional to $1/\sigma$. These Uniform densities for $\log \sigma$ and μ are improper, i.e., cannot be normalized, unless some limits are set on the possible values of μ and $\log \sigma$. Usually, the context of the problem will allow at least rough limits to be set. For instance, if the data values $\{y_n\}$ are mea-sured to precision δ, i.e., with a rounding-off error $\pm \delta/2$, we cannot expect to find σ less than δ.

We will assume the above "colourless" priors for μ and $\log \sigma$, and suppose that prior knowledge limits μ to some range of size R_μ, and $\log \sigma$ to some range of size R_σ. Then

$$h(\mu, \sigma) = \left(\frac{1}{R_\mu}\right)\left(\frac{1}{\sigma R_\sigma}\right) \quad \text{for } \mu, \sigma \text{ in range.}$$

Then, writing v^2 for $\sum_n (y_n - \mu)^2$

$$I_1(x) = N \log \sigma + v^2/2\sigma^2 - 2 \log \sigma + \log \sigma + C$$

$$= (N - 1) \log \sigma + v^2/2\sigma^2 + C$$

where the constant

$$C = (N/2) \log(2\pi) - N \log \delta + \frac{1}{2} \log 2 + \log N + \log(R_\mu R_\sigma) + 1 + \log \kappa_2$$

does not depend on μ or σ.

Choosing μ', σ' to minimize I_1 gives

$$\mu' = (1/N) \sum_n y_n = \bar{y}; \quad \sigma' = \sqrt{v^2/(N-1)}$$

These are the conventional "unbiased" estimates. (Note that σ' differs from the Maximum Likelihood estimate $\sqrt{v^2/N}$, which tends on average to un-derestimate σ slightly.)

With these estimates, the message length becomes

$$I_1(x) = \frac{1}{2}(N-1)\log\frac{v^2}{N-1} + \frac{N-1}{2} + (N/2)\log(2\pi/\delta^2)$$
$$+ \frac{1}{2}\log(2N^2) + \log(R_\mu R_\sigma) + 1 + \log\kappa_2$$

5.6.1 Extension to the Neyman-Scott Problem

The MML analysis of the Normal distribution extends simply to the Neyman-Scott problem of Section 4.2 [13].

Recall that in this problem, the data comprises N instances each of J data values $\{x_{n,j} : j = 1, \ldots, J\}$ where all values in instance n $(n = 1, \ldots, N)$ are selected from a Normal density of mean μ_n and Standard Deviation σ. The N means $\{\mu_n\}$ and the single S.D. σ are unknown. We assume each mean independently to have a prior density uniform over some large range, and σ to have a prior density proportional to $1/\sigma$ in some large range. All $(N+1)$ parameters are to be estimated. The sufficient statistics are

$$m_n = (1/J)\sum_j x_{n,j} \quad (n = 1, \ldots, N)$$
$$s = (1/(NJ))\sum_n\sum_j(x_{n,j} - m_n)^2$$

It easily shown that the Fisher Information is

$$F(\sigma, \mu_1, \ldots, \mu_N) = (2NJ/\sigma^2)\prod_n(J/\sigma^2) = 2NJ^{N+1}/\sigma^{2(N+1)}$$

and hence that the MML estimate is

$$\mu'_n = m_n \ (n = 1, \ldots, N); \quad \sigma' = \sqrt{Js/(J-1)}$$

This is the same, consistent, estimate as was found by the Dowe estimator in Section 4.3, in contrast to the inconsistent Maximum Likelihood estimate $\sigma_{ML} = \sqrt{s}$.

The consistency of the MML estimator is a consequence of the use of the correct "spacing function" in encoding the explanation. The observed value of NJs understates the actual variance of the data around the true instance means, as it shows only the variance around the instance sample means. The straightforward Maximum Likelihood estimator, which in effect fits $\hat{\sigma}^2$ to the observed variance, therefore underestimates σ^2. However, in the MML explanation message, each estimated instance mean is asserted to a limited precision, so the asserted estimates do not exactly equal the instance sample means $\{m_n\}$. The MML estimate of σ^2 in effect fits $(\sigma')^2$ to the variance of the data around these asserted means. The difference between the asserted means and instance sample means increases this variance, so that on average, for any instance,

$$\sum_j (x_{n,k} - \text{asserted mean})^2 \approx J\sigma^2,$$

versus

$$\sum_j (x_{n,k} - m_n)^2 \approx (J-1)\sigma^2$$

The "rounding off" of the asserted means has the effect, on average, of restoring the variance "lost" when the true means μ_n are replaced by the sample means m_n.

5.7 Negative Binomial Distribution

This distribution arises when a series of success or fail trials is continued until a pre-determined number of successes is achieved. The trials are assumed to be independent with unknown success probability p. Thus, the data are just as for the Binomial distribution of Section 5.4, but now the number of successes s is fixed in advance and assumed known to the receiver. The number of trials N is not, being determined by the outcomes of the trials. The likelihood function is exactly the same as in Section 5.4, giving the negative log likelihood

$$L = -s\log p - (N-s)\log(1-p)$$
$$\frac{\partial^2 L}{\partial p^2} = \frac{s}{p^2} + \frac{N-s}{(1-p)^2}$$

For given s and p, $E(N) = s/p$, so

$$F(p) = \frac{s}{p^2} + \frac{(s/p)-s}{(1-p)^2} = \frac{s}{p^2(1-p)}$$

rather than $N/(p(1-p))$ as in the binomial problem. With a Beta (α, β) prior on p

$$I_1(x) = L - \log h(p) + \frac{1}{2}\log F(p) + \frac{1}{2}(1 + \log \kappa_1)$$
$$= -(s+\alpha-1)\log p - (N-s+\beta-1)\log(1-p)$$
$$+ \log B(\alpha,\beta) + \frac{1}{2}\log\left(\frac{s}{p^2(1-p)}\right) + \frac{1}{2}(1 + \log \kappa_1)$$

The MML estimate p'_N of p found by minimizing I_1 is slightly larger than the estimate p'_B for the binomial model with the same N, s and prior:

$$p'_N = \frac{s+\alpha}{N+\alpha+\beta-1/2} \quad \text{versus} \quad p'_B = \frac{s+\alpha-1/2}{N+\alpha+\beta-1}$$

5.8 The Likelihood Principle

The "likelihood principle" has been advocated by several researchers. In essence, it states that all information in some data x relevant to a parameter θ is captured by the likelihood function $f(x|\theta)$. A corollary is that our conclusions about θ should depend only on the data x which occurred, and not on data which did not occur. Suppose two experiments or observations are conducted under different protocols, so that the set X_1 of data which could possibly have been observed in the first experiment is different from the set X_2 which could have been observed in the second experiment, but that the two sets are not disjoint. Now suppose that both experiments are conducted, and in fact yield the same data $x \in (X_1 \cap X_2)$. Then the likelihood principle implies that the inferences about θ drawn from the two experiments should be the same: they both gave the same data, and it is irrelevant that the sets of data which did not occur in the experiments are different.

The MML estimators for the Binomial and Negative Binomial problems show that the MML method violates the likelihood principle. Suppose that the experiments were both series of success-failure trials, the first series being continued until 100 trials had been completed, and the second series being continued until 4 successes had been recorded. The first experiment has a binomial form, the second a negative binomial form. Assume a flat ($\alpha = \beta = 1$) prior on the unknown success probability p in both experiments. The possible results for the first experiment are $\{N = 100,\ 0 \le s \le 100\}$ and for the second, $\{N \ge 4,\ s = 4\}$. If both experiments happened to yield the one result possible under both, viz., $(N = 100, s = 4)$, the MML estimates would be

$$p'_B = \frac{4 + \frac{1}{2}}{100 + 1} = 0.04454; \quad p'_N = \frac{4 + 1}{100 + \frac{3}{2}} = 0.04926$$

These differ, violating the Likelihood principle. The violation is difficult to avoid, given the "message" framework on which MML is based. The receiver of an explanation message is assumed to have prior knowledge of the set X of possible data, and the message is coded on that assumption. It is clear in the SMML construction, and implicit in MML, that the optimum explanation code requires that one assertion or estimate value serve for a range of distinct but similar possible data values. Hence, it seems inevitable that the assertion used to explain the given data will depend to some extent on what distinct but similar data might have occurred but did not.

The violation shown in the Binomial vs. Negative Binomial contrast is innocent enough — a misdemeanour rather than a crime. Whatever the values of $s > 0$ and N, the difference between the two MML estimates of p is of order $1/N$, and so is much less than the expected estimation error (coding region size) which in both cases is expected to be of order $\pm\sqrt{s(N - s)/N^3}$.

The Binomial and Negative Binomial expressions for the Fisher Information

$$F_B(p) = \frac{N}{p(1-p)} = \frac{Np}{p^2(1-p)}, \quad F_N(p) = \frac{s}{p^2(1-p)}$$

have the same value when $s = Np$, which relation is expected to hold approximately for the true value of p, and also for any reasonable estimate of p based on observed s and N. As it is only in the Fisher Informations that the MML treatment of the two problems differ, both lead to almost equal explanation lengths for the same s, N and prior. It is possible that the small differences which do exist between the two MML estimates and between the two explanation lengths would disappear were the MML approximation improved to take proper account of the variation of $F(p)$ with p within a coding region.

6. MML Details in Some Interesting Cases

This chapter develops in more detail some of the techniques for calculating explanation lengths for common components of models. The additional detail is mostly concerned with minor amendments to the "standard" MML formulae needed to cope with extreme and/or exceptional cases, and to ensure reasonable numerical accuracy in less extreme cases. The amendments do not represent any departure from the MML principle of finding the model or hypothesis which leads to the shortest two-part "explanation" message encoding the data. Rather, they are attempts to improve on the Formula I1B used to approximate the explanation length in problems where the rather heroic assumptions of Formula I1B are significantly violated.

Section 6.8 of the chapter, dealing with mixture models, is of more general interest than its title might suggest. It introduces a coding technique useful in many model classes in which there are numerous "nuisance" parameters, and a way of addressing the assertion of discrete parameters to the "limited precision" often required for the optimal coding of assertions. The technique will be used in later chapters describing some fairly complicated applications of MML.

6.1 Geometric Constants

The multi-parameter version of Formula I1B leads to an approximate expression for explanation length derived in Section 5.2.5 as

$$I_1(x) = -\log h(\theta') + \frac{1}{2} \log |\mathbf{F}(\theta')| - \log f(x|\theta') + (D/2) \log \kappa_D + D/2$$

where D is the number of free parameters of the model. Table 6.1 shows for small D the D-dependent constant part of this expression, written as c_D. The constant depends on the lattice constant κ_D. As except for very small D, the optimum quantizing lattice in D dimensions is unknown, the value for the best known lattice has been taken from Conway and Sloane [11]. The value for the constant using a random lattice is also given as z_D, based on Zador's bound (Section 3.3.4). The table also shows the approximation

$$a_D = -\frac{1}{2} D \log(2\pi) + \frac{1}{2} \log(D\pi) - \gamma \approx c_D$$

This approximation is good enough for most purposes, as its errors of order 0.3 nit are small compared with uncertainties in the coding of prior information and (often) the dubious validity of the probability model $f(x|\theta)$. Later, we use the constants in expressions for $I_1(x)$ like

$$I_1(x) = -\log h(\theta') + \frac{1}{2} \log |\mathbf{F}(\theta')| - \log f(x|\theta') + c_D$$

Table 6.1. Constant terms in explanation length vs. number of parameters

D	c_D	z_D	a_D
1	-0.742	0.153	-0.847
2	-1.523	-0.838	-1.419
3	-2.316	-1.734	-2.135
4	-3.138	-2.610	-2.910
5	-3.955	-3.483	-3.718
6	-4.801	-4.357	-4.545
7	-5.655	-5.233	-5.387
8	-6.542	-6.112	-6.239
12	-9.947	-9.649	-9.712
16	-13.471	-13.214	-13.244
24	-20.659	-20.398	-20.393

6.2 Conjugate Priors for the Normal Distribution

For N independent values $x = (x_1, \ldots, x_n, \ldots, x_N)$ all measured to precision $\pm\delta/2$, unknown mean μ, Standard Deviation σ:

$$f(x|\mu, \sigma) = \left(\frac{\delta}{\sqrt{2\pi}\sigma}\right)^N \prod_{n=1}^{N} e^{-(x_n - \mu)^2/2\sigma^2}$$

$$
\begin{aligned}
L &= -\log f(x|\mu, \sigma) \\
&= (N/2)\log(2\pi) + N\log\delta - N\log\sigma + \frac{1}{2\sigma^2}\sum_n (x_n - \mu)^2
\end{aligned}
$$

$$\text{Fisher Information} = F(\mu, \sigma) = 2N^2/\sigma^4$$

For $h(\mu, \sigma) = h_\mu(\mu)h_\sigma(\sigma)$, where

$$
\begin{aligned}
h_\mu(\mu) &= 1/R_\mu \quad (\mu \text{ uniform in some range } R_\mu), \\
h_\sigma(\sigma) &= 1/(R_l\sigma) \quad (\log\sigma \text{ uniform in some range } R_l): \\
I_1(x) &= \log(R_\mu R_l) - N\log\delta + (N/2)\log(2\pi) + \frac{1}{2}\log 2 + \log N \\
&\quad + (N-1)\log\sigma' + \frac{1}{2}\sum_n (x_n - \mu')^2/(\sigma')^2 - \frac{1}{2}\log(2\pi) - \gamma
\end{aligned}
$$

where $\mu' = (1/N) \sum_n x_n; \quad \sigma'^2 = 1/(N-1) \sum_n (x_n - \mu')^2$
Simplifying:

$$
\begin{aligned}
I_1(x) &= \log(R_\mu R_l) - N \log \delta + \frac{1}{2}(N-1)\log(2\pi) + \frac{1}{2}\log 2 \\
&\quad + \log N + (N-1)\log \sigma' + \frac{1}{2}(N-1) - \gamma
\end{aligned}
$$

It is sometimes more convenient to describe a Normal model using the "concentration" parameter $\lambda = \sigma^{-2}$ in place of σ:

$$
N(x_n | \mu, \lambda^{-1}) = \sqrt{\frac{\lambda}{2\pi}} e^{-\frac{1}{2}\lambda(x_n - \mu)^2}
$$

The "uninformative" prior for σ used above over a restricted range R_l of $\log \sigma$ implies an uninformative prior for λ

$$
\begin{aligned}
h_\lambda(\lambda) &= -h_\sigma(\sigma) \frac{d\sigma}{d\lambda} \quad \text{with } \sigma = \lambda^{-\frac{1}{2}} \\
&= -\left(\frac{1}{R_l \sigma}\right)\left(-\frac{1}{2}\lambda^{-3/2}\right) = \frac{1}{2R_l \lambda}
\end{aligned}
$$

or in the unrestricted improper form, simply $1/\lambda$.

Using (μ, λ) as parameters, the Fisher Information is

$$
F(\mu, \lambda) = N^2/(2\lambda)
$$

Then

$$
\begin{aligned}
\frac{h_{\mu\lambda}(\mu, \lambda)}{\sqrt{F(\mu, \lambda)}} &= \frac{(1/R_\mu)(1/(2R_l\lambda))}{\sqrt{N^2/(2\lambda)}} \\
&= \frac{\sigma}{R_\mu R_l \sqrt{2N^2}} = \frac{h_{\mu,\sigma}(\mu, \sigma)}{\sqrt{F(\mu, \sigma)}}
\end{aligned}
$$

so we obtain exactly the same value for $I_1(x)$ and the same MML estimates as above. This is of course just an illustration for the Normal distribution of the general invariance of the MML formulae.

The priors so far assumed for μ and σ (or λ) are the conventional uninformative priors with ranges restricted to allow normalization. If significant prior information is available beyond a simple range restriction, it may be possible to represent it adequately by a prior of invariant conjugate form (Section 1.15.2). This is most conveniently done using (μ, λ) parameters. An invariant conjugate prior is constructed as the posterior density which results from an initial uninformative prior (here $h_0(\mu, \lambda) = 1/\lambda$) followed by "observation" of some real or imagined prior data. This prior data may be represented by its sufficient statistics. Suppose the prior data consists of two independent samples. For the first sample, $(y_1, y_2, \ldots, y_m)$ of size m, only the sample variance v about its mean is recorded:

$$v = \sum_{i=1}^{m} (y_i - \bar{y})^2 \quad \text{with } \bar{y} = \frac{1}{m} \sum_{i=1}^{m} y_i$$

For the second sample, $(z_1, z_2, \ldots, z_{m_1})$ of size m_1, only the sample mean μ_0 is recorded:

$$\mu_0 = \frac{1}{m_1} \sum_{i=1}^{m_1} z_i$$

Then, using the uninformative initial priors, the revised prior given this prior data is

$$
\begin{aligned}
h(\mu, \lambda | m, v, m_1, \mu_0) &= C\, h_0(\mu, \lambda)\, \Pr(v, \mu_0 | \mu, \lambda) \\
&= C\, h_0(\mu, \lambda)\, \Pr(v | \mu, \lambda)\, \Pr(\mu_0 | \mu, \lambda) \\
&= C\, h_0(\mu, \lambda)\, \Pr(v | \lambda)\, \Pr(\mu_0 | \mu, \lambda)
\end{aligned}
$$

where C is a normalization constant, since the two samples are independent and since v is sufficient for λ.

The density of the "sample mean" μ_0 is just the Normal density

$$N(\mu_0 | \mu, (m_1 \lambda)^{-1}) = \frac{1}{\sqrt{2\pi}} \sqrt{m_1 \lambda} \exp\left(-\frac{1}{2} m_1 \lambda (\mu_0 - \mu)^2 \right)$$

The "sample variance" v has the distribution

$$\chi^2(v | m - 1, \lambda^{-1}) = \frac{\lambda^{(m-1)/2} v^{(m-3)/2} e^{-\frac{1}{2}v\lambda}}{2^{(m-1)/2} \Gamma\left(\frac{m-1}{2}\right)}$$

where by $\chi^2(x | d, s)$ we mean the probability density of x when x/s has a Chi-Squared distribution with d degrees of freedom.

After normalization the prior becomes

$$h(\mu, \lambda) = \left(\frac{1}{\lambda}\right) \left[\sqrt{\frac{m_1 \lambda}{2\pi}} e^{-\frac{1}{2} m_1 \lambda (\mu_0 - \mu)^2} \right] \left[\frac{v^{(m-1)/2} \lambda^{(m-1)/2} e^{-\frac{1}{2}\lambda v}}{2^{(m-1)/2} \Gamma\left(\frac{m-1}{2}\right)} \right]$$

This prior is proper provided $m_1 > 0, m > 1$. It may be written as

$$h(\mu, \lambda) = N(\mu | \mu_0, (m_1 \lambda)^{-1}) \chi^2(\lambda | m - 1, v^{-1})$$

The prior data μ_0 and v contribute terms to the length of assertion $(-\log \Pr(v, \mu_0 | \mu, \lambda))$, which depends on the parameters (μ, λ) in just the same way as does the length of the second part of the explanation $(-\log \Pr(x | \mu, \lambda))$. Thus, as described in Section 5.2.10, the "Fisher Information" can be calculated to include the contribution of these terms to the effects of rounding-off the MML estimates to finite precision. For real data sample x of size N as before, the effect is that N is replaced in the expression for $F(\mu, \lambda)$ by "effective sample sizes" which include the prior data. For μ, the effective sample size is $N + m_1$ and for λ it is $N + m$, giving

$$F(\mu, \lambda) = ((N + m_1)\lambda) \left(\frac{N + m}{2\lambda^2} \right)$$
$$= (N + m_1)(N + m)/(2\lambda)$$

The MML estimates become

$$\mu' = \frac{m_1 \mu_0 + \sum_{n=1}^{N} x_n}{N + m_1}$$

$$1/\lambda' = (\sigma')^2 = \frac{m_1(\mu_0 - \mu')^2 + v + \sum_n (x_n - \mu')^2}{N + m - 1}$$

$$I_1(x) = \log \Gamma((m - 1)/2) + \frac{1}{2}[(N + 1)\log(2\pi) + (m - 2)\log 2]$$
$$+ \frac{1}{2}\log((N + m_1)(N + m)/\lambda) - N \log \delta$$
$$+ \frac{1}{2}[(N + m - 1)(1 - \log \lambda') - (m - 1)\log v] + c_2$$

The invariant conjugate prior defined by (m_1, μ_0, m, v) allows considerable freedom in shaping the prior to accord with actual prior beliefs. μ_0 can be chosen to show the mean location thought most likely *a priori*, and v to show what scale of data spread is expected *a priori*. Note that as v is defined as the total variance of a "prior data sample" about its mean, v should be set to about $(m - 1)$ times the expected scale of σ^2. The value of $m > 1$ shows the strength of prior belief in this expected scale, with small m giving a broad prior distribution for λ, and large m a prior tightly concentrated near the anticipated $1/\sigma^2$. Unfortunately, the joint conjugate prior causes the spread of the prior for μ about μ_0 to depend on the estimated value of λ. The width of the spread is about $1/\sqrt{m_1\lambda}$. It is thus difficult to choose the prior to reflect a belief that, even though the data may well prove to be tightly concentrated, the mean may be far from μ_0. The best that can be done is to choose a small value for m_1, perhaps less than one. Alternatively, one may abandon the conjugate prior for μ in favour of one more accurately modelling prior belief, e.g., a Normal form with fixed Standard Deviation or a Uniform density over some finite range.

6.2.1 Conjugate Priors for the Multivariate Normal Distribution

This distribution is a direct extension of the Normal distribution to vector data. The data x comprises N independent K-vectors all measured to within K-volume δ.

$$x = \{y_1, y_2, \ldots, y_n, \ldots, y_N\}$$

where each y-value is a K-vector. Then

$$f(x|\mu, \lambda) = \left(\delta \sqrt{\frac{|\lambda|}{(2\pi)^K}} \right)^N \exp\left(-\frac{1}{2} \sum_n (y_n - \mu)^T \lambda (y_n - \mu) \right)$$

where μ is the mean K-vector and λ is a symmetric positive-definite $K \times K$ concentration parameter, the inverse of the population covariance matrix, and $|\lambda|$ is the determinant of λ.

The quadratic form $\sum_{n=1}^{N} (y_n - \mu)^T \lambda (y_n - \mu)$ may be rewritten as

$$Tr(\lambda S) + N(\bar{y} - \mu)^T \lambda (\bar{y} - \mu)$$

where

$$\bar{y} = (1/N) \sum_n y_n \quad \text{and} \quad S = \sum_n (y_n - \bar{y})(y_n - \bar{y})^T$$

S is the covariance matrix of the sample about its mean. $Tr(M)$ means the trace of matrix M, i.e., the sum of its diagonal elements $\sum_{k=1}^{K} M_{kk}$, so $Tr(\lambda S) = \sum_{i=1}^{K} \sum_{j=1}^{K} \lambda_{ij} S_{ij}$

$$L = -N \log \delta + \frac{1}{2} NK \log(2\pi) - \frac{1}{2} N \log |\lambda| + \frac{1}{2} \sum_n (y_n - \mu)^T \lambda (y_n - \mu)$$

Using the K components of μ and the $K(K+1)/2$ distinct elements of λ as parameters, the Fisher Information is

$$
\begin{aligned}
F(\mu, \lambda) = F(\mu)F(\lambda) &= \left[N^K |\lambda| \right] \left[\frac{N^{K(K+1)/2}}{2^K |\lambda|^K} \right] \\
&= \frac{N^K N^{K(K+1)/2}}{2^K |\lambda|^K}
\end{aligned}
$$

With the same parameters, the uninformative prior for μ is Uniform, and the uninformative prior for λ is $1/|\lambda|$. Both could be normalized by some range restriction, but for λ it is not intuitively clear what this range might be set.

Following the development for the univariate Normal, an invariant conjugate prior can be formed by imagining prior data comprising m data vectors having total sample covariance V about their mean, and m_1 data vectors with mean μ_0. The resulting conjugate prior $h(\mu, \lambda | m, V, m_1, \mu_0)$

$$h_0(\mu, \lambda) \; Pr(V|\lambda) \; Pr(\mu_0 | \mu, \lambda)$$

$$
\begin{aligned}
h_0(\mu, \lambda) &= 1/|\lambda| \\
Pr(\mu_0 | \mu, \lambda) &= N(\mu_0 | \mu, (m_1 \lambda)^{-1}) \\
Pr(V|\lambda) &= W_K(V | m - 1, \lambda^{-1})
\end{aligned}
$$

where $W_K(V|d, A)$ is the Wishart distribution, which gives the joint density of the $K(K+1)/2$ distinct elements of the sample covariance V of a sample of d K-vectors drawn from a multivariate Normal density with mean zero and covariance matrix A. $W_K()$ is the K-variate generalization of the Chi-Squared distribution.

$$W_K(V|m-1, \lambda^{-1})$$
$$= \frac{\pi^{-K(K-1)/4} 2^{-K(m-1)/2}}{\prod_{k=1}^{K} \Gamma(\frac{1}{2}(m-k))} |V|^{(m-K-2)/2} |\lambda|^{(m-1)/2} e^{-\frac{1}{2}Tr(\lambda V)}$$

This distribution is only proper if $m > K$.

After normalization:

$$h(\mu, \lambda) = N(\mu|\mu_0, (m_1\lambda)^{-1}) W_K(\lambda|(m+K-2), V^{-1})$$
$$= \left(\sqrt{\frac{m_1|\lambda|}{(2\pi)^K}} e^{-m_1(\mu-\mu_0)^T \lambda(\mu-\mu_0)} \right)$$
$$\times \left(C|\lambda|^{(m-3)/2} |V|^{(m+K-2)/2} e^{-\frac{1}{2}Tr(V\lambda)} \right)$$

where

$$C = \frac{\pi^{-K(K-1)/4} 2^{-K(m+K-2)/2}}{\prod_{k=1}^{K} \Gamma(\frac{1}{2}(m+k-2))}$$

which is proper if $m_1 > 0, m > 1$.

The interpretation of V as the sample covariance of imagined data is somewhat metaphorical. Were V indeed the covariance of a sample of size m about its mean, V would be singular for $m < K+1$. None the less, one may think of V as being a prior guess at what the average covariance of such samples might be, rather than the covariance of any one set of m imagined data vectors, and any value of m greater than one maybe used. For instance, if there is a prior belief that the kth component of y will have a standard deviation of order a_k, but there is no prior reason to expect any particular correlations among the components of y, one might choose V to be the matrix

$$v_{kk} = ma_k^2 : k = 1, \ldots, K; \qquad v_{jk} = 0 : j \neq k,$$

with $m > 1$ reflecting the strength of the prior beliefs about standard deviations. See also the remarks in Section 6.2 concerning the prior density of the mean and its dependence on the estimate of λ.

As for the univariate Normal, the prior data "sample sizes" m_1 and m may be included in the calculation of the Fisher Information, giving the revised expression

$$F(\mu, \lambda) = \frac{(N+m_1)^K (N+m)^{K(K+1)/2}}{2^K |\lambda|^K}$$

The MML estimates are

$$\mu' = \frac{m_1\mu_0 + \sum_n y_n}{N + m_1}$$
$$(\lambda')^{-1} = \frac{m_1(\mu_0 - \mu')(\mu_0 - \mu')^T + V + \sum_n (y_n - \mu')(y_n - \mu')^T}{N + m + K - 2}$$

$$I_1(x) = \sum_{k=1}^{K} \log \Gamma(\frac{1}{2}(m+k-2))$$

$$+ \frac{1}{2} [K(N+1)\log(2\pi)$$

$$+ \quad \log((N+m_1)^K(N+m)^{K(KH)/2}/m_1)]$$

$$+ \frac{1}{4}K(K-1)\log \pi - N\log \delta$$

$$+ \frac{1}{2}[K(m+K-3)\log 2 - (m+K-2)\log|V|]$$

$$+ \frac{1}{2}(N+m+K-2)(1-\log|\lambda'|) + c_D$$

where D is the number of scalar parameters, $K(K+3)/2$.

6.3 Normal Distribution with Perturbed Data

It is sometimes known that the sample data values

$$x = (y_1, \ldots, y_n, \ldots, y_N)$$

are individually subject to random measurement errors of known expected magnitude. Suppose it is known that the measured value y_n is subject to a Normally distributed error with zero mean and variance ε_n. Then the model distribution for y_n is the convolution of the unknown population distribution Normal(μ, σ^2) and the error distribution Normal$(0, \varepsilon_n)$.

$$f(x|\mu, \sigma) = \frac{\delta}{\sqrt{2\pi}} \prod_n \left(\frac{1}{s_n} e^{-(y_n - \mu)^2/(2s_n^2)} \right)$$

where $s_n^2 = \sigma^2 + \varepsilon_n$

$$L = \frac{N}{2}\log 2\pi - N\log \delta + \frac{1}{2}\sum_n \log(\sigma^2 + \varepsilon_n) + \frac{1}{2}\sum_n \frac{(y_n - \mu)^2}{\sigma^2 + \varepsilon_n}$$

Define $w_n = \sigma^2/s_n^2 = \sigma^2/(\sigma^2 + \varepsilon_n)$. Then

$$\frac{\partial}{\partial \mu}L = \frac{1}{\sigma^2}\sum_n w_n(y_n - \mu)$$

$$\frac{\partial}{\partial \sigma}L = \frac{1}{\sigma}\sum_n w_n - \frac{1}{\sigma^3}\sum_n w_n^2(y_n - \mu)^2$$

Hence, the Maximum Likelihood estimates of μ, σ are

$$\hat{\mu}_{ML} = \sum_n w_n y_n / \sum_n w_n; \quad \hat{\sigma}^2_{ML} = \sum_n w_n^2 (y_n - \hat{\mu})^2 / \sum_n w_n$$

These equations are implicit, as w_n depends on $\hat{\sigma}_{ML}$. In effect, the observed deviation $(x_n - \hat{\mu})$ is given a "weight" w_n: observations whose expected errors are large compared to $\hat{\sigma}$ contribute little to the estimates, but observations with small expected errors are given nearly full weight.

After some algebra, the Fisher Information is

$$F(\mu, \sigma) = \frac{2}{\sigma^4} \left(\sum_n w_n \right) \left(\sum_n w_n^2 \right)$$

Explicit expressions for the MML estimates are not available. From

$$I_1(x) = -\log h(\mu', \sigma') + \frac{1}{2} \log F(\mu', \sigma') + L - \frac{1}{2} \log(2\pi) - c_2,$$

$$\frac{\partial}{\partial \mu'} I_1(x) = -\frac{\partial}{\partial \mu'} \log h(\mu', \sigma') + \frac{\partial}{\partial \mu'} L$$

$$\frac{\partial}{\partial \sigma'} I_1(x) = -\frac{\partial}{\partial \sigma'} \log h(\mu', \sigma') + \frac{\partial}{\partial \sigma'} L$$

$$+ \frac{1}{\sigma'} \left(1 - \frac{\sum_n w_n^2}{\sum_n w_n^3} - 2 \frac{\sum_n w_n^3}{\sum_n w_n^2} \right)$$

As for the Maximum Likelihood estimates, these equations may be solved numerically, e.g., by making some first guess at μ', σ' found by assuming all $w_n = 1$, followed by functional iteration. The numerical solution must be constrained by the condition $\sigma' > 0$, in case the observed variance in the data is less than would be expected from the perturbations $\{\varepsilon_n\}$ alone.

6.4 Normal Distribution with Coarse Data

The Normal distribution and others are sometimes used to model data which has been recorded without error, but only to coarse precision (large δ), e.g., the ages of persons stated in whole years. Here, we treat only the Normal, but the treatment can be adapted to other forms. In an MML model, it is possible that a component of the model is the distribution of such values over a subset of the sample, the subset itself being a feature of the model to be discovered from the data. In such a case it is quite possible for the Normal distribution to be applied to a group of coarse values which comprise only a few distinct values, or may even be all the same. Then the Standard Deviation of the model distribution can be not much greater than the precision δ, and the usual Normal approximation

$$\Pr(x_n | \mu, \sigma) = \frac{\delta}{\sqrt{2\pi}\sigma} \exp(-(x_n - \mu)^2 / (2\sigma^2))$$

is poor, and can lead to "probabilities" greater than one. A correct analysis would use

$$\Pr(x_n|\mu,\sigma) = \frac{1}{\sqrt{2\pi}\sigma} \int_{x_n-\delta/2}^{x_n+\delta/2} \exp(-(x_n-\mu)^2/(2\sigma^2))\, dx$$

Unfortunately, use of this exact expression leads to computational difficulties and is best avoided unless great accuracy is required. A simpler approximation which has been used successfully is to inflate the variance of data about the mean by the variance expected to result from the rounding-off of the true value being measured and/or recorded to the recorded value as it appears in the given data. In effect, we argue that, were the true value available, it would be found typically to differ from the mean by rather more than does the recorded value. The additional variance, assuming unbiased rounding-off, is $(1/12)\delta_n^2$, where we index the data roundoff quantum δ to indicate that different x-values in the given data may have been recorded to different precisions.

The approximated probability for datum x_n recorded to precision $\pm\delta/2$ is

$$\Pr(x_n|\mu,\sigma) = \frac{\delta_n}{\sqrt{2\pi}\sigma} \exp(-[(x_n-\mu)^2 + \delta_n^2/12]\,/\,(2\sigma^2))$$

The maximum value of this expression occurs when $x_n = \mu$ and $\sigma^2 = \delta_n^2/12$ and equals

$$\frac{\delta_n}{\sqrt{2\pi/12}\,\delta_n} \exp(-\frac{1}{2}) = \sqrt{\frac{12}{2\pi e}} = 0.8382\ldots$$

No excessive probability can occur, and for data where $\delta_n \ll \sigma$, little error is introduced. It may be argued that if $\delta_n \gg \sigma$, this formula will give a low probability to x_n even if $x_n = \mu$, but in such a case one could reasonably doubt whether the true quantity recorded as x_n was indeed close to μ.

6.5 von Mises-Fisher Distribution

The d-dimensional von Mises-Fisher distribution describes a distribution of directions in a d-dimensional space. The data are a set of N directions (unit d-vectors represented in some way) and the parameters are a mean direction μ (a unit d-vector) and a concentration parameter ρ. The model density of a datum direction w is

$$f(w|\mu,\rho) = (1/C(\rho)) \exp(\rho w \cdot \mu)$$

For zero ρ, the density is uniform over the d-sphere, and for large ρ the vector $(w - \mu)$ has approximately a $(d-1)$-dimensional Gaussian density $N(0, I_{d-1}/\rho)$ where I_{d-1} is the identity matrix. For a data set $\{w_n : n =$

$1, \ldots, N\}$, the sum $R = \displaystyle\sum_{n=1}^{N} w_n$ is a minimal sufficient statistic. Here we consider the two- and three-dimensional cases.

6.5.1 Circular von Mises-Fisher distribution

For this distribution, the data is a set of N angles in the plane, which we take to be measured in radians, so

$$x = (w_1, w_2, \ldots, w_n, \ldots, w_N)$$

where each w-value is an angle in $[0, 2\pi)$. All w angles are assumed identically and independently distributed. The model distribution is an analogue of the one-dimensional Normal distribution, having a location parameter μ and a concentration parameter ρ. Note that $1/\rho$ does not behave as a scale parameter.

$$f(x|\mu, \rho) = \prod_n g(w_n|\mu, \rho) = \prod_n (1/C(\rho)) \, e^{\rho \cos(w_n - \mu)}$$

where the normalization constant

$$C(\rho) = \int_0^{2\pi} \exp(\rho \cos(\theta) \, d\theta$$

is 2π times a modified Bessel function.

It is convenient to define $y_n = \cos w_n$; $z_n = \sin w_n$. Then a minimal sufficient statistic is the two-element vector

$$R_C = (Y, Z) \quad = \quad (\sum_n y_n, \sum_n z_n) \text{ in Cartesian coordinates}$$

$$R_P = (|R|, m) \quad = \quad (\sqrt{(Y^2 + Z^2)}, \tan^{-1}(Z/Y))$$
$$\text{in Polar coordinates}$$

Define

$$A(\rho) = (d/d\rho) \log C(\rho) = \mathrm{E}(\cos(w - \mu)); \quad A'(\rho) = (d/d\rho) A(\rho)$$

Then

$$L \quad = \quad N \log C(\rho) - \rho \sum_n (y_n \cos \mu + z_n \sin \mu)$$
$$= \quad N \log C(\rho) - \rho(Y \cos \mu + Z \sin \mu)$$
$$= \quad N \log C(\rho) - \rho|R| \cos(m - \mu)$$
$$F_P(\mu, \rho) \quad = \quad \rho N^2 \, A \, A' \text{ in Polar coordinates}$$
$$F_C((\rho \cos \mu), (\rho \sin \mu)) \quad = \quad (1/\rho) N^2 \, A \, A' \text{ in Cartesian coordinates}$$

The Maximum Likelihood estimate for μ is $\hat{\mu}_{ML} = m$. The estimate for ρ satisfies

$$A(\hat{\rho}_{ML}) = |R|/N$$

$\hat{\rho}_{ML}$ tends to overestimate the concentration unless N is large. At the extreme, when $N = 2$ and the two data values are sampled from the Uniform distribution over $[0, 2\pi)$, corresponding to $\rho = 0$, Dowe [53] has shown that $E(\hat{\rho}_{ML})$ is infinite.

A better, Marginal Maximum Likelihood, estimate of the concentration is described by Schou [39]. It equates the observed length of the vector R to the length expected given $\hat{\rho}_{Schou}$ and satisfies

$$\rho_{Schou} = 0 \quad (R^2 < N)$$
$$RA(R\rho_{Schou}) = NA(\rho_{Schou}) \quad (R^2 > N)$$

The Schou estimate of ρ is zero unless $|R|$ is greater than the value to be expected for a Uniform angular distribution, i.e., one with no concentration.

When, as in MML, a Bayesian approach is taken, an "uninformative" prior for μ is the Uniform density $1/2\pi$. No truly uninformative prior for ρ seems possible. Since, for large ρ, the distribution becomes similar to a Normal distribution, a prior in ρ which behaves as $1/\rho$ for $\rho \gg 1$ may be attractive, but is of course improper. Tests comparing the performance of ML, Schou and MML estimators have been conducted using for MML the priors

$$h_2(\rho) = \frac{2}{\pi(1 + \rho^2)} \quad \text{and} \quad h_3(\rho) = \frac{\rho}{(1 + \rho^2)^{3/2}}$$

In Cartesian parameter coordinates $(\rho\cos\mu, \rho\sin\mu)$, the former has a pole at the origin, but the latter is smooth, and is perhaps to be preferred if the data concentration is expected to arise from some physical influence in the plane such as a magnetic field.

The tests, reported in [53], show the MML estimators to give more accurate results than the ML and Schou estimators as measured by mean absolute error, mean squared error, and mean Kullback-Leibler distance.

6.5.2 Spherical von Mises-Fisher Distribution

Similar results have been found for the three-dimensional distribution [12]. Using a prior density which is Uniform in μ over the surface of the 3-sphere and varies with ρ as

$$h_\rho(\rho) = \frac{4\rho^2}{\pi(1 + \rho^2)^2}$$

analogous to the $h_3(\rho)$ used in the circular case, it is found that

$$h(\mu, \rho)/\sqrt{F} = \rho/(\pi^2(1 + \rho^2)^2 A(\rho)\sqrt{N^3 A'(\rho)})$$

where, as before, F is the Fisher determinant and

$$A(\rho) = (d/d\rho) \log C(\rho) = \mathrm{E}(w.\mu); \quad A'(\rho) = (d/d\rho)A(\rho)$$

Minimization of the message length leads to the obvious estimate $\hat{\mu} = R/|R|$ and an estimate for ρ more accurate and less biased than both the Maximum Likelihood and Marginal Maximum Likelihood estimates, especially for small N.

6.6 Poisson Distribution

The data consists of an event count n giving the number of occurrences of a random event within a known interval t. The probability model is

$$\Pr(n) = f(n|r, t) = e^{-rt}(rt)^n/n!$$

where the unknown rate parameter r is to be estimated. An MML analysis has been given in [54].

$$L = -\log f(n|r, t) = rt - n\log(rt) + \log(n!)$$

$$\frac{dL}{dr} = t - \frac{n}{r}; \quad \frac{d^2L}{dr^2} = \frac{n}{r^2}$$

But $E(n) = rt$ so

$$F(r) = E\frac{d^2L}{dr^2} = t/r$$

By way of example, assume a prior density $h(r) = (1/\alpha)\exp(-r/\alpha)$ where α is the mean rate expected *a priori*. This is a conjugate prior obtained from the uninformative prior $h_0(r) \approx 1/r$ and assuming "prior data" consisting of the observation of one event in an internal $1/\alpha$. Then

$$
\begin{aligned}
I_1(n) &= -\log h(r') + \frac{1}{2}\log F(r') + L + c_1 \\
&= \log\alpha + r'/\alpha + \frac{1}{2}\log(t/r') + r't - n\log(r't) + \log(n!) + c_1 \\
&= \log\alpha + (t + 1/\alpha)r' - (n + \frac{1}{2})\log r' - (n - \frac{1}{2})\log t \\
&\quad + \log(n!) + c_1
\end{aligned}
$$

This is minimized by the MML estimate

$$r' = (n + \frac{1}{2})/(t + 1/\alpha) \quad \text{giving a message length}$$

$$I_1(n) = (n + \frac{1}{2})(1 - \log(r't)) + \log(\alpha t) + \log(n!) + c_1$$

These results are invalid if the mean expected count *a priori*, αt, is less than one. In that case, r' is small, and the MML precision quantum $\delta(r') = 1/\sqrt{F(r')/12} = \sqrt{12r'/t}$ may be large compared with α, leading to a "prior estimate probability" $h(r')\delta(r')$ greater than one. The problem is best addressed as described in Section 5.2.10. The Fisher Information is modified to include the observation of the "prior data" of one count in an interval $1/\alpha$, giving

$$F(r) = (t + 1/\alpha)/r$$

This amendment does not alter the MML estimate $r' = (n + \frac{1}{2})/(t + 1/\alpha)$, but the value of $I_1(n)$ becomes, after substituting the MML estimate r' and simplifying

$$
\begin{aligned}
I_1(n) &= (n+1)\log(\alpha t + 1) - n\log(\alpha t) + \log(n!) - \\
&\quad (n + \frac{1}{2})\log(n + \frac{1}{2}) + (n + \frac{1}{2}) + c_1
\end{aligned}
$$

The optimum non-explanation length $I_0(n)$ is found by integrating the posterior density to be

$$I_0(n) = (n + 1)\log(\alpha t + 1) - n\log(\alpha t)$$

The difference $I_1(n) - I_0(n)$ is independent of α and t, and is now positive for all n, increasing from 0.104 at $n = 0$ to 0.173 at $n = 10$, and approaches a limit of 0.1765 as $n \to \infty$. This limit is $\frac{1}{2}\log(\pi e/6)$, as expected for any sufficiently regular single-parameter model (Section 5.2.4).

If convenient, $I_1(n)$ may be approximated by the simpler expression $I_0(n) + 0.176$ with maximum error less than 0.07 nit.

6.7 Linear Regression and Function Approximation

Here, the model concerns how a random variable depends on one or more given variables. We first consider linear dependence among real variables.

6.7.1 Linear Regression

The data comprises N independent cases. For each case n ($n = 1, \ldots, N$), the data consists of given values of K variables y_{nk} ($k = 1, \ldots, K$) and the observed value x_n of a random variable. The probability model is

$$x_n \sim N(\mu_n, \sigma^2) \quad \text{where} \quad \mu_n = a_0 + \sum_{k=1}^{K} a_k y_{nk}$$

The parameters ($a_k : k = 0, \ldots, K$) and σ are unknown.

Without loss of generality, we may insert a dummy variable $(k = 0)$ with dummy values $y_{n0} = 1$ for all cases, and assume the given data to have been standardized so that

$$\sum_{n=1}^{N} y_{nk} = 0, \quad \sum_{n=1}^{N} y_{nk}^2 = N \quad (k = 1, \ldots, K)$$

so any constant term in the expression for μ_n is represented by the co-efficient a_0 of the dummy constant "variable" $y_{.0}$.

Define $\underline{a}$ as the $(K+1)$-vector $(a_0, \ldots, a_k, \ldots, a_K)$. A simple conjugate prior for the parameters is

$$h(\sigma, \underline{a}) = 1/(R\sigma) \prod_{k=0}^{K} Norm(0, \sigma^2/m)$$

That is, we assume a locally uninformative prior for σ, and that each co-efficient a_k has independently a Normal prior with mean zero and variance σ^2/m. Assuming this prior, and treating the y variables as given background information,

$$- \log \left[f(x|\underline{a}, \sigma)\, h(\underline{a}|\sigma) \right]$$

$$= (N/2) \log(2\pi) + N \log \sigma + \frac{1}{2} \sum_{n} \left(x_n - \sum_{k=0}^{K} a_k y_{nk} \right)^2 / \sigma^2$$

$$+ \frac{1}{2}(K+1) \log(2\pi\sigma^2/m) + \frac{1}{2} \sum_{k=0}^{K} m a_k^2 / \sigma^2$$

It can be seen that the conjugate prior has the same effect as extending the N data cases by a further K cases of "prior" data, where in the kth case, $x_k = 0$, $y_{kl} = 0$ $(l \neq k)$ and $y_{kk} = \sqrt{m}$. As in Section 6.2 we will treat the prior data on the same footing as the real data, and so write

$$L = - \log(f(x|\underline{a}, \sigma)h(\underline{a}|\sigma))$$

Define $\mathbf{Y}$ as the $N \times (K+1)$ matrix with elements $(y_{nk} : n = 1, \ldots, N, k = 0, \ldots, K)$, $\underline{x}$ as the N-vector $(x_n; n = 1, \ldots, N)$ and $\mathbf{Z}$ as the symmetric $(K+1) \times (K+1)$ matrix

$$\mathbf{Z} = \mathbf{Y}^T \mathbf{Y} + m\mathbf{I}_{K+1}$$

Then

$$L = (M/2) \log(2\pi) + M \log \sigma - \frac{K+1}{2} \log m + \frac{1}{2\sigma^2} \left[(\underline{x} - \mathbf{Y}\underline{a})^2 + m\underline{a}^2 \right]$$

where $M = N + K + 1$.

The Fisher Information for parameters $(\sigma, \underline{a})$ is

$$F(\sigma, \underline{a}) = \left(\frac{2M}{\sigma^2}\right) \left(\frac{|\mathbf{Z}|}{\sigma^{2(K+1)}}\right)$$

L is minimized with respect to $\underline{a}$ by the MML estimate

$$\underline{a}' = \mathbf{Z}^{-1}\mathbf{Y}^T\underline{x}$$

giving the residual variance

$$v_r = ((\underline{x} - \mathbf{Y}\underline{a}')^2 + m(\underline{a}')^2) = \underline{x}^2 - \underline{x}^T\mathbf{Y}\underline{a}'$$

The MML estimate of σ is

$$(\sigma')^2 = \frac{v_r}{N}$$

$$I_1(x) = (M/2)\log(2\pi) + N\log\hat{\sigma} - \frac{1}{2}(K+1)\log m + \log R$$
$$+ \frac{1}{2}\log(2M|\mathbf{Z}|) + N/2 + c_{K+2}$$

Other priors may be more appropriate in some situations, but unless conjugate will usually require an iterative solution to obtain the MML estimates. With the above prior, a large value of m expresses an expectation that little of the variation in x depends on the y variables, a small value expects accurate prediction to be possible. A neutral value might be $m = 1$.

The above estimates differ little from Maximum Likelihood estimates for $N \gg K$. However, the MML analysis offers a sound method for determining which of the K available given variables should be included in the predictor function $\mu = \sum_k a_k y_k$: that selection of given variables giving the shortest explanation of $\underline{x}$, i.e., the smallest $I_1(x)$, should be preferred.

A recent application of this MML method to the prediction of tropical cyclone intensity change in the Atlantic basin [38] found a regression which predicted rather better than models then in use. Thirty-six seasonal, environmental and meteorological variables, together with all their products and squares, were considered for inclusion in the model. A set of first- and second-order regressors was selected on the basis of message length using data from 4347 cyclone observations. The message length for each model considered was increased by the length of a segment nominating the variables used, leading to a model with $K = 27$.

6.7.2 Function Approximation

Here the data comprise N observed values $(x_1, x_2, \ldots, x_N)$ of a real variable x which varies in some unknown way with an independent real variable t.

The data values are observed values of x corresponding respectively to the independent variable values $(t_1, t_2, \ldots, t_N)$ That is, $x_n = x(t_n)$. It is possible that the observed value x_n is the result of its dependence on t_n, plus some "noise" or "error" ε_n, where it is assumed that $\varepsilon_n \sim N(0, \sigma^2)$ independently for all n, with σ unknown.

Given a set of basis functions $g_k(t)(k = 1, 2, \ldots)$, it is desired to infer a function

$$z_K(t) = \sum_{k=1}^{K} a_k g_k(t)$$

which well approximates $x(t)$. The set of basis functions is ordered in some order of increasing complication, e.g., if the set is a set of polynomials in t, the set would be in order of increasing degree. The inference requires the maximum order of basis function K, the coefficient vector $\underline{a} = (a_1, \ldots, a_k, \ldots, a_K)$ and the "noise level" σ to be estimated.

The mathematics of the message length and MML estimation are identical to the regression problem in Section 6.7.1 with the values of the basis functions replacing the "given variables" of that section: $y_{nk} = g_k(t_n)$. However, in this case the basis functions have no status as real variables, and may without changing the situation be replaced by any other set of basis functions spanning the same space of functions. It appears sensible to choose the basis functions to be an orthonormal set if this is possible. For instance, suppose the range of t were $t_0 < t < t_0 + 1$, and the chosen basis function set were polynomials ordered by degree. If the set were chosen in the naive way $g_1(t) = 1, g_2(t) = t, \ldots, g_k(t) = t^{k-1}$ etc. the coefficients $\{a_k\}$ required to get a good fit to data values $(x_n; n = 1, \ldots, N)$ would depend on the origin of measurement of t, i.e., on t_0, although (usually) the numeric value of t_0 has no significance in the problem. Thus, a "neutral" prior for $\underline{a}$ would have to be constructed to depend on t_0 (in a fairly complicated way) if it were intended that the prior probability distribution over model functions be independent of t_0. Choosing the basis set to be orthonormal over the given values of t removes dependence on t_0 and makes it easier to apply prior beliefs to the choice of the prior $h(\underline{a})$. Whether the set should be chosen to be orthonormal over the discrete set of given t-values, i.e.,

$$\sum_{n=1}^{N} g_k(t_n) g_l(t_n) = \delta_{kl} \quad \text{(the Kronecker delta)}$$

or over a continuous range embracing the given values, i.e.,

$$\int_{t_a}^{t_b} g_k(t) g_l(t) dt = \delta_{kl}$$

might depend on whether the given t-values were a distinguished, irreplaceable set (e.g., the times of successive peaks of the business cycle) or merely a set or randomly or evenly spaced values which happened to be yielded by the experimental or observational protocol.

Assuming $g_k()$ to be an orthonormal set, the prior for fixed K used in Section 6.7.1 might be acceptable,

$$h(\sigma, \underline{a}) = (1/R\sigma) \prod_{k=1}^{K} Norm(a_k|0, \sigma^2/m) \quad (m > 0)$$

The complete prior requires a prior distribution over the order of model, K, and would, lacking relevant prior knowledge, be perhaps chosen as a geometric distribution $h_K(K) = 2^{-K}$ or a Uniform distribution over the integers up to some $K_{max} < N$. In the latter case, the MML inference is the model of order K giving the shortest explanation length $I_1(x)$ as calculated in Section 6.7.1; in the former case the preferred model would be the MML model minimizing $I_1(x) + K \log 2$.

Recent work [48, 16] compared various criteria for selecting the order of an approximating polynomial $z_K(t)$ given a sample of size N of $(t_n, x(t_n))$ pairs $(n = 1, \ldots, N)$ with the $\{t_n\}$ randomly and uniformly selected in $(0,1)$ and $x(t_n)$ generated as $s(t_n) + \varepsilon_n$, where $s()$ is a known "target" function and $\varepsilon_n \sim N(0, v)$. Various target functions (none simple polynomials), sample sizes and "signal to noise ratios" AV $(s^2(t))/v$ $(0 < t < 1)$ were tried. The success of an order-selection method for given $s()$, N and v was assessed by the mean squared error

$$\int_0^1 (s(t) - z_K(t))^2 \, dt$$

In these experiments, the coefficient vector $\underline{a}$ for each order K was chosen as the Maximum Likelihood estimate, i.e., the estimate minimizing $\sum_n (x_n - z(t_n))^2$. The authors concluded that, of the criteria tried, the best in almost all circumstances was one based on the theory of V-C dimension [47].

We replicated their experiments comparing their V-C method (known as Structural Risk Minimization) against MML using the prior described above with $m = 1$, and a uniform prior on K. The results showed a clear advantage overall for MML. The only kind of target function for which MML was (slightly) inferior to the SRM method was functions which deviated slightly from a low-order polynomial, e.g., $s(t) = t + 0.1 \sin^2(2\pi t)$, with high signal/noise ratio. Such a function is strongly at odds with the assumed prior, which implies an expectation that each basis function will contribute about as much to the total variation of x as will the "unexplained" variation σ^2. In retrospect, prior experience of function approximation suggests that the smoothest (low k) basis functions will usually dominate, and that therefore a prior on a_k like $N(0, \sigma^2/K)$ might be a better uninformed choice. Other experiments reported by Baxter and Dowe[3] also support the superiority of the MML method.

The function-approximation problem is one where a probabilistic inter-pretation of the inferred model is not necessarily justified. In some contexts, the $x(t_n)$ values might not be subject to any significant noise or measurement error, and might justly be believed to be deterministic functions of t. The normal Bayesian or non-Bayesian approaches to the problem then have some difficulty in justifying their overtly probabilistic treatment of the model

$$x_n = z_K(t_n) + \varepsilon_n; \quad \varepsilon_n \sim N(0, \sigma^2)$$

The MML approach however can appeal to the Algorithmic Complexity (AC) basis for explanation lengths, rather than a probabilistic basis. In the AC context, the residuals ε_n $(n = 1, \ldots, N)$ need not be considered to be ran-dom quantities in the probabilistic sense of "random". Rather, they are a sequence of values admitting no briefer coding by use of assertions within the space spanned by the basis functions, and hence are "random" in the AC, or Chaitin-Kolmogorov sense.

Historical note. The earliest application of MML to function approx-imation, and one of its earliest applications to any problem, was part of a study by Patrick [29] of the shapes of the megalithic stone circles of Ireland. Here, the model used for the locations (x_n, y_n) of the stones in a distorted circle centred at (x_C, y_C) was (in polar coordinates)

$$r_n = r_0 + \sum_{k=2}^{K} a_k \sin(2\pi k(\theta_n + \phi)) + \varepsilon_n$$

with $(x_n, y_n) = (x_C + r_n \cos\theta_n \, , \, y_C + r_n \sin\theta_n)$ and $\varepsilon_n \sim N(0, \sigma^2)$ The order-1 term in the Fourier series is suppressed by choice of (x_C, y_C). The study showed convincing evidence of two preferred shapes: a simple circle and a symmetric "flattened circle" similar to one previously postulated by Thom [44, 45, 46]. However, more elaborate families of shapes advanced by Thom appeared to have no explanatory power.

6.8 Mixture Models

The data comprises a sample of N independent cases or things drawn from some population. For each thing, a vector of variable values is observed. The variables observed are the same for all cases. Hence, the data can be regarded as the ordered set of vectors $x = (y_1, y_2, \ldots, y_n, \ldots, y_N)$.

The model of the population is that it is the union of, or a mixture of, K different subpopulations. Each subpopulation is modeled by a probability distribution

$$f(y|\theta_k) \quad (k = 1, \ldots, K)$$

where θ_k is a (possibly vector) parameter, and by a relative abundance or proportion

$$a_k \ (k = 1, \ldots, K) \quad \sum_k a_k = 1$$

All probability distributions are of the same, known, form but different sub-populations, or "classes", have different parameter values and proportions.

The class from which each thing in the sample was drawn is unknown. Hence, the model probability for a data vector y_n is

$$\Pr(y_n) = \sum_{k=1}^{K} a_k f(y_n | \theta_k) = P_n \text{ say}$$

If K, the number of classes, is known, the above probability model is just a rather complicated probability distribution with known functional form and parameter vector $\theta = (a_k, \theta_k) \ (k = 1, \ldots, K)$ and in principle any general estimation method, e.g., MML or ML, can be applied to the estimation of θ. However, the complex form of the model creates considerable mathematical difficulties. Further, in practical applications of mixture models, there is usually as much interest in inferring the class $k_n \ (n = 1, \ldots, N)$ to which each thing belongs as in the parameters of each class. The class labels $\{k_n\}$ do not appear in the above model, and hence are not inferred in the estimation of its parameters. However, before proceeding to describe the MML estimation of both class parameters and class labels, it will be useful to review the standard ML (maximum likelihood) approach to the estimation of θ, known as the Expectation Maximization (EM) algorithm.

6.8.1 ML Mixture Estimation: The EM Algorithm

In this approach, the vector $\underline{k} = (k_n; n = 1, \ldots, N)$ of class labels is regarded as conceptually part of the data, which happens not to have been observed. That is, it is treated as "missing data". Were the class labels known, the probability model for the "full" data (k_n, y_n) for thing n could be written as

$$\begin{aligned} \Pr(k_n, y_n) &= \Pr(k_n | \theta) \ \Pr(y_n | k_n, \theta) \\ &= a_{k_n} f(y_n | \theta_{k_n}) = p_{nk_n} \text{ say} \end{aligned}$$

and the log-likelihood of θ on the full data could be written as

$$FLL = \sum_k m_k \log a_k + \sum_k \sum_{\{n: k_n = k\}} \log f(y_n | \theta_k)$$

where m_k is the number of things belonging to class k, and the log-likelihood of θ_k, the parameter of class k, is a function of only the data of things belonging to that class. But of course, the class labels are unknown.

By summing over all classes,

$$P_n = \sum_{k_n=1}^{K} \Pr(k_n, y_n) = \sum_{k=1}^{K} a_k f(y_n | \theta_k) = \sum_{k=1}^{K} p_{nk} \quad \text{say}$$

as before. Using Bayes' theorem,

$$r_k(y_n) = \Pr(k_n = k|y_n) = a_k f(y_n|\theta_k) / \sum_{l=1}^{K} a_l f(y_n|\theta_l) = p_{nk}/P_n$$

where $r_k(y_n) = r_{kn}$ is the probability that thing n belongs to class k, given its data y_n. The rationale of the EM algorithm is to form the expected log-likelihood of θ on the full data, the expectation being taken over the joint distribution of the "missing data" $\underline{k}$, i.e., over the independent distributions $\{r_{kn} : k = 1, \ldots, K\}$ for each thing n.

$$\mathrm{E}(FLL) = \sum_k (Em_k) \log a_k + \sum_k \sum_n r_{kn} \log f(y_n|\theta_k)$$

where $Em_k = \sum_n r_{kn}$.

It is easily shown that maximization of $\mathrm{E}(FLL)$ with respect to $\{(a_k, \theta_k) : k = 1, \ldots, K\}$, with r_{kn} as defined above, results in maximization of LL, the likelihood of θ given the "incomplete" data $(y_n : n = 1, \ldots, N)$.

$$
\begin{aligned}
LL &= \sum_n \log(\sum_k a_k f(y_n|\theta_k)) \\
\frac{\partial}{\partial\theta_k} LL &= \sum_n \left[\frac{\frac{\partial}{\partial\theta_k} f(y_n|\theta_k)}{\sum_l a_l f(x_n|\theta_l)} \right] \\
&= \sum_n \left[\left(\frac{a_k f(y_n|\theta_k)}{\sum_l a_l f(x_n|\theta_l)} \right) \frac{\frac{\partial}{\partial\theta_k} f(y_n|\theta_k)}{a_k f(y_n|\theta_k)} \right] \\
&= \sum_n \left[r_{kn} \frac{\partial}{\partial\theta_k} \log(a_k f(y_n|\theta_k)) \right] \\
&= \frac{\partial}{\partial\theta_k} \mathrm{E}(FLL)
\end{aligned}
$$

Similarly, $\frac{\partial}{\partial a_k} LL = \frac{\partial}{\partial a_k} \mathrm{E}(FLL)$

Maximization of $\mathrm{E}(LL)$ is achieved in a 2-phase iteration by the EM algorithm as follows. Given some initial estimates $(\hat{a}_k, \hat{\theta})$:

Phase 1: For each thing n, compute $\hat{a}_k f(y_n|\hat{\theta}_k)$ for all classes, and hence r_{kn} for all classes. Then accumulate statistics s_k appropriate for the estimation of θ_k, letting the data y_n contribute to the statistics s_k with weight r_{kn}.

Phase 2: For each class re-estimate $\hat{a}_k$ as $\sum_n r_{kn}/N$, and re-estimate $\hat{\theta}_k$ from statistics s_k treated as being accumulated from a sample of size $\hat{a}_k N$.

Phases 1 and 2 are then repeated until convergence of the estimates. Convergence is guaranteed in the sense that each iteration is guaranteed not to decrease $E(FLL)$, but the global maximum is not necessarily found.

The EM algorithm, in the form described, is simply a computational algorithm for maximising the likelihood of θ given the observed data. Although the conventional description of the algorithm, which was followed above, conceives of the class labels $\underline{k}$ as "missing data", this concept is not essential to the derivation of the algorithm, which was in fact in use long before the "EM" name became common.

If we regard our probability models as models of the given data, rather than as models of some hypothetical population from which the data is a sample, then it seems equally valid to regard the vector of class labels as a vector of parameters of the model, rather than as "missing data". We will pursue this view in the next section.

However we regard the class labels, the conventional ML approach to a mixture problem does not estimate them, whether ML is implemented using the EM algorithm or by other means.

There is a further problem with ML estimation of mixture models: the global maximum of the likelihood is usually achieved by an undesirable model. The problem may be illustrated by a simple mixture of two univariate Normal distributions modeling a univariate sample $x = (x_1, x_2, \ldots, x_n, \ldots, x_N)$. The probability model is

$$\Pr(x_n) = \left[aN(x_n|\mu_1, \sigma_1^2) + (1-a)N(x_n|\mu_2, \sigma_2^2) \right] \varepsilon$$

where the five parameters $a, \mu_1, \sigma_1, \mu_2, \sigma_2$ are all unknown, and ε is the precision of measurement of the data values, assumed constant for all n. If ε is small, the highest likelihood is achieved by a model with $\hat{a} = 1/N$, $\hat{\mu}_1 = x_m$, $\hat{\sigma}_1 \approx \varepsilon$, for some m. This is a model in which one class contains the single datum x_m, and the other class contains the rest of the sample. The probability $\Pr(x_m|\hat{a}, \hat{\mu}_1, \hat{\sigma}_1)$ is about $1/N$, and the probabilities for all other data are of order $\varepsilon/\hat{\sigma}_2$. The total log likelihood is thus about

$$LL_1 \approx (N-1)\log(\varepsilon/\hat{\sigma}_2) + \log N$$

If the data were in fact drawn from a population with, say, $a = \frac{1}{2}$, $\mu_1 = 1$, $\mu_2 = -1$, $\sigma_1 = \sigma_2 = 1$, we would have $\hat{\sigma}_2 \approx \sqrt{2}$. The log likelihood of the true parameter values would be of order $LL \approx N \log(\varepsilon/1) + N \log \frac{1}{2}$ for $\varepsilon \ll 1/N$, $LL_1 > LL$, so the incorrect model has the higher likelihood. (The more correct treatment of finite-precision data as in Section 6.4 makes no essential difference to this result.)

In practice, the EM algorithm will not often converge to such an undesirable solution, and in any case, several runs of the EM algorithm from different initial guesstimates are often needed to find the best desirable model. It is however obvious that genuine Maximum Likelihood estimation is not a satisfactory solution for mixture models.

ML also gives no clear method for estimating the number of classes when K is unknown. The likelihood of a model increases monotonically with K.

6.8.2 A Message Format for Mixtures

We consider an explanation message format in which the assertion has the following form. (We omit circumflexes for estimates.)

(a) The number of classes K, if this is unknown.
     The prior assumed has little effect, and can be taken as, for instance, 2^{-K}.
(b) The relative proportions $\{a_k : k = 1, \ldots, K\}$. These are encoded as for the probabilities of a Multinomial distribution (Section 6.3).
(c) For each class k, the distribution parameter estimate θ_k.
(d) For each thing n, the class k_n to which it is assigned.

The obvious format for the detail, encoding the data assuming the model to be correct, is the concatenation of N substrings, each encoding the data for one thing. The substring encoding y_n uses a code based on the distribution $f(y|\theta_{k_n})$, and hence has length $-\log f(y_n|\theta_{k_n})$.

If the second part of the explanation is encoded in this obvious way, the explanation length is not properly minimized, and the parameter estimates $\{\theta_k\}$ will not in general converge to the true values as $N \to \infty$. The problem arises in the estimation and coding of the class label parameters $\{k_n : n = 1, \ldots, N\}$. Clearly, the explanation length using this format is minimized when k_n is chosen to maximize p_{nk_n} (in the notation of Section 6.8.1). That is, thing n is assigned to that class most likely to contain such a thing: k_n is "estimated" by Maximum Likelihood.

$$p_{nk_n} = \text{Max}_k\,(p_{nk})$$

The length of coding for thing n is $-\log(\text{Max}_k\,(p_{nk}))$. However, again using the notation of Section 6.8.1, the probability that a randomly chosen thing would have data y_n is

$$P_n = \sum_k p_{nk}$$

suggesting that an efficient coding should be able to encode thing n with a length of about

$$-\log\left(\sum_k p_{nk}\right) \leq -\log(\text{Max}_k(p_{nk}))$$

and hence suggesting that the obvious coding is inefficient.

Further, using the obvious coding, the estimates of $\hat{a}_k$, θ_k affect only the coding of those things assigned to class k, and so these estimates will depend on the data of only those things. It is easy to show that the resulting estimates are inconsistent.

Suppose the data are univariate and drawn from a mixture in equal proportions of two Normal distributions $N(-a, 1)$ and $N(a, 1)$. Consider the MML estimation of the 2-class model

$$\Pr(x_n) = \frac{1}{2}\{N(x_n|\mu_1, \sigma_1^2) + N(x_n|\mu_2, \sigma_2^2)\}$$

where we assume equal proportions *a priori*. For a sufficiently large sample size N, we expect from symmetry to obtain estimates such that

$$\hat{\mu}_1 \approx -\hat{\mu}_2, \quad \hat{\sigma}_1 \approx \hat{\sigma}_2, \quad k_n = 1 \text{ if } x_n < 0; \quad k_n = 2 \text{ if } x_n > 0$$

Since $\hat{\mu}_1$ and $\hat{\sigma}_1$ affect only the coding of things assigned to class 1, i.e., with $x_n < 0$, we further expect

$$\hat{\mu}_1 \approx \left(\sum_{n:x_n<0} x_n\right) / (N/2)$$

The sum in this expression is not over the true members of class 1. Some members, with $x_n > 0$, are omitted, and a roughly equal number of members of class 2, having $x_n < 0$, have been included. The effect of this substitution is to decrease the sum so $\hat{\mu}_1$ is less than the true class-1 mean $(-a)$. The magnitude of this error depends on a but not on N, so the estimates $\hat{\sigma}_1$, $\hat{\sigma}_2$ are also inconsistent, being too small. Table 6.2 shows the expected estimates $\hat{\mu}_2$, $\hat{\sigma}_1 = \hat{\sigma}_2$ as functions of a. The table also shows the difference in message

Table 6.2. Estimates of 2-component Normal Mixture using MML with "obvious" coding.

a	$\hat{\mu}_2$	$\hat{\sigma}$	$C_2 - C_1$
0.001	0.798	0.603	0.187
1	1.167	0.799	0.123
2	2.017	0.965	-0.147
3	3.001	0.998	-0.460

length per thing between the 2-class model (C_2) and a one-class model (C_1) encoding the whole sample as if drawn from $N(0, a^2 + 1)$. Not only are the two-class estimates seriously in error for $a < 2$, but it appears that MML would favour a one-class model for small a. For a very large sample, the two-class model is not preferred unless $a > 1.52$, although the population distribution becomes bimodal for $a > 1$.

We now show that the coding of the detail can be improved over the "obvious" coding used above, and that the improvement both shortens the explanation length and yields consistent estimates.

6.8.3 A Coding Trick

The difficulties with the obvious format arise because the explanation asserts the estimated class $\hat{k}_n$ of each thing as an unqualified, precise value. It is a cardinal principle of MML explanation coding that no assertion should be stated more precisely than is warranted by the data, but in the "obvious" coding, the class label $\hat{k}_n$ is asserted as if known exactly even if thing n could well belong to any of several classes. We should instead seek a message format which in some sense allows the discrete parameter estimates $\{\hat{k}_n : n = 1,\ldots,N\}$ to be coded to limited "precision", although it is not immediately clear what an "imprecise" value for a discrete parameter might be.

An MML explanation always begins with the assertion of some single model of the data, which is then used to provide a coding scheme for the detail. In this case, the assertion of a single model requires specification of all the parameters $\{(a_n, \theta_n) : k = 1,\ldots,K\}$ and $\{k_n : n = 1,\ldots,N\}$ so we do not intend to change the format for the assertion in Section 6.8.2, and retain all its parts (a) to (d) as before. The difference in format will lie in the coding of the detail. However, we will also reconsider how best to estimate the class label of each thing. As in Section 6.8.1, define for some thing

$$p_{nk} = a_k f(y_n | \theta_k); \quad P_n = \sum_{k=1}^{K} p_{nk}; \quad r_{nk} = p_{nk}/P_n$$

If all but one of $\{r_{nk} : k = 1,\ldots,K\}$ are close to zero, with, say, r_{nl} close to one, then thing n clearly belongs to class l, and we may set $\hat{k}_n = l$ as in the "obvious" code thereby coding y_n with a total message length $-\log p_{n\hat{k}_n}$. However, suppose that the two highest-posterior class probabilities for thing n, say r_{ni} and r_{nj}, are within a factor of two, i.e., that $|\log r_{ni} - \log r_{nj}| < \log 2$. This implies $|\log p_{ni} - \log p_{nj}| < \log 2$, so whether we choose to assign the thing to class i or class j makes at most one bit difference to the message length. Whenever this condition obtains, let us make the choice in such a way as to convey some useful information. Specifically, whenever $n < N$ and $|\log(r_{ni}/r_{nj})| < \log 2$, we choose $\hat{k}_n = i$ if the detail binary code string for thing $(n + 1)$ begins with a binary digit "0", and $\hat{k}_n = j$ if the string begins with "1".

Now consider the decoding of an explanation using this convention. By the time the receiver comes to the detail string encoding y_n, the estimates $\hat{k}_n$, $\hat{a}_{\hat{k}_n}$ and $\hat{\theta}_{\hat{k}_n}$ have already been decoded from the assertion part of the explanation. Thus, the receiver has all the information needed to deduce the code used for y_n, i.e., the Huffman or similar code for the distribution $f(y|\hat{\theta}_{\hat{k}_n})$ and can proceed to decode y_n. But the receiver also at this stage knows $\hat{a}_k$, $\hat{\theta}_k$ for all K classes, and hence can compute $p_{nk} = \hat{a}_k f(y_n | \hat{\theta}_k)$ for all classes, and thence r_{nk} for all k. The receiver can thus observe that the two largest values r_{ni}, r_{nj} differ by less than a factor of two. Knowing the coding convention described above (as the receiver is assumed to know all conventions used in coding an explanation) the receiver can deduce that $\hat{k}_n$ will have been asserted to be i or j according to the first binary digit of the detail for y_{n+1}. Since $\hat{k}_n$ has been given in the assertion, the receiver thus learns the first binary digit of the detail for the next thing. Hence, this digit need not be included in the explanation.

In crude terms, if setting $\hat{k}_n$ to be i or j makes less than one bit of difference to the coding of y_n, we make the choice to encode the first digit of the detail coding of y_{n+1}. Adopting this convention costs at worst one binary digit in coding y_n, but always saves one binary digit in coding y_{n+1}. It is therefore more efficient on average than the obvious coding scheme.

Note that to use this coding trick, the construction of the detail describing the things must commence with the description of the last thing, since its first digit may affect how we describe the second-last thing, and so on.

The above device may be generalized. Suppose that in some explanation (not necessarily a classification) there is some segment of data, not the last to be included in the detail, which could be encoded in several different ways within the code optimal for the "theory" stated in the assertion. Let the lengths of the several possible code segments be $l(1), l(2), \ldots$, etc. In the present case, the data segment is the data y_n of a thing, and the possible codings are those obtained by assigning it to the several classes. The code lengths are the set $\{l(k) = -\log p_{nk} : k = 1, \ldots, K\}$.

The p_{nk} values may be identified with the probabilities of getting the data by each of several mutually exclusive routes all consistent with the "theory". In the present case, they are the probabilities of there being such a thing in each of the classes.

As before, define

$$P_n = \sum_k p_{nk}; \quad r_{nk} = p_{nk}/P_n$$

To choose the encoding for the data segment, first construct according to some standard algorithm a Huffman code optimised for the discrete probability distribution $\{r_{nk} : k = 1, \ldots, K\}$. Note that this distribution is the Bayes posterior distribution over the mutually exclusive routes, given the theory and the data segment. From the standard theory of optimum codes, the length $m(k)$ of the code word in this Huffman code for route (i.e., class) k will be

$$m(k) = -\log r_{nk},$$

the code will have the prefix property, and every sufficiently long binary string will have some unique word of the code as prefix. Now examine the binary string encoding the remainder of the data, i.e., the data following the segment being considered. This string must begin with some word of the above Huffman code, say, the word for route j. Then encode the data segment using route j, hence using a code segment of length $l(j)$. In the mixture case, j is the estimated class $\hat{k}_n$. By an obvious extension of the argument presented for the crude trick, the first $m(j)$ bits of the binary string for the remainder of the data need not be included in the explanation, as they may be recovered by a receiver after decoding the present data segment.

Consider the net length of the string used to encode the data segment, that is the length of string used minus the length which need not be included for the remaining data. The net length is

$$
\begin{aligned}
l(j) - m(j) &= -\log p_{nj} + \log r_{nj} \\
&= -\log(p_{nj}/r_{nj}) \\
&= -\log P_n \\
&= -\log(\sum_k p_{nk})
\end{aligned}
$$

Merely choosing the shortest of the possible encodings for the data segment would give a length of

$$
-\log\left(\max_k(p_{nk})\right)
$$

The coding device therefore has little effect when one possible coding is much shorter (more probable *a posteriori*) than the rest, but can shorten the explanation by as much as log (number of possibilities) if they are all equally long.

When used in the construction of a classification explanation, this coding device has the following effects:

(a) Each thing but the last can be encoded more briefly. The saving is substantial only when the thing could well be a member of more than one class. Since the number of things is not expected to be small, we henceforth ignore the fact that the saving is not available for the last thing.

(b) Although the form of the explanation states a class for each thing and encodes it as a member of that class, the net length of the description is the same as would be achieved by a (much more complex) code which made no assignment of things to classes, but instead was optimised for the sum of the density distributions of the classes.

(c) Since the binary string forming the "remainder of the data" has no logical relationship to the thing being encoded, and since any string resulting from the use of an optimised binary code has the statistical properties of a string produced by a random process, this device effectively makes a pseudo-random choice of class assignment. The probability that the thing will be assigned to a particular class is given by the $\{r_{nk} : k = 1, \ldots, K\}$ distribution, i.e., the posterior distribution for that thing over the classes.

(d) The estimation of the distribution parameters a_k, θ_k of a class becomes consistent.

(e) Given sufficient data, two classes with different distributions will be distinguished by minimisation of the information measure, no matter how much they overlap. Of course, the less the difference in the classes, the larger is the sample size (number of things) needed to reveal the difference, and the existence of two classes rather than one.

(f)   In attempting to find a good model by MML we need not actually construct an optimally coded message in full detail. Rather, we need only consider the length such a message would have if constructed. Where the length of an actual message might depend on essentially trivial details of the coding process (e.g., the exact value to which an estimated parameter value might be rounded in stating it to optimum precision) we calculate an expected message length by averaging over all essentially equivalent encodings. Hence, rather than actually assigning each thing to a single class by constructing a Huffman code and "decoding" some bits of the next thing's description, we treat each thing as being partially assigned to all the classes. Notionally, a fraction r_{nk} of the thing is assigned to class k, and its attribute values contribute with weight r_{nk} to the estimates of the distribution parameters of class k. The net message length required for the description of the thing is computed directly from the sum of the distribution densities of the classes in accordance with the above formula $(-\log P_n)$. Note that the term $-\log P_n$ includes the cost $-\log \hat{a}_{\hat{k}_n}$ of stating the estimated class $\hat{k}_n$.

The upshot of using this improved format is that MML estimation of the mixture model becomes rather similar to the EM process of Section 6.8.1. The major term in the explanation length is

$$-\sum_n \log P_n = -\sum_n \log(\sum_k \hat{a}_k f(y_n|\hat{\theta}_k)),$$

exactly the negative log-likelihood of the "incomplete" data model. However, the MML method differs from EM in that

(a)   In principle, it estimates the class labels k_n.

(b)   The full message length to be minimized includes the assertion of the estimated class parameters $\hat{a}_k, \hat{\theta}_k : k = 1, \ldots, K$, whose coding depends on the Fisher Information and some prior density. The class parameter estimates are therefore MML rather than ML estimates.

(c)   MML allows the number of classes K to be estimated. It is just that number which leads to the shortest explanation.

(d)   Improper solutions do not occur. The explanation length is not in general minimized by models which assign a single thing to some class.

6.8.4 Imprecise Assertion of Discrete Parameters

The mixture model problem introduced the need to devise a coding scheme which in a sense allowed the "imprecise" estimation of discrete parameters (the class labels k_n). Without the ability for imprecise specification, an inconsistency appears in estimating the number of classes and their parameters.

The situation is parallel to that in the Neyman-Scott problem (Section 5.6.1). There, the data comprised N small samples from N different Normal populations, all with the same unknown S.D. σ but with different unknown means. Maximum Likelihood estimation, which is equivalent to exact specification of all estimates, leads to over-precise estimation of the N unknown means ("nuisance parameters") and inconsistent estimation of σ. MML, by optimally choosing the precision of the estimates for the nuisance parameters, gives consistent estimation of σ. Similarly, in the mixture problem, the class labels are "nuisance parameters", and precise (e.g., ML) estimation of them causes inconsistency. "Imprecise" estimation restores consistency, just as in the Neyman-Scott case.

These and other examples suggest that Maximum Likelihood is prone to inconsistency whenever an attempt is made to estimate nuisance parameters whose number increases with sample size. Consistency can be restored to ML by marginalizing-out the nuisance parameters (as in Neyman-Scott) or by integrating or summing over their values (as in Expectation Maximization.) but at the cost of failing to estimate these parameters. MML, on the other hand, remains consistent while estimating all parameters, provided the explanation length is truly minimized.

The coding device described in Section 6.8.3 which has the effect of imprecise estimation of the discrete class labels shows interesting similarities to the behaviour of SMML codes for real parameters. As shown in Section 3.3.4, for models with regular log-likelihood functions, estimation to optimal precision of a set of D real parameters involves use of an "optimal quantizing lattice" in D dimensions. This has the consequence that the estimate for one parameter becomes dependent on data which is irrelevant to that parameter. For large D, the effect is that the SMML estimate of the parameter behaves as a pseudo-random selection from the posterior density of that parameter. The pseudo-randomness is introduced from the values of data not related to the parameter. A further consequence is that the length $I_1(x)$ of the SMML explanation of data x is little more than the length $I_0(x)$ of the optimum non-explanatory message, although the latter contains no estimates. The difference $I_1(x) - I_0(x)$ grows only as the logarithm of the number of estimates asserted. Very similar behaviour is shown by the estimates resulting from the coding scheme of Section 6.8.3. Again, the estimate $\hat{k}_n$ of the class of thing n is pseudo-randomly selected from the posterior distribution over classes $\{r_{nk} : k = 1, \ldots, N\}$, and the pseudo-randomness is introduced from irrelevant other data (y_{n+1}). Also, the length of the MML explanation which estimates the class labels is little more than that of a message which does not. In the mixture model, the additional length is difficult to compute. It arises in part from the fact that the Huffman code constructed over the posterior distribution $\{r_{nk} : k = 1, \ldots, K\}$ will have inefficiencies arising from the integer constraint on code word lengths. These may be reduced by applying the coding trick to groups of things rather than single things, thus using a code over the joint posterior distribution of several class labels (see Section 2.1.6). However, the trick cannot reduce the length required to code the last thing in the sample, and grouping may prevent compaction of the last several things. We conjecture that in the mixture case, the effective increase in explanation length needed to include estimates of the N class labels is not more than $\frac{1}{2}\log N$, and less if the class distributions do not overlap much. In a later chapter, we generalize the coding trick of Section 6.8.3 to real parameters, and show that the "pseudo-random" behaviour, and logarithmic $I_1 - I_0$ increase, still obtains for real parameters.

6.8.5 The Code Length of Imprecise Discrete Estimates

For a single real-valued parameter θ, prior density $h(\theta)$, data x and probability model $f(x|\theta)$, we have shown in Section 5.1 that the expected explanation length using the MML estimate θ' is

$$- \log \left(\frac{h(\theta')}{\sqrt{F(\theta')/12}} \right) - \left(\log f(x|\theta') - \frac{1}{2} \right) + \text{Geometric constants}$$

where the first term represents the code length of the asserted estimate and the second term the expected length of the detail encoding x. The second term includes the expected increase due to the rounding-off of θ' to the imprecise value $\hat{\theta}$ actually stated in the assertion. We now derive a discrete-parameter analogue of the first term, i.e., an expression representing the code length required to encode a discrete parameter to "limited precision". The derivation is based on the example of a class label parameter k_n in the mixture problem, and we use the notation of Section 6.8.3.

The total code length needed to "explain" the data y_n of thing n has been shown to be

$$- \log P_n = - \log \left(\sum_{k=1}^{K} \hat{a}_k f(y_n|\hat{\theta}_k) \right)$$

For brevity, write $f(y_n|\hat{\theta}_k)$ as f_{nk}. This total length will be decomposed as the sum of the length A_n needed to encode an "imprecise" estimate of k_n, and the expected length of the detail used to encode y_n given the estimated class label.

$$- \log P_n = A_n + \mathrm{E}\,(detail\ length)$$

Now, the code construction of Section 6.8.3 leads to a detail where, with probability $r_{nk} = \hat{a}_k f_{nk}/P_n$, y_n is encoded as a member of class k, using a detail length $- \log f_{nk}$. Hence,

$$\mathrm{E}(detail\ length) \quad = \quad - \sum_{k} r_{nk} \log f_{nk}$$

$$A_n \quad = \quad - \log P_n + \sum_{k} r_{nk} \log f_{nk}$$

But $f_{nk} = r_{nk} P_n / \hat{a}_k$. Hence,

$$
\begin{aligned}
A_n &= -\log P_n + \sum_k r_{nk} \log(r_{nk} P_n / \hat{a}_k) \\
&= -\log P_n + \sum_k r_{nk} \log P_n + \sum_k r_{nk} \log(r_{nk} / \hat{a}_k) \\
&= \sum_k r_{nk} \log(r_{nk} / \hat{a}_k)
\end{aligned}
$$

Now, in estimating the class label k_n of a thing, the distribution $\{\hat{a}_k : k = 1, \ldots, K\}$ plays the role of the prior probability distribution of k_n, since it has been asserted to be the distribution of class proportions, and hence the probability distribution of the class of a randomly chosen thing. The distribution $\{r_{nk} : k = 1, \ldots, K\}$ is the posterior probability distribution of k_n, given the data y_n. Hence, A_n, the length needed to encode the "imprecise" estimate of k_n, is just the Kullback-Leibler (KL) distance of the prior from the posterior.

This form for A_n may appear to be quite different from the form

$$
B = -\log \left(\frac{h(\theta')}{\sqrt{F(\theta')/12}} \right)
$$

given above for the length of the encoding of a real-valued parameter. However, there is a close analogy. The form B was derived in Section 5.1 by supposing that the "statistician's" MML estimate θ' was passed to a "coder", who then chose the nearest value $\hat{\theta}$ from a discrete set Θ^* of allowed estimates, and used $\hat{\theta}$ in coding the data x. The "allowed values" in Θ^* were chosen by the coder, but in the neighbourhood of θ, were spaced apart by the amount $w(\theta)$, where $w(\theta)$ was a "spacing function" specified by the statistician. Since the statistician was assumed to have no knowledge of the allowed values apart from the spacing $w(\theta)$, the statistician estimated the explanation length resulting from her choice of θ' on the basis that the estimate $\hat{\theta}$ actually asserted could lie anywhere in the range $\theta' \pm w(\theta')/2$, and should be treated as randomly and uniformly distributed in this range. (The unknown detailed decisions of the coder form the source of "randomness".) Further, since each $\hat{\theta}$ would be used for an interval of width $w(\hat{\theta})$ in the range of θ', the coding probability of $\hat{\theta}$ is set at $h(\hat{\theta})w(\hat{\theta})$, approximately the total prior probability within this interval, and giving an assertion length $-\log(h(\hat{\theta})w(\hat{\theta}))$. The statistician, being ignorant of $\hat{\theta}$, is forced to approximate this by $-\log(h(\theta')w(\theta'))$. Further argument then leads to the choice $w(\theta) = 1/\sqrt{F(\theta)/12}$.

But we may regard this derivation in another light. As far as the statistician is concerned, the asserted estimate $\hat{\theta}$ is randomly distributed with a Uniform density

$$r(\hat{\theta}) = \frac{1}{w(\theta')} \quad (|\hat{\theta} - \theta'| < w(\theta')/2) \quad \text{or 0 otherwise}$$

Since θ' (and in less regular cases the width w) are obtained as functions of the data x, the distribution $r(\hat{\theta})$ depends on x, and should be written as $r(\hat{\theta}|x)$.

Further, the KL distance of $h(\theta)$ from $r(\hat{\theta}|x)$ is (writing w for $w(\theta')$)

$$\int_{\theta'-w/2}^{\theta'+w/2} r(\theta|x) \log \frac{r(\theta|x)}{h(\theta)} \, d\theta = \int_{\theta'-w/2}^{\theta'+w/2} \frac{1}{w} \log \frac{1}{wh(\theta)} \, d\theta$$

which, if $h(\theta)$ is slowly varying as assumed in the MML derivation, is approximately

$$\int_{\theta'-w/2}^{\theta'+w/2} \frac{1}{w} \log \frac{1}{wh(\theta')} \, d\theta = -\log(w(\theta')h(\theta'))$$

With the optimum MML choice

$$w(\theta') = 1/\sqrt{F(\theta')/12}$$

this gives exactly the form B, i.e., the usual MML expression for the length of a single-parameter assertion. Hence, we see that B can be written as the KL distance of the prior $h(\theta)$ from the density $r(\hat{\theta}|x)$, in close analogy to the form A_n for a discrete parameter. Of course, the density $r(\hat{\theta}|x)$ is not the posterior density $\Pr(\theta|x)$, but it is as good an approximation to it as can reasonably be expected of a Uniform density. It is centred on a good guess (the MML estimate θ') and has the same variance as the posterior for probability models having approximately quadratic log-likelihoods about θ'. For multi-parameter problems, it was shown in Section 3.3.4 that the distribution of the asserted estimate of one parameter becomes increasingly close to its posterior as the number of parameters increases.

We conclude that for both real and discrete parameters, the code length required to assert an estimate to optimal precision is close to the Kullback-Leibler distance of its prior from its posterior. This conclusion must be qualified for discrete parameters, as it obtains only when many discrete parameters are estimated simultaneously. For a single real parameter, its MML estimate θ' (together with $w(\theta')$) at least roughly characterizes its posterior, but for a single discrete parameter no such single estimate can serve to capture any more of the posterior than its mode.

In problems with a small number of discrete parameters, all of interest, it does not seem possible to do better than to choose those estimates which minimize the length of an explanation asserting the estimates precisely.

6.8.6 A Surrogate Class Label "Estimate"

Returning to the mixture problem, where there are many "nuisance" class label parameters $\{k_n\}$, consider the univariate two-class model with density

$$(1 - a)f(x|\theta_0) + af(x|\theta_1)$$

We have seen that, for datum x_n, MML will make a pseudo-random assignment of the datum to either class 0 or class 1, with $\Pr(\hat{k}_n = 1)$ being some probability r_n. In some sense, r_n can be regarded as an imprecise specification of the estimate $\hat{k}_n$, and hence as a sort of surrogate MML estimate r'_n which then gets "rounded" by the "coder" to either 0 or 1.

We have argued above that the message length required to encode $\hat{k}_n$ imprecisely, given the prior parameter a, is the KL distance

$$A_n = r_n \log \frac{r_n}{a} + (1 - r) \log \frac{1 - r_n}{1 - a}$$

For datum x_n, dropping the suffix n and writing f_0 for $f(x_n|\theta_0)$ and f_1 for $f(x_n|\theta_1)$, the expected message length required to encode x_n given this random choice of $\hat{k}_n$ is

$$-r \log f_1 - (1 - r) \log f_0$$

The total length needed both to specify $\hat{k}_n$ imprecisely, and then to encode x_n using the "rounded" class label, is thus (in expectation)

$$
\begin{aligned}
l(x) &= r \log \frac{r}{a} + (1 - r) \log \frac{1 - r}{1 - a} - r \log f_1 - (1 - r) \log f_0 \\
&= r \log r + (1 - r) \log(1 - r) - r \log(af_1) \\
&\quad - (1 - r) \log((1 - a)f_0)
\end{aligned}
$$

Regarding r as a "parameter" which may be chosen freely, differentiation gives

$$\frac{d}{dr}l(x) = \log r - \log(1 - r) - \log(af_1) + \log((1 - a)f_0)$$

Equating to zero gives the "MML estimate" r' where

$$
\begin{aligned}
\log \frac{r'}{1 - r'} &= \log \frac{af_1}{(1 - a)f_0} \\
r' &= \frac{af_1}{af_1 + (1 - a)f_0}
\end{aligned}
$$

which is just the value obtained earlier in discussion of the compact explanation format (Section 6.8.3). Thus, whether we regard the values of r as being dictated by use of a coding trick, or as a "parameter" which may be chosen freely, the conclusion is the same: MML chooses to assign datum x_n to class 1 with pseudo-random probability equal to the posterior probability that x_n is indeed drawn from class 1. The result is easily generalized to more than two classes.

6.8.7 The Fisher Information for Mixtures

Calculation of the Fisher Information for a mixture model can become quite difficult. Consider the K-class model distribution (or density) for a variable x given by

$$\Pr(x) = \sum_{k=1}^{K} a_k f(x|\theta_k)$$

where we have omitted the class labels from the model. If x is a vector of M dimensions, the parameter θ_k will often have M or more component scalar parameters, so there are typically of order KM parameters in the model. In some applications of mixture modelling we have seen, $M \approx 100$ and $K \approx 10$, giving well over 1000 parameters. The full Fisher matrix of expected second differentials of the log-likelihood can have around half a million or more distinct elements, few being zero. The cross-differential between some component of θ_k and some component of θ_l ($l \neq k$) will typically be non-zero if there is any region of x-space for which $f(x|\theta_k)f(x|\theta_l) > 0$, i.e., if there is any overlap in the two class distributions. Fortunately, in most practical applications of mixture models with high-dimensional data, the different classes are not expected to overlap closely, so the cross-differentials between parameters of different classes are expected to be small and may even be entirely neglected. If two classes in the model have closely overlapping distributions, a better explanation of the data may well be obtained by combining them into a single class, or by modifying the assumed form of the class distribution function $f(x|\theta)$.

The effect of non-zero cross derivatives between parameters of two different classes is to reduce the Fisher Information below the value obtaining if these derivatives were neglected. Similarly, overlap between classes causes non-zero cross derivatives to appear between the "proportion" parameters $\{a_k\}$ and the distribution parameters $\{\theta_k\}$ which again act to reduce the Fisher Information. In the extreme of exact overlap, i.e., if for two classes k and l, $\theta_k = \theta_l$, then the data cannot provide information about their individual proportions a_k and a_l, but only about the sum $(a_k + a_l)$, and the Fisher Information becomes zero. Note that in such a case, the "standard" approximation for the length of the assertion,

$$-\log \frac{h(\text{parameters})}{\sqrt{\text{Fisher Information}}} + \text{constant}$$

breaks down, and some amended expression such as suggested in Section 5.2.9 should be used in preference.

As the algebraic difficulties in obtaining the expected second differentials are formidable, the use of the empirical approximation (Section 5.3) based on the sample data is recommended.

So far, this section has omitted mention of the class label parameters $\{k_n : n = 1, \ldots, N\}$. That is, we have tacitly followed the EM approach

(Section 6.8.1) in treating the class labels as "missing data" rather than as parameters of the model, and have discussed the Fisher Information based on a marginal log-likelihood from which the class labels have been eliminated by summing over their possible values. That is, we have considered an "unclassified" model. However, we consider the class labels to be true parameters of the mixture model, and in Section 6.8.2 *et seq.* have developed an explanation format allowing estimates of these parameters to be asserted. We now revise our consideration of the Fisher Information to reflect the inclusion of these parameters, giving a "classified" model.

6.8.8 The Fisher Information with Class Labels

In Section 6.8.3, we show that the combined code lengths needed to assert a class label estimate $\hat{k}_n$ and then to encode the datum y_n according to the estimated distribution of class $\hat{k}_n$ can be reduced by a coding trick to an effective value of $-\log \sum_{k=1}^{K} \hat{a}_k f(y_n | \hat{\theta}_{\hat{k}_n})$. Hence, the length needed to name the class labels and encode the data for all N data in the sample is numerically equal to the negative log-likelihood, or detail length, of an unclassified model in which the class labels do not appear (apart from a small term probably about half the log of the sample size). It might therefore be concluded that there is no essential difference between the two models, and the Fisher Information relevant to the assertion of the estimates $\{(a_k, \theta_k) : k = 1, \ldots, K\}$ in an unclassified model would apply equally to the classified model of Section 6.8.2. However, this is not the case.

At first sight, treating the $\{k_n\}$ class labels as parameters would appear to make the already difficult calculation of F totally hopeless, by adding a further N to the dimension of the Fisher matrix. Indeed, a wholly satisfactory analysis remains to be done, but the following outline is suggestive.

The Fisher Information is used in MML as an indication of how sensitive the detail length is expected to be to the effects of "rounding-off" parameter estimates to values useable in the assertion. We will regard the MML "estimate" of a class label $\hat{k}_n$ in the two-class mixture of Section 6.8.6 as being represented by the surrogate value r'_n of Section 6.8.6, which gives the probability that $\hat{k}_n$ will be pseudo-randomly rounded to class 1 rather than class 0. The detail length expected after "round-off" of r'_n values to $\hat{k}_n$ estimates either 0 or 1 is

$$L = -\sum_{n=1}^{N} (r'_n \log f(x_n | \hat{\theta}_1) + (1 - r'_n) \log f(x_n | \hat{\theta}_0))$$

First, note that the mixture proportion parameter a does not appear in this expression. It now has the role of a hyper-parameter in the prior for class labels rather than as a direct parameter of the data model. Second, note that the second derivative of L with respect to θ_1 is given by

$$\frac{\partial^2 L}{\partial \theta_1^2} = -\sum_n r' \frac{\partial^2}{\partial \theta_1^2} \log f(x_n | \theta_1)$$

Substituting the MML estimate

$$r'_n = \frac{a f_1(x_n | \theta_1)}{a f_1(x_n | \theta_1) + (1 - a) f(x_n | \theta_0)}$$

the expectation is easily shown to be

$$\mathrm{E}\frac{\partial^2 L}{\partial \theta_1^2} = -aN \int f(x | \theta_1) \frac{\partial^2}{\partial \theta_1^2} \log f(x | \theta_1) \, dx$$

which is just the ordinary expression for the Fisher Information for parameter θ_1 and a sample size of aN. Similarly, $\mathrm{E}\dfrac{\partial^2 L}{\partial \theta_0^2}$ is the ordinary Fisher Information for θ_0 and a sample size of $(1 - a)N$. Also, $\dfrac{\partial^2 L}{\partial \theta_1 \partial \theta_0} = 0$. The effect of introducing the class-label parameters is to decouple the parameters of the two classes so cross-derivatives do not appear, and the diagonal Fisher matrix elements for θ_0 and θ_1 assume the simple forms which they would have were the class labels known.

This line of analysis may be continued to show that the optimum precision for the assertion of the proportion parameter a is also unaffected by interaction with other parameters.

There is a much more direct, and to my mind more persuasive, demonstration that the Fisher for a classified model with class labels is simply the product of the Fisher Informations of the class proportion parameters and of the class distribution parameters $\{\theta_k\}$ considered separately. It follows from remembering that what we are really about is calculating the length of an efficient explanation of the data. Recall the explanation format for a classified model as described in Section 6.8.2.

- It begins by asserting the number of classes K (if this is not known *a priori*).
- Next comes the assertion of the class proportions $\{a_k\}$. The receiver, on reading this, has no reason to expect these proportions to be other than the K state probabilities of an unknown Multinomial distribution, so encoding them as such is efficient.
- Next come the class labels of the N things, $\{k_n\}$. The receiver at this stage knows only the probabilities of each label, i.e., the proportions in each class, and so must expect the sequence of N labels to behave as a sequence of random draws from a K-state Multinomial distribution with probabilities as asserted. So again a simple sequence of coded labels, label k being coded with length $(-\log a_k)$, is efficient.
- Having received the class labels, the receiver now knows exactly which (and how many) things in the sample are asserted to belong to each class. He

knows that the members of class 1 will be encoded using a code efficient for the distribution $f(y|\hat{\theta}_1)$, and he knows how many members there are. He also knows that the distribution parameters for other classes will have no bearing on the detail code used for members of class 1. So there is no question as to how precisely $\hat{\theta}_1$ should be asserted: the precision is that dictated by the Fisher Information for the distribution $f(y|\hat{\theta}_1)$ with sample size equal to the number of members of class 1. The same is true for all classes: there is no interaction between the parameters of different classes. It follows that the assertion can continue by asserting $\hat{\theta}_k$ for each class in turn with no inefficiency, using for each class a Fisher Information depending only on that class.

 − In effect, once the assertion of the class labels $\{k_n\}$ has been decoded by the receiver, the descriptions of the class distributions become logically independent of one another.
 − The coding of the details for the N things are not mutually independent, since the receiver, using the data y_n decoded for thing n, uses the coding trick to recover some bits of the string encoding y_{n+1} which have been omitted from the detail.

We conclude that the message-length effects of overlapping class distributions are completely absorbed by the "coding trick" used to economize the detail part of the explanation, and have no effect on the length of the assertions of the class parameters.

6.8.9 Summary of the Classified Model

To summarize and generalize this approach to the classified mixture model with MML parameter estimates $\{a_k, \theta_k : k = 1, \ldots, K\}$, $\{k_n : n = 1, \ldots, N\}$:

(a)   The detail length is

$$L = -\sum_{n=1}^{N}\sum_{k=1}^{K} r_{kn} \log f(y_n|\theta_k)$$

where

$$r_{kn} = \frac{a_k f(y_n|\theta_k)}{\sum_{l=1}^{K} a_l f(y_n|\theta_l)}$$

(b)   Class label estimate k_n is chosen pseudo-randomly from the distribution

$$\Pr(k_n = l) = r_{ln}$$

(c)   The assertion length for k_n, after allowing for the coding efficiency introduced in Section 6.8.3, is

$$A_n = \sum_{k} r_{kn} \log \frac{r_{kn}}{a_k}$$

(d) Datum y_n contributes with weight r_{kn} to the statistics S_k used to estimate θ_k.

(e) θ_k should be chosen as the MML estimate of θ_k for the distribution $f(.|\theta_k)$ based on statistics S_k, "sample size" $N_k = \sum_n r_{kn}$, and some assumed prior $h(\theta_k)$. (We do not mean to imply that the priors for the parameters of the K components are necessarily independent, although this might commonly be the case.)

(f) The assertion length for θ_k may be calculated as

$$A_{\theta_k} = \frac{-\log h(\theta'_k)}{\sqrt{F_k(\theta_k)}} + \text{constant}$$

i.e., under the same conditions as for (e). That is, $F_k(\theta_k)$ is the determinant of expected second differentials

$$-N_k \int f(y|\theta_k) \left(\frac{\partial^2}{\partial \theta_k^2} \log f(y|\theta_k) \right) dy$$

(g) The class proportions (assuming a uniform prior) are estimated as

$$a_k = \frac{N_k + \frac{1}{2}}{N + K/2}$$

(h) The assertion length A_a for the class proportions is the usual form for a Multinomial distribution with sample size N, data $\{N_k : k = 1, \dots, K\}$

(i) If K is unknown, the assertion length for K is based on its assumed prior. For instance, if $h(K) = 2^{-K}$, $A_k = K \log 2$.

(j) The total message length is

$$I_1 = A_k + A_a + \sum_{k=1}^{K} A_{\theta_k} + \sum_{n=1}^{N} A_n + L + c_D + \frac{1}{2} \log N - \log K!$$

where the constant c_D is the constant from Section 6.1 for D real parameters; $D = K - 1 + K \times$ (Dimension of θ_k); the term $\frac{1}{2} \log N$ is explained in Section 6.8.4; and the $-\log K!$ term arises because the numbering of the classes is arbitrary.

(k) If K is unknown, choose K to minimize I_1.

(l) In computing I_1, use may be made of the identity

$$\left(\sum_{n=1}^{N} A_n \right) + L = \sum_{n=1}^{N} (A_n - \sum_{k=1}^{K} r_{kn} \log f(y_n|\theta_k))$$

$$= - \sum_{n=1}^{N} \log \sum_{k=1}^{K} a_k f(y_n|\theta_k)$$

6.8.10 Classified vs. Unclassified Models

Despite the logic of the preceding sections, the reader might well be skeptical of the results. Whether the class labels are treated as missing data (the unclassified model) or as parameters (the classified model with class labels asserted for the things) seems to have a more dramatic effect on the Fisher Information, and hence on estimates, than would be anticipated from the close correspondence between the main parts of the message length in the two cases. Some numerical experiments have been conducted to discover how different the two models turn out in practice. In one set of experiments, values were sampled from the univariate density

$$aN(x|-c, 1) + (1-a)N(x|c, 1)$$

with various sample sizes, proportions (a) and mean separation ($2c$). The data were then analysed to estimate the parameters, with a, the two means μ_1 and μ_2, and the standard derivations σ_1, σ_2 all being treated as unknown. For each data set, an "unclassified" model was fitted by numerical minimization of the message length, with the Fisher matrix approximated by the empirical form of Section 5.3.4. A classified model was also fitted as summarized in Section 6.8.9, using an obvious variant of the EM algorithm. The same priors were assumed in both models, namely, that the means have independent Uniform priors in some range of size $4R$, the standard deviations have independent Uniform priors in the range $(0, 2R)$, and the proportion parameter a has a Uniform prior in $(0, 1)$. The range scale R was set equal to the Standard Deviation of the true population distribution. Of course this is an unknown as far as the estimation problem is concerned, but we sought to mimic a situation in which prior information imposed fairly strong limits on the credible ranges of the means and Standard Deviations relative to the overall variability of the data.

We also fitted a "one class" model to the data set, viz., a Normal distribution of unknown mean and SD, and calculated its message length. The mean and SD were assumed to have the same priors as assured for the means and SDs of the two-class components. The reason for choosing fairly tight prior ranges for these was to ensure that two-class models would not be unduly prejudiced by a prior expectation that, if two classes existed, they would be very distinct ($|\mu_1 - \mu_2| \gg \sigma_1, \sigma_2$).

The results showed, as expected, that the explanation length of the classified 2-class model always exceeded that of the unclassified model. However, the difference was usually quite small, less than 2 nits, and the parameter estimates also insignificantly different. The unclassified model tended to estimate a slightly smaller separation between the means and slightly greater standard deviations. Again, this is to be expected because the Fisher Information of an unclassified model decreases as the two components become more similar, whereas that of a classified model does not. The only circumstances when the message lengths and estimates for the two models diverged significantly was when the one-class explanation length was the shortest. Whenever the explanation lengths showed a clear preference for the unclassified 2-class model over the one-class model, say, 3 nits or more corresponding roughly to a 5% "significance level", the classified model gave essentially identical results.

The close agreement between classified and unclassified approaches whenever the data justified a 2-class model was found whether the two component classes were distinguished by different means and/or by different standard deviations. Table 6.3 shows a few example results for $a = 0.5$, $\sigma_1 = \sigma_2 = 1.0$, $\mu_1 = -c$, $\mu_2 = c$ and sample size 100.

Table 6.3. Two-class classified and unclassified mixture models

True	Unclassified		Classified		One-Class
$2c$	$\hat{\mu}_2 - \hat{\mu}_1$	I_1	$\hat{\mu}_2 - \hat{\mu}_1$	I_1	I_1
2.2	0.01	194.8	2.51	199.0	192.4
2.6	2.70	207.3	2.81	210.0	204.3
2.8	2.66	206.1	2.96	208.4	207.2
3.0	2.88	203.6	2.94	205.8	204.3
3.2	3.07	206.1	3.12	207.7	208.0
3.6	3.89	224.2	3.92	225.5	228.3
4.0	3.92	215.6	3.95	215.8	226.9

With the exception of the one case ($2c = 3.0$) right on the borderline of significance, both classified and unclassified methods agree on where one or two classes are to be preferred. The only great difference in parameter estimates was in the case $2c = 2.2$, where the unclassified model collapsed the two classes together to give effectively only one class. In this case, neither two-class model was competitive with a one-class model.

Other experiments with other class distribution functions and more classes also lead to the conclusion that, in a mixture model with an unknown number of classes, it makes no practical difference whether a classified or unclassified MML model is used. As the classified model is far easier to solve, gives class labels, and is arguably sounder, it is the better choice.

There remains the possibility that for some data, either the number of classes K is known certainly *a priori*, or the probability model has no concept of a mixture of classes behind it, but just happens to have the form of the sum of a known number of parameterized density functions. In the first case, the divergence of the classified and unclassified models when classes overlap greatly is not "censored" by the alternative of a simpler model with fewer classes, and so may be of some concern. However, even if K is known certainly, it is reasonable still to entertain models with fewer than K classes and to prefer one of these if it gives a shorter explanation. The simpler model can be viewed, not as a hypothesis that only $K - 1$ classes exist, but rather as expressing the conclusion that two or more of the K classes are indistinguishable given the volume and nature of the data. To use an invidious example, it is beyond much doubt that humans come in two classes which are easily distinguished, male and female. But it would be a rash statistician who insisted on fitting a two-class model to the examination results of a mixed-gender group of students where the data did not identify the sex of each student.

In the second case, where the conceptual probability model is not a mixture, use of a classified model is inadmissible, and the full computational complexity of the unclassified MML model must be faced. The decrease in the Fisher Information as the components become more similar causes the MML parameter estimates to tend towards similarity, or even to converge as shown in the $2C = 2.2$ example in Table 6.3. This is not a "bias" or defect in the estimator: it simply reflects the fact that there are many formally distinct models with similar component parameters which cannot be expected to be distinguished by the data. If the prior gives about equal prior probability to all of these, the total prior attaching to an MML estimate with strongly overlapping components is much higher than that attaching to weakly overlapping models, and MML is indeed estimating the model of highest posterior given the data.

It may well be that in such cases where the population model is not a conceptual mixture of distinct classes, prior information will suggest a prior density on the component parameters $\{\theta_k : k = 1, \ldots, K\}$ which becomes very low if the parameters are closely similar, rather than the independent priors assuming in the above experiments. The tendency of the MML estimates to converge may then be weakened.

6.9 A "Latent Factor" Model

This section discusses a "factor analysis" model of a multivariate real-valued distribution. Like the Mixture problem, it illustrates the graceful MML handling of nuisance parameters, and also uses a modification to the Fisher matrix to resolve an indeterminism in the parameterization of the model.

The data are N independent observations from a K-dimensional distribution

$$x = \{\underline{x}_n : n = 1, \ldots, N\}; \quad \underline{x}_n = (x_{nk} : k = 1, \ldots, K)$$

The assumed model is

$$x_{nk} = \mu_k + v_n a_k + \sigma_k r_{nk}$$

where the variates $\{r_{nk} : k = 1, \ldots, K, n = 1, \ldots, N\}$ are all i.i.d. variates from $N(0, 1)$. In concept, each of the K variables measured is regarded as having a value for the Nth case determined by a mean, a dependence on an unobserved variable v having a value v_n for the case, and a specific variability with variance σ_k^2. Hence, the model models a multivariate Normal density with a special covariance structure.

We wish to estimate the unknown parameters $\{v_n : n = 1, \ldots, N\}$ (the "factor scores"), $\{\mu_k : k = 1, \ldots, K\}$ (the means), $\{\sigma_k^2 : k = 1, \ldots, K\}$ (the specific variances), and $\{a_k : k = 1, \ldots, K\}$ (the "factor loads"). Define for all n, k

$$
\begin{aligned}
w_{nk} &= x_{nk} - \mu_k \\
y_{nk} &= w_{nk}/\sigma_k, \quad \underline{y}_n = (y_{nk} : k = 1, \ldots, K) \\
b_k &= a_k/\sigma_k, \quad \underline{b} = (b_k : k = 1, \ldots, K) \\
b^2 &= \underline{b}^2 = \sum_k b_k^2, \quad v^2 = \underline{v}^2 = \sum_n v_n^2 \\
\mathbf{Y} &= \sum_n \underline{y}_n \underline{y}_n^T
\end{aligned}
$$

It is clear that the model as shown is indeterminate. The parameters $\underline{b}$, $\underline{v}$ enter the distribution model only via their Cartesian product, so the data can give no information as to their absolute sizes b^2 and v^2. It is conventional to assume that the factor scores $\{v_n\}$ are of order one in magnitude. We assume a prior on the scores such that each score is independently drawn from the Normal density $N(0, 1)$. For $\underline{b}$ we assume all directions in K-space to be equally likely, and a prior density in K-space proportional to

$$(1 + b^2)^{-(K+1)/2}$$

This prior on $\underline{b}$ is proper and expresses an expectation that in each dimension, a_k will be of the same order as σ_k but could be considerably larger. The prior on the mean vector μ is assumed Uniform over some finite region large compared with $\underline{a}$.

The negative log-likelihood is

$$L = \frac{1}{2} KN \log(2\pi) + N \sum_k \log \sigma_k + \frac{1}{2} \sum_n \sum_k (x_{nk} - \mu_k - v_n a_k)^2 / \sigma_k^2$$

If the Fisher matrix is evaluated in the usual way, its determinant is zero, because the log-likelihood is unchanged by the simultaneous substitution $\underline{a} \to \alpha \underline{a}$, $\underline{v} \to (1/\alpha)\underline{v}$ for any α. We resolve the singularity by noting that while the data provide only K values relevant to the estimation of each factor score v_n, the score is subject to a fairly tight $N(0, 1)$ prior. Following the Section 5.2.10 on curved priors, we add the second differential of $- \log h(v_n)$ (namely one) to the second differential of L with respect to v_n in the Fisher matrix. We could similarly take account of the curvature of the prior $h(\underline{b})$, but as in the usual factor problem $K \ll N$ and the curvature is small, it makes little difference and is omitted.

With this modification, the Fisher Information is found as

$$F(\underline{\mu}, \underline{\sigma}, \underline{a}, \underline{v}) = (2N)^K (Nv^2 - S^2)^K (1 + b^2)^{N-2} / \prod_k \sigma_k^6$$

where $S = \sum_n v_n$. The explanation length is then

$$I_1 = (N-1) \sum_k \log \sigma_k$$
$$+ \frac{1}{2} \left[K \log(Nv^2 - S^2) + (N + K - 1) \log(1 + b^2) \right]$$
$$+ \frac{1}{2} \left[v^2 + \sum_n \sum_k (x_{nk} - \mu_k - v_n a_k)^2 / \sigma_k^2 \right] + \text{constant terms}$$

Minimization of I_1 leads to the implicit equations for the parameters shown in Table 6.4 in the column headed MML. Note that all parameter symbols here stand for the estimates of these parameters. The table also gives corresponding equations for Maximum Likelihood (ML) and for a marginalized Maximum Likelihood model in which the nuisance vector score parameters are eliminated by integration over their $N(0,1)$ priors and hence not estimated (ML*).

Table 6.4. Defining equations for the Factor Model

Parameter	MML	ML	ML*
v_n	$(1 - 1/N)\underline{y} \cdot \underline{b}/C^2$	$\underline{y} \cdot \underline{b}/C^2$	not estimated
S	0	0	not calculated
$\underline{\mu}$	$\sum x_n / N$	$\sum x_n / N$	$\sum x_n / N$
$\underline{b}$	$\dfrac{1 - K/((N-1)b^2)}{(N-1)(1+b^2)} \mathbf{Y} \cdot \underline{b}$	$\dfrac{1}{Nb^2} \mathbf{Y} \cdot \underline{b}$	$\dfrac{1}{N(1+b^2)} \mathbf{Y} \cdot \underline{b}$
σ_k^2	$\dfrac{\sum_n w_{nk}^2}{(N-1)(1+b^2)}$	$\dfrac{\sum_n w_{nk}^2}{N(1+b^2)}$	$\dfrac{\sum_n w_{nk}^2}{N(1+b^2)}$

The form of the equation for $\underline{b}$ in all three estimators shows that it is an eigenvector (actually the one with the largest eigenvalue C^2) of the sigma-scaled covariance matrix $\mathbf{Y}/N$. The defining equations for the estimates can be readily solved by an iterative scheme.

The ML solution is unsatisfactory. The maximum likelihood is usually reached when $\underline{a}$ is parallel to some data axis and the corresponding σ_k approaches zero. It is thus inconsistent.

The ML* estimator is that usually described as the "Maximum Likelihood" estimator in the literature and derives from the work of Joreskog [21]. While consistent, it is badly biased for small factors, tending to overestimate

b^2, underestimate the specific variances $\{\sigma_k^2\}$, and to bias the direction of $\underline{b}$ towards a data axis. When tested on artificial data drawn from a population with no factor, it tended to give an estimated b^2 of order 1 with $\underline{b}$ nearly parallel to an axis. It does not directly give any estimate of the factor scores. Mardia et al. [27] suggest the estimator $v_n = y_n \cdot \underline{b}/(1 + b^2)$, but this does not maximize the likelihood for given ML estimates of the other parameters.

The MML estimator gave more satisfactory results, showing no sign of bias. On artificial data with a known factor, the estimation errors of MML were consistently smaller than those of ML*, as measured by $(\hat{a}-a)^2$, $(\hat{v}-v)^2$, or by the estimation of the specific variances. The MML defining equations do not have a solution unless the correlation matrix of the data shows a dominant eigenvalue greater than would be expected in a sample from an uncorrelated (zero-factor) population. A fuller derivation and test results are given in [56].

6.9.1 Multiple Latent Factors

The MML latent factor model has been extended to the discovery and estimation of multiple latent factors. The general framework, data, prior assumptions and notation follows the single-factor case, but for a J-factor model, we now assume that there are J true latent factors whose "load" vectors, when scaled by the specific variances of the observed variables, have, independently, Uniform priors over their directions, and, also independently, identical length priors similar to that assumed for the single load vector $\underline{b}$ in the single-factor model. We also assume that the factor scores associated with the different factors have uncorrelated unit Normal priors.

However, with the obvious data distribution model

$$
\begin{aligned}
x_{nk} &= \mu_k + \sum_{j=1}^{J}(v_{jn}a_{jk}) + \sigma_k r_{nk} \\
&= \mu_k + \sigma_k\Big(\sum_j(v_{jn}t_{jk}) + r_{nk}\Big)
\end{aligned}
$$

where $j = 1, \ldots, J$ indexes the J true scaled factors

$$
\{\underline{t}_j = (t_{jk} : k = 1, \ldots, K) : j = 1, \ldots, J\}
$$

the data covariance matrix is a sufficient statistic, and we cannot hope to distinguish among the infinitely many sets of J true factor load vectors which would give rise to the same population covariance. This indeterminacy is additional to the indeterminacy between the lengths of the load and score vectors. It can be resolved by accepting that the true latent load vectors cannot be estimated, and instead estimating a set of J load vectors

$$
\{\underline{b}_j = (b_{jk} : k = 1, \ldots, K) : j = 1, \ldots, J\}
$$

which are constrained to be mutually orthogonal. Similarly, one accepts that the J factor score vectors $\{\underline{v}_j\}$ be estimated as mutually orthogonal. It then follows from the prior independence assumed for the true latent vectors that the orthogonal latent vectors $\{\underline{b}_j\}$ do not have independent priors. Their directions (subject to orthogonality) remain Uniformly distributed, but the joint prior density of the lengths of the orthogonal $\underline{b}_j$ and $\underline{v}_j$ vectors now contains a factor

$$\prod_j \prod_{i<j} |v_j^2 b_j^2 - v_i^2 b_i^2|$$

showing that, with independent priors on the true factors, it is most unlikely that any two orthogonal factors will have similar "strengths" as measured by their load-score products $v_j^2 b_j^2$. The full form for the joint prior density of load and score vector lengths is (writing b_j for $|\underline{b}_j|$ etc):

$$h(\{b_j, v_j \text{ all } j\})$$

$$= G_{NKJ} \prod_{j=1}^{J} \left(T_K(b_j) H_N(v_j) \prod_{i<j} \frac{|v_j^2 b_j^2 - v_i^2 b_i^2|}{v_j b_j v_i b_i \sqrt{(1+b_j^2)(1+b_i^2)}} \right) 2^J \, j!$$

where

$$T_K(b) = \frac{2\Gamma(\frac{1}{2}(K+1))}{\sqrt{\pi}\Gamma(\frac{1}{2}K)} \frac{b^{K-1}}{(1+b^2)^{(K+1)/2}}$$

$$H_N(v) = \frac{N}{2^{N/2}\Gamma(1+N/2)} v^{N-1} e^{v^2/2}$$

the 2^J factor occurs because the sign of load vectors is immaterial, the $J!$ factor occurs because the ordering of the factors is arbitrary, and G_{NKJ} is a normalization constant.

When the Fisher Information is calculated for the orthogonal-factor parameterization (adding as in the single-factor case the log-prior curvature 1 to the second derivatives with respect to scores), it is found to contain the factor

$$\prod_j \prod_{i<j} (v_j^2 b_j^2 - v_i^2 b_i^2)^2$$

because if any two orthogonal factors have almost equal load-score products, the log-likelihood is almost unaltered by a rotation of the pair in the plane containing them. Hence, when the prior and Fisher are combined in the MML explanation length term

$$-\log\left(\frac{\text{prior}}{\sqrt{\text{Fisher}}}\right)$$

the interaction terms disappear, and the resulting defining equations for the J-factor model are little more complex than for the single-factor model.

Introducing variables R_j; Q_j for each factor, the MML estimates satisfy

$$\mu_k = (1/N) \sum_n x_{nk}$$

$$w_{nk} = x_{nk} - \mu_k$$

$$y_{nk} = w_{nk}/\sigma_k$$

$$\mathbf{Y} = \sum_n \underline{y}_n \underline{y}_n^T$$

$$\sigma_k = \frac{\sum_n w_{nk}^2}{N - 1 + \sum_j R_j b_{jk}^2}$$

$$Q_j = 1 + b_j^2 + (K - J + 1) / v_j^2$$

$$v_j^2 = \underline{b}_j^T \mathbf{Y} \underline{b}_j / Q_j^2$$

$$R_j = v_j^2 + \frac{N + K - 1}{1 + b_j^2} \quad \text{or} \quad N + J - 2$$

$$\underline{b}_j = \mathbf{Y} \underline{b}_j / (Q_j R_j)$$

The J factor load vectors are multiples of the J eigenvectors of $\mathbf{Y}/N$ with the largest eigenvalues. As in the single-factor case, a solution with J latent factors will exist only if all these J eigenvalues are larger than would be expected to arise with a smaller number of latent factors.

The message length, which is required to compare models with different numbers of latent factors, is given by

$$I_1(x) = (K/2 - 2J) \log 2 + K \log N + \frac{1}{2} JN \log(2\pi) - \log(J!)$$

$$- J \log S_N + \sum_{i=0}^{J-1} \log(S_{N-i} S_{K-i} - J \log(S_K/S_{K+1}$$

$$+ \frac{1}{2}(K - J + 1) \sum_j \log v_j^2 + \frac{1}{2}(N + K - 1) \sum_j \log(1 + b_j^2)$$

$$+ (N - 1) \sum_k \log \sigma_k + \frac{1}{2} \sum_j v_j^2 + \frac{1}{2} \sum_n \sum_k y_{nk}^2$$

$$+ \frac{1}{2} \sum_j v_j^2 b_j^2 - \sum_j \underline{b}_j^T \mathbf{Y} \underline{b}_j / Q_j - \frac{1}{2} D \log(2\pi)$$

$$+ \frac{1}{2} \log(D\pi) - \log G_{NKJ}$$

where $D = 2K + J(N + K - J + 1)$ is the number of parameters, and

$$S_m = \frac{m \pi^{m/2}}{\Gamma(m/2 + 1)} \quad \text{(The surface of a unit m-sphere)}$$

The finite ranges of the Uniform prior densities on μ_k and $\log \sigma_k$ have been omitted from the message length, as they affect all models equally.

The constant G_{NKJ} is the normalization constant for the prior density of the orthogonal factor load and factor score vectors $\{\underline{b}_j \, \underline{v}_j : j = 1, \ldots, J\}$ derived from the independent prior densities assumed for the true factor load and score vectors. An exact closed-form expression for G_{NKJ} is not known, but for $N > 40$, $\log G_{NKJ}$ depends little on N and is well approximated by

$$- \log G_{NKJ} = A_J + \frac{1}{2} J(J - 1) \log(K - B_J)$$

where the constants A_J, B_J are shown below for $J \leq 6$.

$J:$	1	2	3	4	5	6
$A_J:$	0	0.24	1.35	3.68	7.45	12.64
$B_J:$	$-$	0.58	0.64	0.73	0.86	1.21

Tests on artificial data sets having up to five latent factors have shown the MML estimator to be less biased and more accurate than the Maximum Likelihood (ML*) estimator. The latter does not directly choose the number of factors, so in the testing it was given the benefit of being given J.

When J is unknown, it may be estimated using the ML* estimator by comparing the log-likelihoods of models with different numbers of factors, and choosing the model with the highest "penalized" log-likelihood. Two well-known "penalty" functions were tried, Akaike Information Criterion (AIC) and the so-called Bayes Information Criterion (BIC). These penalties, which depend on the number of free parameters in each model, are subtracted from the log-likelihoods of the models to gibe the penalized log-likelihoods. For a model with D parameters fitted to a data sample of size N, they are

$$\text{AIC penalty} = D; \quad \text{BIC penalty} = \frac{1}{2} D \log N$$

Using the MML estimator, there are two ways in which J may be estimated. One may try models with various assumed J and choose the one of shortest explanation length, or one may choose the model with the largest assumed J for which an MML solution exists.

When AIC and BIC using ML* models were compared with MML estimates of J, it was found that AIC was much inferior to BIC, and generally MML was superior to both. For some tests with weak factors, BIC appeared to find the true J more frequently than MML, which did not find the weakest factor. However, when the true-J ML* model selected by BIC was inspected in these cases, its estimate of the weakest factor was found to have no clear relation to the true weakest factor, and the Kullback-Leibler distance of the ML* model from the true population was greater than that of the MML model even though the latter omitted the weakest factor. The two MML methods for choosing J generally agreed, but there was some indication that choosing the highest possible J permitting an MML solution was slightly the more reliable of the two.

7. Structural Models

In this chapter we describe some applications of MML to the inference of some simple discrete structures. Some use is made of the expressions developed in Chapter 6 for the message lengths for statistical distribution models, which appear as components in the inferred structural model, but the main emphasis is on the additional assertion components needed to describe the discrete structure.

7.1 Inference of a Regular Grammar

Here, the data comprises a sequence of strings, or "sentences", in a finite alphabet of $K + 1$ symbols. The hypothesis to be inferred from the strings is a regular grammar defining a language to which all the strings belong. We will use Roman letters (a, b, c, etc.) as the symbols of the alphabet, save for a distinguished symbol ("#") which is a punctuation mark occurring at the end of each sentence, and only there. The number K of ordinary symbols is known.

There are many formalisms in which a regular grammar may be specified, and an efficient assertion code could be defined for each of them. All are equivalent in the sense that any regular grammar may be described in any formalism, but the prior (or coding) probability implied for a particular grammar may vary considerably between one formalism and another. In this example we have chosen a formalism based on finite-state machines as it is relatively easy to work with, but we do not argue that it provides universally the most appropriate assertion code in all situations where a regular grammar might be inferred.

7.1.1 A Mealey Machine Representation

A regular grammar on a K-symbol alphabet may be represented by an abstract finite-state machine (FSM). An FSM has a set of S "states" numbered $1, 2, 3, \ldots, S$ and indexed by s. Each state may have from one to $(K + 1)$ "transition arcs" (or more simply, "arcs") which are directed arcs leaving the state and going to some state. Each arc is labelled by a symbol of the alphabet. No two arcs from the one state have the same label symbol. The destinations of arcs labelled with ordinary symbols are unrestricted: an arc may go to the state it leaves, and two or more arcs from one state may lead to the same state.

State 1 is distinguished, and called the "start state". Any arc labelled with # must go to state 1.

An FSM generates a sentence from the grammar it represents as follows. The machine starts in state 1. It then moves to some state by following one of the arcs leaving state 1. In so doing, it generates as the first symbol of the sentence the symbol labelling that arc. Having reached the destination state of the arc, the machine then follows some arc leaving that state, and in doing so generated the next symbol of the sentence. This process of moving from state to state by following arcs, and generating a symbol on each move, continues until an arc labelled # is followed. This arc necessarily generates the # symbol terminating the sentence, and leads to state 1, leaving the machine ready to start another sentence. Two simple examples of FSMs on the binary alphabet a, b are shown in Tables 7.1 and 7.2. In these figures, a next-state entry "−" for some symbol means the symbol cannot be produced from the current state.

Table 7.1. FSM M1, grammar S ::= b | aS

Current State	Next State for Symbol		
	a	b	#
1	1	2	−
2	−	−	1

Table 7.2. FSM M2, grammar S ::= aFb; F ::= Λ | abF

Current State	Next State for Symbol		
	a	b	#
1	2	−	−
2	4	3	−
3	−	−	1
4	−	5	−
5	6	3	−
6	−	5	−

The first machine defines the grammar whose sentences begin with any number of "a"s and end with b (or b# if you prefer). In Bakus-Naur Form (BNF):

S ::= b | aS

The second machine generates sentences which are either ab# or aab followed by any number of ab pairs followed by b#. In BNF:

$\qquad$ S ::= aFb

$\qquad$ F ::= Λ | abF

Note that the grammar generated by the second machine is also generated by the simpler machine M3 shown in Table 7.3. However, as we now describe, we will not necessarily regard M2 and M3, nor their grammars, as equivalent.

Table 7.3. FSM M3, grammar S ::= aFb; F ::= Λ | abF

Current State	Next State for Symbol		
	a	b	#
1	2	–	–
2	4	3	–
3	–	–	1
4	–	2	–

7.1.2 Probabilistic FSMs

In most contexts where a collection of symbol strings (sentences) in a finite alphabet might be "explained" by a hypothetical grammar, the collection will contain regularities not fully captured by a formal grammar. In particular, we usually find regularities expressible in terms of the frequency or probability of occurrence of certain structures. For example, the formal grammar for the "C" computer language places no restriction on the length (in symbols) of a mathematical expression, yet as the language is used in practice, most expressions contain less than 20 symbols, and expressions containing more than 1000 symbols are very rare. Similarly, most attempts to formalize a grammar for the English language admit sentences like:

"The man died."

"The man the dog bit died."

"The man the dog the boy kicked bit died."

"The man the dog the boy the teacher hit kicked bit died."

etc.

All but the first two constructions are in fact uncommon, and the fourth is arguably no longer English as she is spoke.

The formalism discussed here extends the simple FSM model to allow for inclusion of some probabilistic structure in the grammar. The extended model is called a probabilistic finite-state machine (PFSM). In a PFSM, each arc is labelled with a "transition probability" as well as a symbol. The probabilities on the arcs leaving a single state sum to one. The probability on an arc from state s_1 to state s_2 shows the probability that, if the machine is in state s_1, it will follow the arc to state s_2.

The grammar represented by a PFSM is called a probabilistic regular grammar. It defines not just a set of possible sentences but also a probability distribution on that set. That is, it defines a population of sentences.

When machines M2 and M3 above are given transition probabilities, they can no longer be regarded as equivalent. For instance, M2 can represent a grammar such that 70% of all generated sentences are "ab", 20% of all sentences are "aabb" but sentences beginning "aaba..." have an average length of 100 symbols. The simpler machine M3 cannot represent such a population of sentences.

7.1.3 An Assertion Code for PFSMs

We will first consider a code for the discrete structure of the FSM, and defer consideration of the assertion of transition probabilities.

(a)   The code starts with the number of states S, using some sensible prior. This part of the assertion is short, and its length would usually have little effect on the comparison of two competing hypothesized FSMs. We will ignore its length.

(b)   For each state, the code describes the arcs leaving the state as follows. First, the number of arcs leaving the state, say, a_s for state s, must be between 1 and $(K+1)$, so can be coded in $\log(K+1)$ nits.

(c)   Following a_s, the code specifies the symbols labelling the arcs. This set of symbols is some selection of a_s symbols from the alphabet of $(K+1)$ symbols and can be coded with $\log \binom{K+1}{a_s}$ nits.

(d)   For each of the a_s exiting arcs, in lexographic order, the code names the destination state. If none of the arcs is labelled with #, this takes $(a_s \log S)$ nits. If some exit arc is labelled #, the cost is $((a_s - 1) \log S)$, since an arc labelled # necessarily leads to state 1.

The above code can describe any FSM, but contains some inefficiencies. First, the numbering of all states other than state 1 is arbitrary, so the code permits $(S-1)!$ different, equal-length descriptions of the same FSM. This inefficiency is easily allowed for by subtracting $\log(S-1)!$ from the calculated assertion length, or, equivalently, determining some convention for a canonical numbering of the states. Second, the code permits the description of FSMs some of whose states cannot be reached from the start state. No such FSM could sensibly be asserted in an explanation, so the code is somewhat redundant. In practice, we have found the redundancy of this code to be small, and minimal-length explanations using it are not seriously affected. Finally, there is a small inefficiency arising from the fact that at least one arc must be labelled #, but this condition is not enforced by the code.

There is an important and essentially arbitrary choice in this coding scheme. Although the number of arcs leaving a state can have any value between 1 and (K+1), it is not obvious that all these values are equally likely *a priori*. Hence, although the number of exit arcs can be coded in $\log(K+1)$ nits, such a coding may not be optimal given prior domain knowledge. For instance, in computer languages the number of symbols which may follow a given string is usually quite restricted, so we might expect the average number of exit arcs from a state to be significantly less than $(K+1)/2$. We will assume henceforth all exit arc numbers to be equally likely, but this is just for purposes of illustration.

Using this coding scheme, the length of the description of machine M2 is as shown in Table 7.4 (remembering $K=2$).

To the total cost of $(12\log 3 + 7\log 6)$ shown in Table 7.4, we must add the cost of specifying the number of states (6) according to whatever prior has been assumed, e.g., add $(\log^* 6)$, and subtract an allowance for arbitrary state numbering $(\log(5!))$.

Table 7.4. Length of the description of machine M2.

State	a_s	Cost	Label(s)	Cost	Destn(s)	Cost
1	1	log 3	(a)	log 3	(2)	log 6
2	2	log 3	(a,b)	log 3	(4,3)	2 log 6
3	1	log 3	(#)	log 3	(1)	0
4	1	log 3	(b)	log 3	(5)	log 6
5	2	log 3	(a,b)	log 3	(6,3)	2 log 6
6	1	log 3	(b)	log 3	(5)	log 6

7.1.4 A Less Redundant FSM Code

A more complex coding scheme can remove most of the inefficiencies of the simple code. One possibility is the following coding scheme.

The transition table row for a state in a table such as 7.2 can be represented by an ordered list of $(K+1)$ state numbers. We first agree on a lexical ordering of the alphabet symbols, e.g., {a,b,#} and will present the transition table row entries for a state in this order. Next, we augment the possible state numbers which may appear in such a row by the integer "0". Here, "0" in the entry for some state and alphabet symbol will mean that emission of the symbol from this state is impossible, as was shown by a "−" entry in Table 7.2. Note that the entry for the symbol "#" can only be 0 or 1.

The coding process will use an "open list" of state numbers and a "highest state number" variable H. Initially, the open list contains the single number "1" and $H = 1$. Also, no state is assigned a state number except the start state, which is assigned the number "1". The code construction consists of repeated execution of the following step:

− If the open list is empty, exit.
− Remove the first state number from the open list. Let this be C.

– For each of the K alphabet symbols in lexical order, if the destination state of state C has not been assigned a number, add 1 to H, assign it the number H and place H on the end of the open list. In any case, output a code string for the destination state number, using a code length of $\log(H+1)$ since the number must be in the range $\{0 \ldots H\}$.

– For the symbol #, output a code for 0 if state C cannot emit "#", or 1 if it can (code length one bit).

The number of states is not explicitly encoded. It is determined by the emptying of the open list, whose operation is easily recovered by the receiver. This code improves on the original in two ways. First, it avoids the redundancy arising from the arbitrary numbering of states by imposing a "canonical" numbering scheme. Second, it cannot encode a machine containing states inaccessible from the start state. Certain other redundancies remain. It permits the coding of machines which never emit "#", or which can become trapped in a cycle of states. To illustrate the code, the code produced for machine M2 of Table 7.2 is, in numeric form with punctuation added for legibility:

$$2,0,0; \quad 3,4\ 0; \quad 0,5,0; \quad 0,0,1; \quad 6,4,0; \quad 0,5,0$$

The complications of a non-redundant code are illustrated here to show that devising a truly non-redundant code for discrete structures can be difficult, and the implications of the code for the implied prior can be quite obscure. For instance, the less-redundant code for FSMs outlined here implies some joint prior over the numbers of states and arcs, but this prior is not obvious from the structure of the code, and may be quite unreasonable (e.g., the expected number of states may be infinite).

7.1.5 Transparency and Redundancy

At least in the case of FMSs, an attempt to devise a code scheme less redundant than the simple code of Section 7.1.3 has led to a scheme in which the implied prior is not obvious, and depends in obscure ways on the number of alphabet symbols and the topography of the transition table. There is in our view a strong case for tolerating some degree of redundancy in the code schemes adopted for discrete structures if the redundancy allows a simple interpretation of the details of the scheme in terms of prior probability. The use of a hypothesis assertion code which is redundant in the sense of allowing the assertion of nonsense hypotheses and/or allowing more than one coding of the same hypothesis will have the effect of inflating the length of the assertion. If the amount of inflation is bounded by a constant factor independent of the hypothesis, the amount of data (in this case the number of sentences) required for MML to discover a complex hypothesis will be increased by about this factor. Thus, a small percentage increase in assertion length arising from redundant coding simply reduces slightly the efficiency of the MML estimation, but is not expected to have more serious effects. Serious distortion of the estimates will occur only if the inflation factor of assertion lengths increases with the complexity of the assertion. It has been shown [2] that MML inference remains consistent provided the inflation factor is less than two.

In Bayesian terms, replacing the "correct" prior distribution $h(\theta)$ by a sub-normalized approximation $k(\theta)$ will not result in inconsistency provided $h(\theta) \geq k(\theta) > h(\theta)^2$ for all θ.

In the present case, the simple code of Section 7.1.3 is not very redundant. The lengths it gives for the assertion of a FSM are less than 50% longer than a bound on the minimal lengths calculated by enumerating all acceptable FSMs of a given number of states, at least for up to 6 states. The inflation ratio appears to decrease with the size of the alphabet. The advantages of using the simple code are considerable, in that variants of it allow prior beliefs about the number of states and the average number of transition arcs per state to be incorporated in the coding in an obvious way. We have used the simple code in all experiments on the inference of regular grammars.

In the Algorithmic Complexity approach to induction, where the asserted hypothesis may be any computable function and is represented by a program for a Universal Turing Machine, redundancy of the assertion code is ineluctable. As there is no general method for deciding whether a program halts, there is no general method for identifying nonsense assertions. Similarly, there is no general method for determining whether two programs, and the hypotheses they assert, are equivalent. Although most applications of MML will deal with less general hypothesis spaces where questions of nonsense and equivalence are in principle decidable, the decision processes may often be so computationally difficult that the practical situation resembles that of Algorithmic Complexity: strictly non-redundant assertion codes may not be reasonably achievable.

This position may be philosophically objectionable, but it does not undermine the MML approach, nor is it in conflict with observation of the progress of scientific theory. If redundancy exists in our assertions even in an in-principle decidable domain, it simply means that our understanding of the domain does not, as yet, allow us to express that understanding as a properly normalized prior probability distribution. If we attempt a nonsense assertion, we can confidently expect the data to reveal our folly, since the assertion will not explain the data. If we fail to recognise the equivalence of two distinct assertions, we may require more data to reach confidence in the assertion we make, but should eventually realize the equivalence empirically because the data never seems to distinguish between the assertions.

Even when we are aware of the redundancy of our assertion language, there may be good reason to tolerate it. An explanation of the operation of a synchrotron may well use assertions framed partly in classical terms (when describing the generation of the accelerating and beam-focussing fields), relativistic terms (when describing the relationship of field strength to particle speed), and quantum-mechanical (in describing the formation of antiparticles). These assertions are partly redundant, in that the quantum electrodynamics of the particles implies their response to applied fields, and partly nonsense, in that the classically framed assertions rely on classical theory strictly incompatible with relativity. Yet we understand, accept and use such explanations routinely.

7.1.6 Coding Transitions

Having described the discrete structure of a hypothesized PFSM, the assertion should next encode the transition probability distribution over the transition arcs leaving each state. For each state, this distribution is a Multinomial distribution, and it seems natural to assume a uniform prior over all possible sets of probabilities. Thus, the analysis of Section 5.4.1 is applicable. However, there is a technical problem. In that section it is assumed that the sample size is known *a priori*, and the code used for asserting a probability distribution depends on the sample size. The relevant sample size for the coding of the transition probabilities out of some state of a FSM is the number of times the state is visited by the available data sentences. However, this number is not known *a priori*, and cannot be known to the receiver of an explanation until the detail encoding the sentences is decoded.

One possible way round this problem is to base the coding of the transition probabilities for a state on the *expected* number of visits to the state. This expected number is a function of the structure of the PFSM (already asserted) and the transition probability distributions of all of its states. In principle, one could devise a code encoding all the transition distributions simultaneously, with coding precisions based on the transition distributions themselves. In effect, the set of asserted distributions could be treated as being a single vector parameter of the model PFSM, and the Fisher Information expressed in terms of this vector. Unfortunately, the expected frequencies of visiting the various states, and hence the Fisher Information, are complex functions of the transition probabilities, making this approach computationally difficult.

A much simpler approach is to use an incremental code for the transitions out of each state (Section 2.1.10). In this approach, the assertion asserts only the discrete structure of the PFSM, and not the transition probabilities. In the detail, the code fragment encoding the next transition to be made in generating the data sentence is encoded with a coding probability $\frac{n_{sk}+1}{v_s+a_s}$ where s is the current state, a_s is the number of arcs leaving it, v_s is the number of times the state has already been left (including transitions directly to itself), k is the symbol labelling the arc to be followed, and n_{sk} is the number of times this arc has already been followed. The length of the code fragment encoding the next transition (and hence the next symbol) is of course the negative log of this coding probability. Use of this incremental code means that the total detail length needed to encode all transitions out of some state s is $\log \frac{(N_s + a_s)!}{\sum_k (N_{sk}!)}$, where N_s is the total number of transitions out of state s, and N_{sk} is the number of transitions generating symbol k.

Use of an incremental coding of transitions in the detail means that no estimates are actually asserted about the transition probabilities. This is probably not a matter of concern, as interest is usually focussed on the discrete structure. If it is desired to treat the probabilities as an essential feature of the inferred model, a small correction (from Section 5.2.13) can be added to the above expression, given by $\frac{1}{2} \log(\pi(a_s - 1)) - 0.4$ for each state.

Note that for any state having only one exit arc ($a_s = 1$), there is no message length required to encode the transition probability or the transitions.

7.1.7 An Example

Some of the earliest work on the inference of regular grammars was by Gaines [17], from whom the PFSA representation used here is adopted. He considered this set of sentences in the alphabet {a, b, c, #}:

```
c a a a b #
b b a a b #
c a a b #
b b a b #
c a b #
b b b#
c b #
```

He then found PFSA structures of one to six states, choosing for each number of states the PFSA with the highest likelihood (essentially the shortest detail length as given by an incremental code). The likelihood of course increased monotonically with increasing number of states, but after an informal consideration of the amount of increase given by each additional state, he suggested the best choice to be the 4-state machine in Table 7.5.

Table 7.5. 4-state machine.

Current State	Next State for Symbol			
	a	b	c	#
1	—	2	3	—
2	—	3	—	—
3	3	4	—	—
4	—	—	—	1

Using the assertion code of Section 7.1.3 and the incremental detail code, the MML process found just this machine, with an explanation length of 37.5

nits. The nearest competitor was the 3-state machine in Table 7.6 giving 39.8 nits. This machine combines states 2 and 4 of the best machine.

Table 7.6. 3-state machine.

Current State	Next State for Symbol			
	a	b	c	#
1	−	2	3	−
2	−	3	−	1
3	3	2	−	−

Numerous experiments with artificial data sets generated by PFSMs with up to 9 states have shown that the MML criterion reliably identifies the source grammar, given sufficient data. To avoid confusion between the merits of the MML criterion and the effects of the search method used to find the PFSM giving the shortest explanation, a search algorithm was used which was guaranteed to find the global minimum. It was therefore very slow and impractical for learning grammars of any complexity. The most complex data used in these experiments was a set of 42 sentences in the binary alphabet $\{0, 1, \#\}$. Writing each sentence as

$$x_0 y_0 z_0 x_1 y_1 z_1 \ldots x_n y_n z_n \#$$

the grammar can be described by the rules:

(a)   All sentences have lengths which are a multiples of 3.
(b)   If the symbols $x_n x_{n-1} \cdots x_2 x_1 x_0$ are regarded as the digits of a binary integer X, and similarly $y_n \cdots y_0$ and $z_n \cdots z_0$, then in each sentence, $Z = X + Y$ if overflow is neglected.

On this data set, the search algorithm found a 9-state machine which implements serial binary addition in the triplet representation defined by the grammar.

Much faster heuristic search algorithms to perform MML inference of regular grammars have been devised by Raman and Patrick [33] and used in several applications, e.g., learning the "rules" of historic phonetic changes in Chinese, and the temporal patterns of events in the play of Australian Rules Football.

7.2 Classification Trees and Nets

Consider data comprising N independent cases, where each case is characterized by values of K attributes, and by a classification into one of C classes. A Classification or Decision function is a function of the K attribute values which more or less accurately predicts the class to which a case belongs. If the attribute values of case $n(1 \leq n \leq N)$ are regarded as a vector

$$v_n = (v_{n1}, v_{n2}, \ldots, v_{nk}, \ldots, v_{nK})$$

then the function is a function of v_n with range the integers $\{1 \ldots C\}$.

Many forms of classification function have been studied. A Classification Tree is a classification function representable as a finite singly rooted tree. Each non-leaf node of the tree represents a test of one of the K attributes, and the branches from the node represent possible results of the test. Typically, a test of a discrete attribute having m possible values is represented by m distinct branches from the node, each corresponding to one possible value. A test of a real-valued attribute is typically represented by a binary branching, one branch for values $v_{nk} \geq t$ and the other for values $v_{nk} < t$, where t is some threshold value which may vary from node to node.

The leaves (leaf nodes) of the tree are called "categories". Each case can be assigned to a category by starting at the root node, applying the test defined for that node, and following the branch corresponding to the result of the test. If this branch leads to another non-leaf node, the test at that node is applied and another branch followed, and so on until a leaf or "category" is reached. In traditional decision trees, each category is labelled with a class, which is the class predicted for those cases assigned to the category.

Given N "training" cases for which the true class is known, the inference of a decision tree consists of constructing a tree which correctly predicts the class of many training cases, while not producing an over-elaborate tree. The inferred tree may then be used to predict the classes of new cases where the true class is unknown.

There is an extensive literature on decision trees. The most widely known and used algorithms are probably those developed by Quinlan [32]. Quinlan and Rivest [31] introduced an MML approach to learning decision trees. Their coding techniques had some technical deficiencies, and the treatment here follows the amendments suggested by Wallace and Patrick [58].

7.2.1 A Decision Tree Explanation

Following Quinlan and Rivest, the attribute values of the training cases are not regarded as subject to explanation. That is, the inferred hypothesis makes no assertion about why or how the training cases come to have their observed attribute values. Since this part of the data is not to be explained, it may be assumed to be transmitted to the receiver in any convenient code, or alternatively assumed to be already known to the receiver. Either way, the length of the representation of the training data attribute values will not enter into the explanation length. These assumptions are analogous to those made in regression problems (Section 6.7.1) which may also be regarded as requiring the construction of predictor functions.

The data to be encoded in the explanation are the true classes of the training cases, and the explanation is intended to be decodable by a receiver who already knows or has received full information about the number and nature of the attributes, the attribute values of all training cases, and the number of possible classes. The assertion part of the explanation describes a decision tree, but departs from traditional practice in that each leaf of the tree, rather than being labeled with the most probable class of cases assigned to the leaf, specifies a Multinomial probability distribution over classes. The detail part of the explanation encodes the actual class of each case in the training sample using a code based on the distribution found in the leaf to which the case is assigned.

On receipt of the assertion, a receiver can construct the decision tree. He can then decode the detail, as for each case, he can assign the case to a leaf using his knowledge of the cases' attributes. Then, knowing the asserted distribution for that leaf, he can decode the detail for that case to recover its class.

7.2.2 Coding the Tree Structure

A reasonably efficient way of encoding the tree structure will now be described. The assertion coding of the Multinomial class distributions for each leaf is deferred for the moment.

The tree structure proceeds node by node, beginning with the leaf node. The code string for a node can take one of two forms.

- If the node is a leaf, the code string is just a code for "leaf". This could be followed by a string specifying the leaf's class distribution, but this part will be left till after the structure is asserted.
- If the node is a non-leaf, the code string begins with a code for "non-leaf". Being a non-leaf, the node will represent a split of the cases reaching the node into two or more subsets, each of which will be directed along a different branch from the node to a different further node. The code string for the current node therefore continues with a specification of which attribute determines the split, and if the attribute is real-valued, the threshold t against which the attribute value is tested. Finally, the code string concludes with the code strings for each of these different "child" nodes. The order in which the children are listed follows an agreed ordering of the possible values of discrete attributes and of the two possible results of the threshold test of a real-valued attribute. These orderings are assumed to be known to the receiver.

The code for the structure of a tree is just the code string for its root node. It will be seen that the form of the code resembles the binary "tree code" of Section 2.1.14 with a modification to allow for multi-way branching when a discrete attribute with more than two possible values is tested, and the addition of the specification of which attribute to test at non-leaf nodes.

In the binary tree code, the leaf/non-leaf specification for a node was a single bit, implying a prior probability of one half for a split. This prior had the effect that the expected size of the tree was infinite, but the probability assigned to the set of all infinite trees was zero. To achieve the same effect in trees with different numbers of branches at different nodes (different "arity") we chose to code the "non-leaf" specification with a prior probability of one over the arity of the node's parent, e.g., $\Pr(\text{non-leaf}) = 1/3$ if the node is the child of a three-way split.

The code to specify the tested attribute at the root node (if it is not a leaf) must choose among all K attributes, giving a code length of $\log K$. However, at a lower non-leaf node, all discrete attributes which have been tested in the path from the root to this node will have the same, known, value in all cases reaching the node, and are therefore not candidates for testing at this node. The number of candidate attributes is thus less than K and a shorter code is needed to specify the one to test.

7.2.3 Coding the Class Distributions at the Leaves

As was the case for the inference of PFSMs (Section 7.1.6), it is awkward to encode the class distribution probabilities in the leaves as part of the assertion, as the receiver, before decoding the data, cannot know how many cases reach each leaf category and hence how precisely the class probabilities will be stated. Instead, we use an incremental code for stating the detail for each case, i.e., its class. The effect is that the explanation message does not explicitly assert the leaf class distribution probabilities, and the total detail length for the cases assigned to a category becomes the I_0 (non-explanatory) message length for the category's Multinomial distribution (Section 2.1.10). The total message length of the explanation may be corrected to reflect the I_1 message lengths for the category distributions and the details of their cases, the correction being, for each leaf, about $\frac{1}{2}\log(\pi(C-1)) - 0.4$ where C is the number of classes.

The length of the incremental detail coding of the classes of the cases reaching a leaf depends on the prior probability distribution assumed for the class probabilities. Because the decision tree is designed to have leaf categories highly informative about the class of cases assigned to the category, we do not a priori expect the class probabilities in a category to be similar, but hope and expect one class to be dominant. A Uniform prior on the class probabilities is not appropriate, and we assume a conjugate multi-state Beta prior over the class probabilities $\{p_c : c = 1, \ldots, C\}$ of the form (Section 5.4.1)

$$\Pr(\{p_c : c = 1, \ldots, C\}) = \frac{\Gamma(C\alpha)}{\Gamma(\alpha)^C} \left(\prod_{c=1}^{C} p_c\right)^{\alpha-1}$$

where $0 < \alpha < 1$. The lower the value of α, the more the prior expects the probabilities to be extreme. A value of 0.5 seems to give generally adequate results.

7.2.4 Decision Graphs and Other Elaborations

The performance of decision trees based on the MML criterion appears to be competitive with those using other methods, but strict comparison is difficult because the search space is so large that exhaustive search for a global minimum of the explanation length is infeasible. The same problem affects other principles for constructing decision trees, so a comparison is clouded by the unknown efficiency of the search heuristics employed.

An elaboration of the basic tree model does offer a demonstrable improvement with many classification problems. Instead of a simple tree of branching attribute tests, the model space is enriched by allowing branches to join. That is, a leaf or split node may receive cases from more than one branch. The classification function is therefore represented by a single-rooted acyclic directed graph, and cases may reach a leaf node or category by more than one route. A tree is still a possible model, as it is just a graph which happens to have no joining branches.

Apart from (perhaps) having joining branches, a Decision Graph is otherwise identical to a Decision Tree. At each non-leaf ("split") node, one attribute is tested and cases directed along one of several arcs or branches leaving the node according to the result of the test. Each leaf node ("category") is described by a Multinomial class probability distribution. The only difference is that in a graph, two or more branches, from the same or different split nodes, may have the same destination node. A quite minor change to the assertion part of the explanation allows acyclic graphs to be encoded, and no change need be made to the detail part which encodes the classes of cases using a separate incremental code for each category. A possible assertion code for the structure of a decision graph is described below.

The assertion of the structure of a decision graph commences with the description of a decision tree structure essentially as described in Section 7.2.2. However, the resulting "leaf" nodes are not yet determined to be categories. This assertion code lists the split and leaf nodes of the tree structure in a depth-first order, and implicitly allows the nodes to be numbered in the order in which they appear in the coded structure. The ordering is such that the descendants of a node appear after the node, i.e., if we consider the arcs to be directed from a split node to its children, all arcs go from lower-numbered to higher-numbered nodes.

The assertion of a graph structure is completed by a further code string for each leaf of the tree, listed on descending node number order. The string for a leaf labels it as either a category (i.e., a terminal node of the graph) or a dummy. If it is a dummy, the string continues with the node number of some non-dummy (split or category) node with a higher node number, and the string is interpreted as meaning that the single branch reaching the dummy node should in fact lead to the named non-dummy. Note that the descending order in which these strings are listed means that the receiver, on receiving a "dummy" label code, already knows which of the higher-numbered nodes are dummies. Note also that the redirection of the branch leading to a dummy towards a higher-numbered non-dummy cannot create a cycle in the graph, since all ancestors of the dummy have lower node numbers than the dummy.

In the structure code for a simple tree, the assertion of the attribute to test at a split node could usually be shortened to less than $\log K$ because discrete attributes tested en route to the split node would not be candidates for testing. In the graph structure, during the assertion of the initial tree, this economy is not available, since the possibility of redirection of additional branches to the split node, which are not known to the receiver at the time he must decode the identity of the test attribute, means that the receiver cannot assume that all cases reaching the node will have the attribute values implied by the single route to the node which is apparent in the tree structure. All K attributes must be considered possible candidates for testing, so the assertion of the one to test will have a length of $-\log K$.

In coding the category/dummy labels, an incremental code for these labels can be used, based on a Uniform prior probability distribution. However, a simple incremental coding is unlikely to be ideal, since low-numbered leaves would seem *a priori* to be more likely to be dummies than high-numbered leaves, having more possible redirection destinations.

The above code for the assertion of a graph structure is fairly efficient, but is awkward to employ during the search for a good graph, which usually proceeds by incremental elaboration of the graph by splitting and joining. The effects of an incremental change to the graph on the form and length of its structure assertion are not simple, so recalculation of the explanation length following a change is a little slow. More convenient but more complicated codes have been described by Oliver [28] and Lee [24].

The Advantages of a Graph over a Tree. A decision graph can give a shorter explanation message than a decision tree whenever the tree structure required to represent the classification function contains identical subtrees. This situation may arise just at the category nodes: the tree for a two-class classification function commonly has several category nodes in which the first class is dominant and several in which the second is dominant. If the class probability distributions in two or more categories with the same dominant class are similar, a graph structure allows them to be replaced by a single category, saving the cost of asserting some class probability distributions. (This saving appears in the detail of the explanation if incremental codes are used for the classes of cases in each category.)

A more substantial improvement is possible when a Boolean expression for probable membership of a class is most briefly expressed as a disjunction of two or more conjunctions involving different attributes. The well-known artificial data set "XD6" has two classes and ten binary attributes labelled "a" to "j". A case probably belongs to class 1 if its attributes satisfy

$$a.b.c + d.e.f + g.h.i = 1 \quad \text{where } + \text{ means OR}$$

A tree for this classification function needs 49 split nodes and 40 categories. A graph needs only 9 split nodes and two categories. While this is an extreme case, it is not uncommon for disjunction to appear in the best "rule" for a class in real data sets.

A graph model does not extend the set of possible classification functions beyond what is possible with tree models. However, when the best explanation of the data within this set involves disjunction, the graph algorithm will need less data to find the best function, and becomes only slightly less effective than a tree algorithm in finding the best function when the best function is efficiently represented by a tree.

Regression in Categories. The set of possible classification functions may be extended by allowing a regression model to be asserted for each category. It is possible that some class-relevant information remains in the attribute values of the cases assigned to a category, but this information cannot be readily represented by further splitting of the category. For example, it could be that for some category in a two-class tree or graph model, the sum of three real-valued attributes has predictive value. A subtree refining the category would at least have to have three test nodes, one per attribute, and could only give a rough approximation to a test of their sum. The assertion code for tree or graph models could easily be modified to provide the option, at each category, of asserting a regression predictor function using $(C-1)$ linear functions of some selection of attributes to define a probability distribution over the classes of the form

$$\Pr(c) = \frac{\exp\left(f_c(x)\right)}{\displaystyle\sum_{j=1}^{C} \exp\left(f_j(x)\right)} \quad (c = 1, \ldots, C)$$

where $f_1(x) = 1$ and for $c > 1$, $f_c(x)$ is a linear function of the vector x of selected attributes.

Given sufficient data, regression predictors of this form might well be able to give shorter detail lengths for the classes of cases in the category, sufficiently shorter to outweigh the additional assertion cost of specifying the attribute selection and linear function coefficients.

Of course, the search for the shortest explanation of a data string would become much slower, but it would then be possible for the search to discover, for some data strings, that a single regression predictor gave a better explanation than any tree or graph.

7.3 A Binary Sequence Segmentation Problem

The inference problem addressed in this section is wholly artificial, and unlikely to be encountered in real life, at least in the form considered here. It is included because it introduces in a conceptually simple form a class of models of real importance, and also because it illustrates how a naive attempt to apply MML principles can give poor results.

The problem was introduced in a paper by Kearns *et al.* [22] which attempted to compare the performance of three model selection criteria on a problem which was not trivial but which permitted exact optimization of each criterion. This was a worthwhile enterprise, since many model selection problems present such mathematical difficulty that comparison of different model selection criteria is clouded by doubt whether each criterion had actually been optimized by whatever search or optimization method is employed. The three criteria compared were Cross-Validation (CV), Structural Risk Minimization (SRM) and Minimum Description Length (MDL). Of these, MDL is, for the problem considered, essentially identical to MML.

The problem as stated supposes the data to be a set of N pairs $\{x_n, y_n : n = 1, \ldots, N\}$ where x_n is a real value in $(0 \ldots 1)$ and y_n is binary, either 0 or 1. Let the pairs be indexed in order of increasing x-value. The y-values then form a binary sequence. The model for the data is that the interval $[0, 1]$ is divided into $K + 1$ sub-intervals by K "cut points" $\{c_k : k = 1, \ldots, K\}$ indexed in order of increasing value, so that the kth subinterval ends at c_k and the $(K + 1)$th subinterval ends at 1, and that if x_n lies in the kth subinterval, $\Pr(y_n = 1) = p$ if k is odd, but $(1 - p)$ if k is even. The inference problem addressed in [22] is, given the data, but no knowledge of p of K, to estimate the number of cuts K, and incidentally, the cut points $\{c_k\}$ and the probability p. The results presented concerned the success of the three criteria in estimating K, and their success in estimating the cut points and probability was not explicitly reported.

Their experiments were all based on applying the criteria to artificial data sets all generated from basically the same "true" model, which had $K = 99$, generating 100 subintervals, and evenly spaced cut points $\{c_k = k/100 : k = 1, \ldots, 99\}$. Each experiment used 100 data sets randomly generated with a specified sample size N and probability p. For each of several values of p, a series of experiments were done with N increasing from low values to a few thousand. The chosen values of p were all less than 0.5, so that y values were likely to be zero in odd subintervals and one in even subintervals. As the problem posed is essentially unchanged if the "true" p is replaced by $(1 - p)$, this restriction does not affect the validity of the experiments. The average estimated K was found for each experiment using each criterion. For each data set in an experiment, a dynamic-programming algorithm was used to find, for every assumed K, the cut-point set which minimized the number of "errors" in the y values, where an "error" is a zero in an even subinterval or a one in an odd subinterval. This choice of cut points for a given K is a maximum-likelihood estimate of the cut points. For the cross-validation criterion, a randomly chosen 10% of the N data were withheld for validation use, and the maximum-likelihood cut points chosen using the remaining 90%.

For each data set, the maximum-likelihood models with differing numbers of cuts were compared according to each criterion. For the CV criterion, the number of cuts giving the fewest errors in the validation data was chosen. (Since the dynamic programming algorithm can localize a cut only to within the interval between two data points, the mid-point of this interval is taken as the cut point.) For the SRM criterion, the number of cuts giving the lowest probabilistic bound on the generalization error was chosen. For the MDL criterion, the number of cuts giving the shortest message encoding the data was chosen, where the message length computation method is described below.

7.3.1 The Kearns *et al.* "MDL" Criterion

The form for the message length used in [22] is:

$$
\begin{aligned}
KMDL \;=\; & -N\left[R_K \log R_K + (1 - R_K) \log(1 - R_K)\right] \\
& -N\left[R_E \log R_E + (1 - R_E) \log(1 - R_E)\right]
\end{aligned}
$$

where E is the number of "errors" in the maximum-likelihood model with K cuts, $R_K = K/N$ and $R_E = E/N$. It is intended as an approximation to

$$
\log \binom{N}{K} + \log \binom{N}{E}
$$

which would give the length of a message specifying the location of the cut points (to within adjacent data points) and which y-values are "in error". The form adopted neglects the need to specify K and E to the receiver, and the "entropy" approximation to the combinatorial form is in error by the order of $\log N$.

The experimental results reported showed reasonable behaviour and performance using either the CV or SRM criteria. For instance, for $p = 0.1$, as the sample size N was increased starting from 100, the average estimate of K gradually rose to about 100 at $N = 500$, remained a little above 100 until $N = 1500$, then stabilized at 99 ± 1. However the "MDL" criterion choosing K to minimize $KMDL$ showed rather bizarre behaviour. For sample sizes up to about 1200, it usually chose a model with sufficient cuts to reduce the "error" count to zero, with the average estimated K rising to almost 300 at $N = 1200$. As N was increased further, the estimate dropped sharply, but did not stabilize at 99 ± 1 until $N = 1600$.

Similar behaviours were found with the higher error probability $p = 0.2$, but of course at larger sample sizes. The CV criterion gave estimates in the range 100 to about 130 for $500 < N < 2300$, stabilizing at about 100 for $N > 2400$, but the *KMDL* criterion gave estimates rising to about 670 at $N = 2000$ and did not stabilize near the true value until $N > 2600$. With still higher "noise" $(p = 0.3)$, CV's estimate hovered above 100 until $N > 2800$, but *KMDL*'s was about 1300 and still rising at $N = 2900$!.

The reader of this paper might well conclude that MDL (or MML) was a quite poor method of selecting among competing models, but in fact its purported implementation of MDL is no such thing.

7.3.2 Correcting the Message Length

Viswanathan *et al.* [49] have presented an analysis of the flaws in the "MDL" criterion used in the above work, and comparative tests of it and a more correct MML criterion. This section is adapted from their work.

Correcting the obvious flaws in *KMDL* to include the message length costs of stating the number of cuts K and the number of errors E, and to use the exact combinatorial forms for the cost of locating the cuts and errors, resulted in a slight improvement in behaviour, but the corrected criterion still tended strongly to choose "models" with many cuts and no errors up to quite large sample sizes, especially with high "noise" levels.

The crucial flaw in the *KMDL* criterion is that it uses a method of encoding the data which simply does not minimize the message length. It is not an MDL or MML criterion. The flaw is that it assumes the message encodes the locations of the cut points to within the interval between adjacent data points. The reader by now should be aware that minimizing the length of an "explanation" encoding of data requires that the parameters of the asserted model be stated to an optimally chosen precision. In the present problem, it is easily seen that the optimum precision for stating the position of cuts is not necessarily to within adjacent data points. Except in data generated with very low noise (p close to zero or one) a lower precision will give a shorter message.

For example, suppose we choose to allow the asserted model to have cut points only preceding odd-numbered data points. Then there are only $N/2$ rather than N possible cut points, and we save roughly K bits in asserting a model with K cuts. Of course, restricting possible cut points may increase the number of "errors". For each cut, with probability $1/2$ it will be in its maximum-likelihood position, resulting in no additional error, and with probability $1/2$ it will be displaced by one data point, so the number of errors will increase by one. Hence, (assuming $p < 0.5$) the detail part of the message, which locates the errors, will be on average longer by $(K/2) \log_2((1-p)/p)$ bits. Overall, the total message length is expected to change by about

$$(K/2) \log_2((1-p)/p) - 1 \quad \text{bits}$$

which is negative for $p > 0.2$. With a noise level above $p = 0.2$, the lower precision of cut-point location makes the message shorter.

To extend this argument, we suppose the cut points to be specified, not by which data point it precedes, but by its position in the real interval $[0, 1]$ in which the x-values of the data set lie, and that the position of each cut is specified to a precision $\pm(\delta/2)$. So constraining the positions at which cuts may be asserted means that the asserted position of a cut may differ from its ideal position by up to $\delta/2$. The expected number of data points lying between the ideal and asserted position is $N\delta/4$. The expected increase in detail length, per cut, is this expected number of data points times the expected increase per point due to encoding it with the "incorrect" probability. This gives (for $p < 0.5$)

$$
\begin{aligned}
\text{E(increase in detail)} &= (KN\delta/4)\left[(1-p)\varepsilon - p\varepsilon\right] \\
&= (KN\delta/4)(1-2p)\varepsilon
\end{aligned}
$$

where $\varepsilon = \log((1-p)/p) = \log(1/p - 1)$ assuming $p < 0.5$. The cost of specifying the positions of K cut points to precision δ is $K \log(1/\delta) - \log(K!)$ since the order in which they are stated is immaterial.

Differentiating the sum of the excess detail cost and the cut-point specification cost shows the optimum δ to be

$$
\delta = \frac{4}{N(1-2p)\log(1/p - 1)}
$$

With this choice of δ the expected increase in detail length is just one nit per cut.

The effective precision of $1/N$ used in $KMDL$ is justified only for very low noise rates with $p < 0.02$.

The MML estimate of p must now be based on the expected total number of errors, which exceeds the error count E of the maximum-likelihood model found by dynamic programming by the extra "errors" arising from the imprecise cut-point locations. Since this expected number depends on δ, which depends on the estimated p, a few iterations of the equations are needed for convergence. The resulting estimate of the error rate p exceeds the maximum-likelihood estimate (E/N), and roughly corrects for the overfitting of the maximum-likelihood model, which led the $KMDL$ criterion to underestimate the number of errors.

This derivation of an MML criterion is still imperfect, as it assumes that the number of data points within $\pm(\delta/2)$ of an ideal cut position is $N\delta$. This is the average value which might be expected by a receiver who is ignorant of the data x-values, but these are in fact known to the receiver, so the precision for a cut-point location should take into account the actual data point locations in the neighbourhood of the ideal cut position.

7.3.3 Results Using the MML Criterion

When the Kearns et al. experiments were replicated using the MML criterion, it showed no bizarre behaviour and a performance comparable to CV's. At high noise ($p = 0.3$) it required a rather larger sample size to reach an estimate of K around 90, but whereas CV's estimate rose to about 150 and did not return to near 99 until $N \approx 3000$, the MML estimate rose smoothly to near 99 by $N = 2500$ and then stayed there. At lower noise rates of 0.1 and 0.2, there was little to choose between the criteria.

These replicated experiments used the single true model of 99 equally spaced cut points used by Kearns *et al.* This model is highly atypical of the model class proposed by them, and assumed in all criteria, in which the true cut points are supposedly randomly distributed in $[0, 1]$. One consequence is that in their experiments, and in the replicated trial of MML, the numbers of data points in each subinterval were far nearer equality than would be expected in typical models of the proposed class. The later work [49] also conducted experiments in which the data were generated using a different set of randomly chosen cut points in each replication. The results showed MML to be generally superior to CV at all noise levels, being perhaps a little conservative in estimating K, but giving models which diverged less than CV's from the true pattern of $\Pr(y_n = 1 \mid x_n)$.

7.3.4 An SMML Approximation to the Sequence Problem

Section 4.10.2 introduced an approximation to the SMML explanation length for some data x, which finds a region in hypothesis space such that the log-likelihood of all models in the region is not less than the prior-weighted average in the region. The assertion length is taken as the negative log of the total prior probability in the region, and the detail length as the prior-weighted average negative log-likelihood.

Following the work of [22] and [49], I decided to apply this approximation to the binary sequence problem. As the presented problem was primarily concerned with estimating the cut number K, the model region was required to contain only models of a specified K, with the models in the region having all the same prior probability

$$\text{Prior}(K) = (1/N)\frac{1}{\binom{N}{K}}$$

but different cut positions. The dynamic programming algorithm was modified to yield, for each K, not only the minimum number of "errors" E, but also the number of different K-cut models giving exactly $(E + m)$ errors for $m = 0, 1, \ldots$. The "detail" length of a model with $(E + m)$ errors was taken as the "I_0" form, which does not actually involve an assertion of the error probability p, and is hence smaller than the correct "I_1" form by a small constant about 0.17 nit.

$$Detail = \log(N) + \log\binom{N}{E + m}$$

Given this information for any K, it is easy to construct the collection of K-cut models required by the SMML approximation. The message length for an SMML K-cut model is then given by an assertion length the negative log of the total prior probability of the collection, plus a detail length equal to the average detail length in the collection. The value of K minimizing the total length was then chosen.

The results obtained were similar to those given by the MML approximation in [49], but were somewhat more accurate, suggesting that the explanation-length approximation of Section 4.10.2 is fairly good. Although testing was not as extensive, a couple of points which the previous work had not made clear emerged.

- With a sample size barely large enough to reveal the data subintervals for a given noise rate, there can be a wide range of K estimates all giving explanation lengths within a few nits of the minimum. That is, the estimation of K is quite difficult.
- The number of different K-cut models in the collection can be very large.
- The occasional data set generated with evenly spaced cuts yielded a model with many cuts and zero errors, but when this occurred, the explanation length found with this model was almost always longer than N bits. As we consider an "explanation" acceptable only if it is shorter than the original data string, these cases represented a failure to find any acceptable explanation with a precisely asserted number of cuts.

7.4 Learning Causal Nets

A "Causal Net" is a representation of known or inferred causal relations among a number of variables. It is conventionally represented in part by a directed acyclic graph (DAG) in which each variable is shown as a node, and an arc from one node (the parent) to another (the child) implies that the parent variable has a direct causal effect on the value of the child. A node may have several parents, all of which affect its value, and a node may have many children, all of which are affected by it. Since the graph is acyclic, no variable can have a direct or indirect effect on itself. The acyclic condition further implies that some variables have no parents. They are "autonomous". Similarly, some nodes have no children.

The graph only partially represents the known or inferred network of causal effects. Each node is annotated with a function whose arguments are the values of its parents and whose value is a probability distribution over the possible values of the node variable. That is, the annotation for a variable V is a conditional probability distribution for V, conditioned by the parents of V (if any).

Different flavours of causal net can be defined, depending on the nature of the variables concerned (e.g., real or discrete) and the form or forms assumed for the conditional probability distributions. For instance, if all variables are real-valued, the causal net model may assume linear influences of parents on children and Gaussian unexplained variation. If we have a set of K real variables, and N independent instances of data values for the set, i.e., if we have data $\{x_{nk} : n = 1, \ldots, N, k = 1, \ldots, K\}$, such a causal model of the data source specifies, for each variable v_k, a possibly empty "parent set" D_k of variables and a conditional probability distribution

$$x_{nk} = \sum_{v_j \in D_k} a_{kj} x_{nj} + r_{nk}$$

where the coefficients $\{a_{kj}, v_j \in D_k\}$ give the linear effects of the parents on v_k, and the residual r_{nk} is assumed to be a random variate from the Gaussian density $N(0, \sigma_k^2)$. (For simplicity, we assume that the data values have been standardized to zero mean.) Given K, N and the data values $\{x_{nk}\}$, we address the problem of inferring a causal model of the above form. This requires inference of the structure of the DAG, the parameters of each probability distribution, and possibly a choice of the form of each distribution.

Causal Nets have been widely used for prediction and diagnosis, and are common components of Expert Systems. In early applications, the structure of the Causal Net model was usually elicited from human domain experts. The coefficients of the causal relations could be also elicited, or could be estimated from known data sets. However, the difficulty of eliciting consistent and reliable models from "experts", and the desire to apply causal modelling in domains where there is no acknowledged expertise, has led to interest in the automatic inference of causal nets from data. For the simple linear case described above, the most widely known work is due to Spirtes, Glymour and their collaborators, and has led to a commercially available program TETRAD II, which served as a benchmark for the MML method described here.

7.4.1 The Model Space

The model used to explain the data is intended to be a representation of the real-world situation giving rise to the data, insofar as that situation can be discovered from the data and prior knowledge. In the real world, the situation of a set of variables which may be causally related is characterized by (at least):

- A temporal order among the variables. The implication of this order is that early variables may affect later ones, but not vice versa.
- The existence, mathematical form and strengths of the direct effects of one variable on another. The mathematical form may be fixed *a priori* or be assumed to be one of a usually very limited set of forms.
- The magnitude of endogenous or inexplicable variation of each variable.

The full situation may involve effects due to variables outside the set of variables observed, whose values and (perhaps) existence are unknown. We will ignore the possible existence of such hidden variables, while admitting that a more complete treatment should include them.

Any prior knowledge which is available may constrain the model. We may have partial or probabilistic knowledge of the temporal order, the existence or impossibility of direct effects between certain pairs of variables, etc., and our method allows the use of some of these forms of prior knowledge. In the absence of useful prior knowledge, we will assume all temporal orders to be equally likely *a priori*, and assume that the existence of a direct effect between any pair of variables has a fixed prior probability P_a independent of the identity and temporal rank of the variables. Of course, if the effect exists, its direction is dictated by the temporal order. We may also assume priors over the choice of form of probability distribution (if there is a choice) and for the parameters of these distributions.

To summarize, a hypothesized real-world situation among the K variables is specified by an ordering of the K variables (prior probability $1/K!$), a set of direct effect arcs (prior probability $P_a^{N_a}(1-P_a)^{M_a-N_a}$ where N_a is the number of effect arcs asserted to exist, and $M_a = K(K-1)/2$ is the maximum possible number of arcs), and a set of K conditional probability distributions, each giving the distribution of one variable conditioned on its parents.

7.4.2 The Message Format

A naive format for the explanation message first describes what we shall call a "Totally ordered model" or TOM. It begins by specifying an ordering of the variables, and then which of the M_a possible arcs are present. The direction of each effect is implied by the order. Rather than specifying next all the conditional probability distribution forms and parameters, it then deals with each variable in turn in the stated order, with a section of code for each variable. Let j index the variables in the chosen order, so v_1 is autonomous.

The message section S_j for variable v_j comprises two parts, in effect a mini-explanation of the data values $\{v_{nj} : n = 1, \ldots, N\}$. The first part states the form of the conditional distribution for v_j (if there is a choice), and then the parameters of the distribution. The second part encodes all N values of v_j according to the asserted distribution. At the point where S_j is to be decoded, the receiver will already know the identity and values of the parents of v_j, and so can properly compute the conditional probability distribution for each value of v_j. Message section S_j has exactly the form and structure of a simple regression model explanation message asserting the regression of v_j on its parents, such as described in Section 6.7.1 for linear regression, and encoding the values of v_j according to this model. For a reason to be shown later, the parameters of the regression model are estimates by the Maximum Likelihood rather than by minimizing the total message length of S_j, but for the simple regression models used, the difference is insignificant. Both the total length I_k of S_j and the maximized likelihood L_k of the asserted regression model are computed.

The re-arrangement of the TOM explanation to intersperse assertions and details does not violate MML principles, and has the useful effect of making the precision of parameter estimation depend on the actual, rather than expected, distributions of parent values. The explanation length thus follows the more accurate Formula I1A of Section 5.1 rather than Formula I1B.

7.4.3 Equivalence Sets

MML considers as "equivalent" models which cannot be expected to be distinguished on the basis of the available data, the code it in concept uses to assert a model allows for the assertion only of members of a subset Θ^* of models chosen so that no two members are equivalent in this sense. The coding probability of an assertable model is taken as the sum of the prior (or coding) probabilities of all those models equivalent to it. Part of the aggregation of equivalent models has already been incorporated in the calculation of the length of a TOM model, in that the parameters of the regression models are specified only to a precision based on the relevant Fisher Information.

Other equivalences among causal net models also exist and lead to further aggregation of the naive TOM models into equivalence sets each represented by a single assertable model. The joint probability of a TOM and the data is given by

$$
\begin{aligned}
&- \log \Pr(\text{TOM,Data}) \\
=\ & \log(K!) - N_a \log P_a - (M_a - N_a) \log(1 - P_a) + \sum_k L_k \\
=\ & -\log(K!) - M_a \log(1 - P_a) + \sum_k C_k = I_T
\end{aligned}
$$

where $C_k = L_k - d_k \log(P_a/(1 - P_a))$ and d_k is the number of parents of v_k. The latter form is convenient for rapidly computing the change in I_T consequent on the addition or removal of an arc. When two or more TOMs are aggregated into a single model, not necessarily a TOM, their single replacement acquires the sum of their joint probabilities with the data. The various forms of aggregation performed are now described.

7.4.4 Insignificant Effects

Suppose two TOMs A and B differ only in that B contains an effect arc not present in A. If the strength of this effect is sufficiently small, the explanation length may be reduced by grouping A and B together, and using A as the representative model. This shortens the assertion, because the prior probability of the aggregate is the sum of the priors of A and B. The data may have a shorter detail length if encoded using model B rather than model A, but if the extra effect is small, the increase in detail length resulting from using model A instead of B may well be less than the reduction in assertion length, resulting in a reduction in C_k for the affected variable if B is deleted from Θ^*. We therefore consider any TOM with such a weak effect to be grouped with the representative TOM obtained by deleting the small effect arc. A TOM without weak effects will be called "clean". The details of the test to discover "weak" effects depends on the form of the regression model. For linear regression a satisfactory test requires only a comparison of the effect coefficient and the "unexplained" variance.

It may be thought that this is an unimportant refinement of the MML process. TOM B must have an assertion length exceeding that of A by an amount of order $\frac{1}{2}\log N$, because B's assertion is lengthened by the inclusion of an estimate for the coefficient of the "weak" effect. Hence, the coding probability of B will be less than that of A by a factor of order $1/\sqrt{N}$. However, for any A there will typically be about $K^2/4$ other TOMs differing from A only by the addition of an insignificant effect. The total coding probability (and hence posterior probability) of all such TOMs may therefore be about $(K^2/(4\sqrt{N})$ times that of A. Unless N is greater than K^4 there may be more coding probability on these TOMs than on A itself, so inclusion of them in the set represented by A adds significantly to the posterior probability of the aggregate.

7.4.5 Partial Order Equivalence

A set of TOMs which differ only in the order of the variables, and have the same direction for all arcs, are clearly indistinguishable given the data, and so will be grouped. All TOMs in the group are linear extensions of the DAG representing the directed effect arcs. All will have the same regression parameter estimates. Their aggregate is the DAG model.

7.4.6 Structural Equivalence

Two DAGs with the same skeleton, i.e., with the same effect arcs but some differences in arc directions and coefficients, can still imply exactly the same probability distribution over the set of possible data values. The strongest form of such equivalence is sometimes called "structural equivalence". If the two DAGs are structurally equivalent, then whatever regression coefficients are specified in the first DAG, there is a set of coefficients for the second DAG which will produce the same data distribution. The conditions on the DAGs required for structural equivalence have been described by Chickering [10]. In particular, if two DAGs are structurally equivalent, their maximum likelihoods on any data set will be equal. Hence, the data cannot distinguish between them, and MML will group the DAGs together. A rigorous test for structural equivalence between two DAGs can be done in polynomial time but is not quick. We adopt the heuristic of grouping together DAGs which have the same skeleton and the same maximum likelihood on the given data. This test will certainly aggregate all structurally equivalent DAGs, but may (rarely) include in the set a DAG which is not structurally equivalent to the other members of the set. Two DAGs with the same skeleton may have identical maximum likelihoods on the given data even if they are not structurally equivalent, but such coincidences require special and unlikely relations among the data, and the possibility may be ignored. The necessary relations have probability measure zero under our assumed priors.

7.4.7 Explanation Length

MML groups together "equivalent" TOMs, representing each group by a representative DAG with prior probability equal to the sum of the priors of all TOMs in the group. The assertion of the inferred representative is therefore coded with an assertion length of minus the log of this total prior probability. The detail length is the detail length given by the representative DAG. The MML process ensures that the detail lengths for all TOMs in the aggregate are approximately equal, so the choice of representative is not crucial. Hence, we may take the total explanation length of a causal net explanation asserting a DAG R which is the aggregate of all TOMs T in a set T_R as given by

$$I_1 = -\log\left[\sum_{T \in T_R} \Pr(T, \text{ Data})\right] = -\log\left[\sum_{T \in T_R} \exp(-I_T)\right]$$

The exact form of the code used to assert a hypothesized representative TOM would presumably be quite complex, since the length of the code for the TOM must depend not just on the topology and parameters of the TOM, but also on those of all TOMs which it represents. Fortunately, the precise nature of this code need not concern us, since we need to know only the length of the assertion, not its actual binary digits.

7.4.8 Finding Good Models

Given a data set, we would ideally like to find the DAG giving the shortest explanation. Even for modest K, the number of possible TOMs (or more correctly, representative TOMs in the MML code scheme) is so large that exhaustive search is infeasible. Further, computing the explanation length for some DAG is not simple. The detail length is easy to compute, but to compute the prior would require that every TOM structurally equivalent to the given DAG be enumerated, and that every TOM derivable from these by addition of "weak" arcs be enumerated and its prior computed. Even for $K = 10$, the number of TOMs to be enumerated in order to determine the MML equivalence set of the given DAG could run into hundreds. Instead, we have developed an algorithm based on a Monte Carlo sampling from the Bayes posterior distribution over ungrouped, individual TOMs. To be precise, the grouping over similar regression parameters is done, so the joint probability of a TOM T and data is as given by I_T as defined in Section 7.4.3, but no grouping according to order, structural, or insignificant-effect equivalence is done.

To perform the sampling, we use a version of the Metropolis algorithm for sampling from a discrete distribution. The current TOM is represented by a vector $V = (v_j : j = 1, \ldots, K)$ holding the variable indices in the current total order, together with a $K \times K$ incidence matrix of bits showing which effect arcs are present in the model. The ordering of V is such that each arc goes from a variable earlier in V to a variable later in V. In addition, the program maintains a K-vector of the current C_k values of the variables, a step count and the current value of I_1. After initializing these structures to some initial TOM, the algorithm enters a sampling phase.

Sampling from the posterior over TOMs is done by a process which steps from TOM to TOM in such a way that the number of visits to a TOM is proportional to its posterior probability. Three types of step are used:

- S1: (Temporal order change) Pick an integer j uniformly in $(2 \ldots K)$. Examine the variables v_j and v_{j-1}. If there is no arc between them, swap the two vertices in V. This swap merely moves to an order-equivalent TOM and does not change I_1. If an arc exists (necessarily from v_{j-1} to v_j, attempt to swap the vertices and reverse the arc. To make the attempt, the C values of both variables must be recomputed, since one will lose a parent and the other gain one. The new value of I_1 can then be computed from the new and old C values.
- S2: (Skeletal change) Pick two distinct integers j and k uniformly in $(1 \ldots K)$. Suppose $j < k$. If an arc exists from v_j to v_k, attempt to remove it. Otherwise, attempt to add such an arc. In either case, the value of C_k must be recomputed, and the resulting change in I_1 found from the new and old values.
- S3: (Double skeletal change) Pick three distinct integers i, j, and k uniformly in $(1 \ldots K)$. Suppose $i < k$, $j < k$. If v_i has an arc to v_k, attempt to remove it, and if it does not, attempt to add such an arc. Similarly, and simultaneously, attempt to remove or add an arc from v_j to v_k. To calculate the change in log joint probability I_1, only C_k need be recomputed. Type S3 steps are strictly speaking unnecessary, since the full set of TOMs can be explored using only types S1 and S2. However, it is included in the hope of accelerating transitions between TOMs which differ in one parent of some variable, where the two alternative parents are correlated. It is

also possible that adding two parents to a variable reduces its C_k, whereas adding either alone does not.

When a step is attempted, it is accepted if the change in I_1 is negative, or if

$$\exp(I_1(\text{old}) - I_1(\text{new})) > U$$

where U is a pseudo-random value drawn uniformly from the range $(0, 1)$. If the step is accepted, the changed TOM becomes current, and I_1, C, V, and the incidence matrix are updated. After any step, accepted or rejected, the step count is incremented.

This Monte Carlo process meets sufficient conditions for the Metropolis algorithm to apply, viz., the number of possible transitions from each TOM is constant, and every transition is reversible. The process will therefore visit every TOM with a frequency proportional to its joint probability with the data, and hence proportional to its posterior.

We are of course not directly interested in the posterior probabilities of individual TOMs, but rather in the posterior probabilities of sets of TOMs which will be aggregated together and represented by a single DAG in an MML explanation code. The total posterior given by MML to the representative of such a set is the sum of the posteriors of the member TOMs. We therefore do not count visits to TOMs, but visits to MML models. To enable visits to be counted in this way without explicit enumeration of all members of a set, we characterize a set of TOMs, i.e., an MML model, by the skeleton and likelihood of one of its clean members.

Whenever the current TOM is changed by a step, a "clean" version of the TOM is constructed by deleting all insignificant effect arcs, and the skeleton and maximized likelihood of the clean TOM found. Recall that all structurally equivalent TOMs share these properties. We then attempt to count visits to (skeleton-likelihood) pairs. In a problem with many variables, far too many such pairs may be visited to make exact visit counting easy. Instead, we form a hash of the skeleton and the log likelihood and use it to index a 65,536-entry vector of visit counts. The program uses a $K \times K$ symmetric matrix of 32-bit constant integers initialized with pseudo-random values, one for each undirected pair of variables. (The diagonal entries are unused.) A "skeleton signature" is formed for the current TOM as the sum modulo 2^{32} of the entries corresponding to the arcs of the TOM, and a hash index formed as the sum modulo 2^{16} of the upper and lower halves of the skeleton signature and an integer representing 100 times the current log likelihood. Thus, cleaned TOMs with different skeletons or different maximized likelihoods are unlikely to yield the same hash indices. The visit count indexed by the current hash index is incremented after each sampling step. The "cleaned" TOM is used only to generate the hash code and does not replace the current TOM from which it is derived. The latter remains the current TOM for the Monte Carlo process.

At the end of the sampling phase (which currently makes $200K^3$ steps) the final visit count in a cell of the hash vector, divided by the number of steps, estimates the total posterior probability of all MML models hashing to that cell. If posterior probability is concentrated in a few "good" MML models, the highest cell count will give an over-estimate of the posterior of the "best" model, as it may also contain some counts from visits to other models. To minimize the chance of serious over-estimation, we actually maintain two different signatures and use them to increment two different visit count vectors. The posterior probability of a model is then estimated by the smaller of the two final visit counts to which it hashes.

During sampling, the program accumulates a list of up to 50 MML models, being the ones of highest estimated posterior. No model with a posterior less than 1/200 of the highest posterior is retained. Each retained MML model is represented by its highest-posterior DAG. The DAG posterior is estimated using a further two visit count vectors for which the hash indices are calculated from "DAG signatures". These signatures are formed like the set signatures, but using a different $K \times K$ matrix of pseudo-random constants. This matrix is not symmetrical, so the DAG signature captures effect directions as well as the skeleton. The DAG hash indices do not involve the likelihood. The DAG visit count vectors are updated after each sampling step.

The retained models are further cleaned using a more accurate form of the test for "weak" effects. Models which become the same after cleaning are merged. Finally, a very general test for MML equivalence is applied. If two models A and B have prior probabilities P_A and P_B, and the Kullback-Leibler distance of B from A is D, then the expected change in message length if A is merged with B is

$$\frac{P_A N D + P_A \log P_A + P_B \log P_B - (P_A + P_B) \log(P_A + P_B)}{P_A + P_B}$$

If this expression is negative, model A is merged with model B. To apply the test, unnormalized model prior probabilities are estimated from

$$P_A = \frac{\text{Posterior}(A|\text{Data})}{\text{Pr}(\text{Data}|A)} \text{Pr}(\text{Data})$$

The unknown value $\text{Pr}(\text{Data})$ cancels out in the test expression. The selected models are then displayed.

The program described above gives rather more useful information about possible models of the data than does a simple greedy search. It gives several (up to 50) distinct models, and estimates of their posterior probabilities. These should help judgements of the confidence to be placed in the hypothesized models. During the sampling phase, the program notes and records the DAG of highest posterior, which may be helpful if no MML model was found with a convincingly high posterior. This model can also be displayed. Also during the sampling phase, the program accumulates the frequency with which each possible directed arc appears in the current (unsimplified) TOM. These frequencies provide estimates of the marginal posterior probabilities of each arc, and so can be used for judgement of the plausibility (given the data) of a postulated causal relation between a pair of variables, without a commitment to any particular model for the entire net.

7.4.9 Prior Constraints

Knowledge of the data domain may imply prior constraints on the hypothesized temporal order of some variables. For instance, weight-at-birth cannot be later in the model's partial order than income-in-present-job for the same person. Such a constraint can be incorporated in the sampling by forbidding any step which would place the former variable after the latter in the total order. The program allows any self-consistent set of such constraints to be imposed. It also allows the inclusion of specific arcs to be forbidden or required. Softer prior knowledge, in the form of different prior probabilities for the presence of each possible directed arc, could also be incorporated without much trouble.

7.4.10 Test Results

Three major versions of this MML algorithm for learning causal nets have been implemented. The first deals with real-valued variables and uses a simple linear regression model for the dependence of a variable on its parents. The second deals with discrete variables and models the dependence of a variable by a set of Multinomial probability distributions, one for each combination of parent values which occurs in the data. The third extends the second by allowing both the unrestricted dependence model of the second version and a "logit" dependence model, which has fewer parameters when there are many possible combinations of parent values. The "logit" model in effect assumes that the parent variables do not interact in their effects on the child. This version chooses either the unrestricted or the logit model for each non-autonomous variable independently, on the basis of which form gives the shorter message length.

Various tests of the first version are reported in [57] and compared with results obtained by TETRAD II. In general, the MML method was the more accurate in recovering the DAG structures from which the artificial data were generated, making fewer errors about the presence and directions of arcs. The tests involved data with up to 30 variables.

An interesting test case is an artificial data set with 27 variables. A causal net model was constructed where the variables may be thought of as indexed by a triple (i, j, k), with each of i, j, k taking values 0, 1, or 2. The net contained all possible arcs of the forms $(i, j, k) \rightarrow (i+1, j, k)$, $(i, j, k) \rightarrow (i, j+1, k)$ and $(i, j, k) \rightarrow (i, j, k+1)$, giving 54 direct effect arcs in all. Most effects were given coefficients of 0.4, but 11 had -0.5. A random sample of 1000 cases was generated. No prior precedence information was used. TETRAD II was unable to find any causal link in this data using default run settings. Perhaps the fact that it relies on finding significant patterns in very high-order partial correlations is the culprit, as orders of over 20 would be involved in this problem, making accurate estimation difficult.

According to the MML algorithm with P_a automatically estimated, the best model was the true model. The posterior of this model was estimated to be 97%. No other model was retained. The data was re-run with a crippled version of the MML algorithm, which followed exactly the same Monte Carlo steps, but did not account for "small effect" equivalence. The best model it found had an estimated posterior of only 0.1% and 27 false arcs. This result shows that almost all of the posterior of the best MML model was contributed by "unclean" equivalents. In fact all the estimated 97% came from unclean equivalents: it was found that the Monte Carlo process had never visited the best model.

8. The Feathers on the Arrow of Time

The fundamental laws of Physics appear to be essentially time reversible at the microscopic level: anything that can happen can equally well happen backwards. The "laws" applying at the everyday macroscopic level seem quite different, describing irreversible changes inevitably leading to a decrease in order. In particular, the Second Law of Thermodynamics says that the "entropy" of a closed system (a measure of its disorder) never decreases, and (almost) all changes to the macroscopic state of the system with the passage of time lead to an increase of entropy. Thus, the macroscopic laws define the so-called "Thermodynamic Arrow of Time", which points unambiguously from a more ordered past to a less ordered future. The emergence of this directed Arrow at the macro level from fundamental micro laws which are symmetric with respect to time has been well discussed. It has been shown to follow naturally from the micro laws and the way "order" is defined in our descriptions of the world we see. It is not the paradox which it first appears. I will briefly rehearse these arguments, but add little to them: they are no longer matters of controversy. However, there remains an aspect of the Arrow which does not appear to have been much discussed, and which seems worth some exploration.

If we observe a closed system at present, and find it to be partially ordered, there are indeed excellent reasons to suppose that some little time in the future it will be less ordered than at present. It has also been well noted that these same reasons would also lead us to suppose that, at some little time in the past, it was also less ordered than at present. That is, the standard (and valid) arguments showing that entropy, or disorder, of the closed system will almost certainly increase in the future also show that its entropy will almost certainly increase as we peer further into its past. Thus, while our arguments yield us the point of the Arrow pointing to a less ordered future, they also imply a point at the other end of the Arrow pointing to a less ordered past. However, as a matter of common sense, if we see a closed system to be partially ordered at present, we will usually infer that it was more ordered in the past, consistently with the Second Law of Thermodynamics which tells us that the system's orderliness should have decreased from the past time to the present.

For instance, if at present we open a well insulated box and find within it a metal bar, one end of which is hotter than the other, we feel entitled to predict that if we quickly shut the box, in ten minutes' time the two ends will be closer in temperature (a less ordered state) than they are now. But we also feel entitled to infer that ten minutes ago, the hot end was even hotter, and the cold end colder, than we now observe. This latter, commonsense, inference of a more ordered past is the kind we invariably make in discussing the physical history of closed systems. We assume that the Arrow does not have a point at both ends, but rather has a point at the Future end and feathers at the Past end. The point points to higher entropy in the future, the feathers trail from lower entropy in the past.

The substantive question I wish to address is why the Arrow has feathers at its rear rather than another point? In other words, why do we accept the arguments that entropy will increase in the future, but reject their equal implication that entropy decreased in the past? Why do we use and accept a certain, valid, logical apparatus in reasoning about the future destiny of a system, yet reject it in favour of something else in reasoning about the past history of the same system? And if we do use some other reasoning apparatus for the past, what is it and what validity does it have? I will attempt to show that we do indeed reason about future and past in different ways. In essence, I will conclude that our reasoning about the future, when rational, is a deductive process using the standard logic of statistics, but that our reasoning about the past is usually an inductive process not deducible from the standard logic of statistics. Note that I am not criticising this inductive process, nor the "commonsense" equivalent we informally and habitually employ. Rather, I will attempt to show that inductive reasoning about the past is well motivated, and, given certain assumptions about the real world, emerges from Minimum Message Length principles.

Most previous discussion of the asymmetry of the Arrow have treated the problem in different terms. Here, I draw heavily on a recent work by Huw Price [30].

Stefan Boltzmann "proved" that in an ideal gas, whatever its current distribution of particle velocities, a collision between particles would be expected to change the velocity distribution towards one of higher entropy. He noted that the proof did not rely on the direction of time, and hence that an entropy increase is to be expected when going from the time after the collision back to the time before it, just as much as in going from before the collision to after it. There is a technically valid objection to this argument. Boltzmann assumed that the velocities of his colliding particles were uncorrelated before the collision. But after the collision, they will inevitably be correlated. When retracing the history back in time, the final velocities become (when negated) the "initial" velocities of the reversed collision, and since they are correlated, Boltzmann's assumption is violated, and we cannot conclude from his "proof" that as the history is retraced into the past, entropy will increase into the past.

Actually, the objection to Boltzmann's proof applies in both directions. If a body of gas has a past, there will have been ample opportunity for its molecules to interact, both directly and via chains of collisions, so if two molecules collide at the present instant, even if they have never met before, their initial velocities cannot validly be assumed to be uncorrelated (or more precisely, statistically independent). Thus, the "proof" fails to prove rigorously that entropy will probably increase into the future. But in fact the conclusions of Boltzmann's argument can be shown to hold by more robust arguments, which again imply probable entropy increase into both the future and the past.

Price goes on to say that, even if Boltzmann's statistical argument is correct, it is overridden by the fact that we *know* the past to have been more ordered, and this acts as a "boundary condition" enforcing an unbroken entropy increase from past to present and beyond. For Price, therefore, the only question is why the past was so ordered. But (as Boltzmann already knew) we cannot rely on any memory or record of the past, of the body of gas or anything else, to deduce the probability, let alone certainty, of a low entropy past. Only *after* we have established some reason to accept a low entropy past can we consider a low entropy past as a boundary condition on system behaviour, or more weakly, a prior on the past favouring low entropy.

So, rather than accepting a more ordered past as a given, I have attempted to start from what I must take as given, namely my *present* senses and mental state, and try to elucidate why Price, I, and most others conclude that the past was more ordered than now.

8.1 Closed Systems and Their States

To simplify discussion, the question of what we can deduce about the future and past of some part of the world will start with the treatment of "closed systems". By a closed system I mean a physical, real world collection of matter and energy which is totally isolated from any outside influence. No matter or energy may escape from the system, nor may any new matter or energy enter it. While no known totally isolated system exists (except maybe the Universe), the concept of a closed system is a standard tool of Physics, as for practical purposes many real systems of interest are sufficiently isolated to be treated as being closed. For this discussion, I make the simplifying assumptions that the closed system is of fixed, finite spatial extent. In effect, our closed systems will be some matter and energy confined within a fixed, impenetrable box. The amounts of matter and energy are assumed to be known. (In one example, energy will be allowed to enter or leave the system, but discussion of this is deferred till that example. Until then, assume the energy of the system cannot change.) The "state" of a system at some instant, e.g., the present, is a collection of numbers which together completely specify all of the information about the system which is relevant to its future behaviour. For instance, in the case of a closed system comprising a box containing some inert gas, the present state would comprise specification of the position and velocity of every atom of the gas. As time goes by, the state of the system will change, the changes being governed by some "laws of physics".

A "view" of a system at some instant is a usually incomplete and/or imprecise specification of its state at that instant. For our box of gas, a view might specify every atom's velocity exactly, but its position only with an accuracy of one millimetre. A view could be much broader: it might only specify the temperature of the gas, and the observation that the gas appeared to be in equilibrium.

The "state space" of the system is the set of all states which can be specified, subject to the given confinement and the amounts of matter and energy. For simplicity, I will assume that these givens, or *invariants*, are known exactly (and of course do not change).

The "possible states" of a system is the set of all those states into which any present state consistent with the invariants and the present view could evolve at any time in the future, or could have obtained at any time in the past. Clearly, given a present view of the system, any state rationally asserted to have obtained in the past, or to obtain in the future, must lie in the set of possible states.

I make the further simplifying assumption that the state space of the system is discrete. That is, no numbers with infinitely many decimal places are needed for an exact specification of a state. This assumption has some justification in quantum mechanics, where the state space of a confined system is indeed a discrete set.

Finally, I assume that time changes in discrete, equal steps. Thus, the history of the system during some period is fully specified by the sequence of states it reaches at the discrete instants within the period. The discrete instants are indexed with the integers, and unless otherwise noted, the present instant is $t = 0$. This discretization of time is of course unrealistic, but I hope its assumption is not critical to the argument. With these assumptions, the "Laws of Physics" obeyed by the system are described by a mapping of the state space into itself.

8.2 Reversible Laws

I assume the Laws of Physics to obey the conservation of energy, momentum and angular momentum and, like the real Laws as we know them, to be essentially reversible. Let $s(t)$ denote the state of the system at time t. If the system is deterministic, the Laws form a one-to-one mapping $M()$ where $s(t + 1) = M(s)$. Time reversibility requires that if q and r are states, and $r = M(q)$, then $q' = M(r')$ where s' signifies a state identical to s, save that all velocities (and any other state variables which are odd derivatives with respect to time) are negated. Then the laws of physics appear exactly the same whether going forwards or backwards in time. I assume time reversibility.

If the system is non-deterministic, the mapping is a probabilistic mapping described by the function $M(r|q)$ where r and q are states and $M(r|q)$ is the probability that $s(t+1) = r$, given that $s(t) = q$. In this case, we require for reversibility of the laws that $M(r|q) = M(q'|r')$. Here and elsewhere, all probabilities and conditional probabilities are implicitly further conditioned on the known invariants of the system, that is, those properties of the system which are known and constant, e.g., the amount and type of matter, the spatial confinement and the total energy.

8.3 Entropy as a Measure of Disorder

In general, the calculation of the entropy of a closed system given its invariants and some view of its macroscopic situation can be complicated. In the special case, typified by a confined ideal gas, of a system of N identical tiny particles the state of each being fully determined by just a position and a velocity, a simple expression applies. Let z denote a (discrete) position-velocity pair which a particle may have, and Z the set of all possible such particle states. I use z to emphasize that a particle state is not to be confused with a system state. In this simple case, the system state s of the gas is the N-tuple of the particle states of all its N particles.

The known invariants and view may suffice to allow us to estimate the fraction of particles in each particle state z, or equivalently, the probability $\Pr(z)$ that a certain arbitrarily chosen particle has particle-state z. Then the entropy, here denoted by the symbol H, is classically defined by the equation

$$H = - \sum_{z \in Z} [\Pr(z) \log \Pr(z)]$$

The probability distribution $\Pr(z)$ is conditioned only on the known invariants and view. H measures the expected amount of information needed fully to specify the instantaneous state of every particle, and hence the system state, given only the invariants and whatever information is contained in the view. (Note that the particle-state distribution is constrained by the requirement that the sum of the particle energies equal the known invariant total system energy.)

Informally, a system is regarded as being in *equilibrium* if its present view is expected to persist forever with no significant change (assuming no external intervention). More formally, the system is in equilibrium if its entropy, as calculated from the present view, is close to the maximum value it could obtain if constrained only by the invariants. That is, the information in the present view does not imply an entropy lower than the maximum possible. As the probability distribution $\Pr(z)$ which maximizes H subject to the invariants is unique, the system appears to be in equilibrium if the present view is consistent with the maximum entropy particle-state probability distribution.

If the present view implies that $\Pr(z)$ differs from the maximum entropy distribution, the system is not in equilibrium, and its entropy is less than the maximum. Since the entropy is a measure of the amount of information needed fully to specify the system state given the present view, a non-equilibrium view conveys some information about the present state, and the system may be said to be partially ordered. The nature of its order is described by the non-equilibrium features of the present view.

Strictly speaking, for a partially ordered system, the above expression for its entropy should have an additional term, showing the amount of information needed to describe the view features which condition the probability distribution $\Pr(z)$. However, this term is negligible compared with the main term except for very small systems and those views including highly detailed features, and is usually ignored.

The classical definition of entropy defines the entropy of a view, i.e., a collection of macroscopically similar states. Since it shows the amount of information needed to specify the exact system state given the view, and there is no information available about the state save what is contained in the view and invariants, the entropy of a view is essentially the log of the number of system states consistent with the view. It is the amount of information needed to specify one of this number. If instead of a view we consider a state, no further information is needed to define it: the "collection" has only one member. Hence, the definition is not directly applicable to a single state. However, the definition can be extended to apply to a state if we allow the particle-state probability distribution $\Pr(z)$ to be re-interpreted, in the case of a single system state, as the frequency distribution of particle states in the system state. Then a "high entropy" state is one in which the particle-state distribution approximates the particle-state probability distribution of an "equilibrium" or non-informative view. Some low entropy states would present a low entropy macroscopic view to an observer whose measurements revealed the state's departures from the equilibrium particle-state distribution, but for some low entropy states, the departures might be too subtle to show up in macroscopic view having a realistic level of detail. For example, a state in which only every second possible particle-state was occupied (according to some arbitrary enumeration of particle states) would have low entropy according to this definition, but would probably show no macroscopic evidence of order. This difficulty in the definition of the entropy of a state fortunately is of no consequence to our arguments, as almost certainly partial order visible only at a microscopic level would be destroyed in a very short time.

It follows from this definition of the entropy of a state that the number of states with entropy H increases exponentially with H. The great majority of states in the state space have entropies close to the maximum possible value. If the system is in equilibrium (no partial order and no informative current macroscopic view) we may expect that as time goes by the state entropy will jitter around just a little below its maximum value.

8.4 Why Entropy Will Increase

If we are presented with a closed system which appears to be in equilibrium, there is nothing which can be usefully predicted of its future, save that it will probably remain in equilibrium. We can hope to make interesting predictions about the future of a closed system only if it is now partially ordered, i.e., has entropy significantly below the maximum. I will now argue that we may confidently predict that its entropy will increase. Several cases will be treated.

(a) Deterministic Laws, Exact View.

Assume the Laws of Physics are deterministic, and that we have exact knowledge of the present state $s(0)$ at time $t = 0$. Suppose $s(0)$ is a state of low entropy H_0. The deterministic laws allow us to deduce the state $s(f)$ at some future time $f > 0$. The repeated application of the time-step mapping $s(t+1) = M(s(t))$ will no doubt have changed the state and its entropy. Since the mapping is based on the microscopic laws, which make no reference to entropy and are reversible, there is nothing inherent in the mapping which can lead us to expect that the entropy of a state will be related in any obvious way to the entropy of its successor, although, since a single time step will usually produce only a small change in the macroscopic view of the system, we may expect the resulting change in entropy to be small. After the f steps leading to time f, therefore, we can expect the initial entropy to have changed by some amount, but otherwise to be almost a random selection from the set of possible entropy values. Recall that in a closed system, the number of states of entropy H is of order $\exp(H)$. There are therefore far more high entropy states than low entropy ones, so the chances are that $s(f)$ will very probably have an entropy greater than the initial state $s(0)$.

In a deterministic closed system of constant energy, the sequence of states from $t = 0$ into the future is a well-defined trajectory through the state space. The trajectory may eventually visit all states in the space, or only some subset. Given that the state space is discrete, it can only contain a finite, albeit huge, number of states. The trajectory from state $s(0)$ thus must eventually return to state $s(0)$, and since the mapping is one-to-one, must form a simple cycle in which no state is visited more than once before the return to $s(0)$. Since it is cyclic and produced by time reversible laws, the trajectory must show just as many steps which decrease entropy as increase it. However, because the trajectory can be expected to visit far more high entropy states than low entropy ones, if we start from some

low entropy state and advance for some arbitrary number of steps, we
must expect to arrive at a state of higher entropy. If this argument is
accepted, the Second Law will in general apply even given deterministic,
time reversible laws. No paradox is involved.

(b) Deterministic Laws, Inexact View.

If, rather than knowing the precise initial partially ordered state $s(0)$,
we have only an imprecise view of the initial state, the argument for
expecting entropy to increase can only be strengthened. We may well
feel more justified in regarding the entropy at future time f as being in
some sense a random selection from a range of values, since we no longer
have any certainty about the future state $s(f)$. Given this uncertainty, the
larger number of high entropy than low entropy states becomes perhaps
a sounder basis for expecting $H_f > H_0$.

Because each time step has but a small effect on the macroscopic view,
we also expect the macroscopic behaviour of the system to be insensitive to the fine detail of the initial state, and to be well predicted by a
macroscopic, or at least inexact, initial view.

(c) Nondeterministic Laws

If we suppose the fundamental microscopic laws to be nondeterministic
(but still reversible) the state-to-state mapping $M()$ is replaced by a
probabilistic law

$$\Pr\left[s(t+1) = r\right] = M(r|s(t))$$

Given $s(0)$, we can at best compute a probability distribution over the
possible states at time $f > t$, and are the more justified in expecting $s(f)$
to have higher entropy than $s(0)$.

8.5 A Paradox?

The preceding arguments, backed up by the results from simulations of a
small but, I hope, plausible thermodynamic closed system described in a
later section, show that macroscopically irreversible behaviour not only can,
but must be expected to, emerge from the operation of time reversible microscopic laws on an initially partially ordered system, especially but not only if
these laws are non-deterministic. In particular, we can confidently expect the
entropy of the system to increase. The arrow of time unambiguously points
to a future less ordered than the present.

But a paradox of sorts remains. Re-reading the arguments of cases (a),
(b) and (c), it can be seen that, although it was assumed that the predicted
situation occurs at a time f in the future ($f > 0$), this assumption actually
played no part in the argument. If f is replaced by a time b in the past
($b < 0$), all that was argued to the effect that $H_f > H_0$ becomes an equally
valid argument that $H_b > H_0$. Thus, I conclude that while simple statistical
deductions show that entropy will very probably increase in future, these

same deductions show that entropy very probably decreased from the past. The entropy shown by the present state or view of the system is therefore almost certainly close to the lowest it will ever have, or ever did have, at least for times not hugely remote from the present. This is precisely the conclusion reached by Boltzmann. However, it is not what we would normally conclude about the past of a presently partially ordered closed system, and many would like to avoid it.

8.6 Deducing the Past

Before dismissing this deduction of past disorder as nonsense, let us consider what means might be used in the real world to deduce something about the past state T (for "Then") from exact or partial knowledge of the system's present state N (for "Now"). As before, I assume the microscopic laws (whether deterministic or probabilistic), the exact system energy, other invariants, and the state space all to be fully known. I will consider three cases: non-deterministic laws given an inexact view N^V of N; deterministic laws given full knowledge of N; and deterministic laws given an inexact view N^V. There are several approaches which might be taken to the deductive task.

8.6.1 Macroscopic Deduction

The most direct approach would be to assume the universal validity of the macroscopic laws of motion, Thermodynamics, Chemistry, etc., and attempt to use them to deduce the past state from the present state. While sometimes successful, this approach has severe limitations, as will appear.

These macro laws appear, as normally presented, to be deterministic: they are expressed in equations with no random terms. Some of them, e.g., the second law of Thermodynamics and the laws of viscous flow and diffusion, are not time symmetric, and describe "dissipative" behaviour which reduces order and increases entropy as time advances. Today, the dissipative "laws" are not seen as fundamental, but rather as emerging from the large-sample statistics of reversible micro processes. However, they have been well tested and found to be excellent models and predictors of macroscopic behaviour, so it is not unreasonable to assume that they held in the past period from state T to state N. The empirical evidence supporting them is so strong that we may be prepared to treat these laws as fundamental, and to ignore the argument that the large-sample statistics of reversible micro processes do not support them except for the prediction of future behaviour. If so, we may proceed to set up the differential equations embodying these laws, set N (or some macroscopic view N^V of the present) as boundary conditions at $t = 0$, and integrate the equations back to $t = -k$.

This approach may work adequately for small k, i.e., for the recovery of the very recent past, but in general is likely to fail. For instance, let our

system be a long thin pipe closed at both ends and filled with water, with
the present view showing some dye colouration in a short stretch of the pipe.
If there is no bulk motion of the water and no significant temperature or
gravity gradient, one of the applicable macro equations will be the equation
for one-dimension diffusion of the dye concentration $v(x,t)$:

$$\frac{\partial v}{\partial t} = K \frac{\partial^2 v}{(\partial x)^2}$$

Taking the present-view dye distribution $v(x,0)$ as boundary condition, this
differential equation can be integrated to negative times. Unfortunately, if the
integration is carried too far back, it is very likely to yield negative values of
dye concentration at some points of the pipe, which are of course impossible.
For any negative time t, the function $v(x,0)$ is the convolution of $v(x,t)$ with
a Gaussian function of x, so recovering $v(x,t)$ from $v(x,0)$ is a deconvolution
calculation which is very ill-conditioned and may have no positive solution.

   Another sort of possible failure arises from the fact that a purely deductive
process based on backward integration of the macro laws has no room for
our background knowledge. For example, suppose our system is contained
in a strong, insulated box, and our present view shows the box to contain
a hot mixture of nitrogen, various gaseous oxides, some sooty deposits of
carbon and potassium compounds on the walls of the box, a small wooden
stick charred at one end, and some fragments of discoloured paper. I cannot
believe that integration of the macro laws could lead us to a past state where
the box contained a firecracker and a lighted match.

   More generally, if in the original state T some part of the system was
in a low entropy metastable state, and the actions of some other part of
the system triggered the collapse of the metastability, it is hard to see how
deduction from the present N using the macro laws can lead to a recovery of
the metastable state. In summary, the use of macro laws to deduce the past
seems to have limited utility, and be unlikely to reach many "commonsense"
inferences that we make about the past, except in cases such as the motions
of the major bodies in the solar system, where dissipative processes and hence
entropy increases are very small.

8.6.2 Deduction with Deterministic Laws, Exact View

If the laws are deterministic and we have exact knowledge of the state N,
there is no problem. We simply trace back the history leading to N by using
the inverse of the one-to-one mapping of states $M()$. Of course, we recover T
without error. I will term this backwards tracing "devolution", and say that
N devolves to T.

   The case is wholly unrealistic. In the real world, we never have exact
knowledge of the present state of a closed system.

8.6.3 Deduction with Deterministic Laws, Inexact View

Assuming our knowledge of the present state N is an inexact view N^V, we can at best enumerate the collection of states compatible with N^V and trace each back to its only possible ancestor at the "Then" time. This will give us a set T^P of possible ancestors exactly as numerous as the collection of possible present states, and one of this set must be the true ancestor T. Deduction can at best lead to a probability distribution over the members of T^P. If all "Then" states were considered equally likely *a priori*, the distribution over T^P will be uniform, since each state in T^P has the same probability (one) of evolving into a "Now" state in N^V. We consider a non-uniform prior over "Then" states in the next section, but find the idea unsound.

We are left with the problem of deducing some conclusion about T knowing only that it is some member of T^P. One possible approach is to form a view of the past by averaging over the members of T^P. This seems not to be very helpful. The true "Now" state N devolves to the true "Then" state T, which may well have lower entropy than N. However, the precise configuration of N which implies its ancestor has lower entropy is extremely fragile. A state N^* differing only microscopically from N will very probably devolve to an ancestor of higher entropy. If we imagine the state space to be arranged so that states which are microscopically similar are close together, we find that in general, two trajectories which pass through neighbouring states in state space will diverge rapidly (whether traced forwards or backwards). While the (past) trajectory through N may pass through a lower entropy T, it is probable that (past) trajectories through close neighbours of N will diverge to higher entropy "Then" states.

Simulation experiments support this assertion. If a simulation is run from a low entropy state T, it will almost always reach a higher entropy state N which, providing the elapsed time is not too great, will still be partially ordered. In many hundreds of trials, it was found that if N was then perturbed by the smallest possible displacement of just one of the typically 10^5 particles involved, or by the reversal of a single particle velocity component, the resulting perturbed state devolved to a "Then" state with higher entropy than N and of course T. Further, the "Then" state reached presented a macroscopic view very different from T. It appears that for any realistic view N^V of the present state, the vast majority of states in T^P differ greatly from T, and in particular have entropies greater than N, not less.

It follows that trying to deduce the nature of T by averaging over T^P will lead to the conclusion that the past was less ordered than the present, and very different from what common sense would suggest. Simply averaging T^P to estimate some property of the past state corresponds to the (well justified) Bayesian "minimum loss" process if the "loss function" assumed is the square of the difference between the true and estimated value. One might consider whether some other loss function might yield estimates more in line with commonsense inferences about the past.

The only candidate loss function I have thought of which might work to discriminate among the members of T^P is a function which shows how much the evolution of some member of T^P to a member of N^V does violence to the usual (but not fundamental) irreversible physical laws such as the laws of diffusion, viscosity, friction, and in particular the Second Law. It is highly likely that the true evolution of T to N has pretty much conformed to these laws, but the evolution of a high entropy past state to some state in N^V will be in criminal violation. There are, unfortunately, two flaws in this suggestion. First, we are likely to infer a past state of *lower* entropy than T, showing fine-scale macroscopic structure which dutifully is attenuated in strict accordance to the irreversible laws until it matches the faint irregularities visible in the present view which common sense would consider the result of statistical fluctuations such as thermal noise, Brownian motion, etc., and in no way informative of the past. Second, and more importantly, we have no warrant for assuming such a loss function. We have as yet found no deduction proving that these "laws" held sway in the past, and hence no reason to suppose that their past violation is in any way a "loss".

8.6.4 Deduction with Non-deterministic Laws

I will assume an inexact view N^V of the present state N. Since the laws are non-deterministic, every state in N^V may have several possible immediate ancestors at time $t = -1$, and each of these several possible ancestors at $t = -2$, and so on. In general, we expect the set T^P of possible ancestral states at $t = -k$ to have cardinality much greater than N^V, and in some cases T^P may include the entire state space. Clearly, any deduction must deal with the inherent uncertainty of the past by using the logic of probability. The reversibility of the microscopic laws is easily shown to lead to the following equality, where r is a state occurring at some time t_1 and s is a state occurring at some later time $t_1 + k$.

The evolution probability $P_e(s|r, k)$ that state s will occur at time $t_1 + k$, given that state r occurred at time t_1, equals the probability $P_e(r'|s', k)$ that state r' will occur at time $t_2 + k$, given that state s' occurred at some time t_2.

(Recall that s' is a state identical to s save that all velocities and other odd derivatives with respect to time are negated.) In what follows, r and s will always refer to states separated by k time steps. We can therefore abbreviate the equality to the symbolic form:

THE RETRACE EQUALITY: $P_e(s|r) = P_e(r'|s')$

The probability that r evolves into s after k time steps along some particular trajectory from r to s is the product of the probabilities of all the random choices made along the trajectory. For every such trajectory from r to s there exists a trajectory from s' to r' along which exactly the same random choices are made, in reverse order. Thus, s' has just the same probability of evolving into r' along this trajectory as r has of evolving into s along its trajectory. Summing probabilities over all trajectories from r to s to get $P_e(s|r)$, and over all trajectories from s' to r' to get $P_e(r'|s')$ yields the retrace equality.

The retrace equality gives us a simple way to investigate the Bayesian posterior probability distribution over T^P given a present view N^V. Let s be some state in N^V. The posterior distribution over states in T^P given present state s is given by Bayes' theorem as:

$$P_f(r|s) = P_e(s|r)P_p(r)/P_m(s)$$

where r is a state in T^P at time $-k$, and $P_p(r)$ is the "prior" probability that the system would be in state r at time $-k$. The prior probability is the probability we would ascribe to r given only knowledge of the system invariants, i.e., before seeing the present view.

$P_m(s)$ is the implied marginal distribution over present states given by:

$$P_m(s) = \sum_r [P_p(r)P_e(s|r)]$$

Let us assume that $P_p(r)$ is the same for all states in the state space. This is a reasonable assumption given that in a closed system, all states in the space have equal probability unless some non-equilibrium feature of the state is known, which is of course not the case before we have observed the present view. If $P_p(r)$ is uniform over the state space, so is the marginal distribution $P_m(s)$. In fact, for every state in the state space, its marginal probability at time $t = 0$ equals its prior probability at time $t = -k$, and $P_e(s|r) = P_f(r|s)$.

Given this uniform prior, a Monte Carlo approximation to the posterior could be obtained by repeatedly selecting a state s in N^V at random and using the retrace equality to trace a possible trajectory backwards from that state to time $-k$. The tracing backwards is performed by first replacing s by s', then letting s' evolve for k steps according to the microscopic laws, and finally replacing the resulting state r' by the time reversed state $r = (r')'$. The retrace equality guarantees that the probability that a state r at time $-k$ would evolve to a state s at time zero equals the probability that tracing backwards from s will give r at $t = -k$. I will say that s devolves to r.

More generally, we may say that a state r at an early time evolves to a probability distribution over states at some later time, and that a state s at the later time devolves to a probability distribution over states at the earlier time. The collection of states found by the repeated devolution of states randomly selected from N^V is an unbiased sample from the posterior over T^P. States in T^P in general will not have equal likelihoods of evolving into states in N^V. Hence, assuming equal prior probabilities for all states at $t = -k$, the posterior probabilities given N^V of states in T^P will be unequal. The devolution of s in N^V to r in T^P is equivalent to the evolution of s' for k time steps into r'. As argued, and as confirmed by simulation, any evolution of a par-tially ordered state is expected with very high probability to conform to the normal macroscopic laws of thermodynamics, and in particular, to result in a state of higher entropy than the original. Almost certainly, $H(r') > H(s')$. Since time reversal does not alter the entropy of a state, almost certainly, $H(r) > H(s)$. That is, almost all states in T^P will have higher entropy than those in N^V, and the Monte Carlo sampling of states from the posterior probability distribution over T^P will contain a great majority of states with higher entropy than N.

Both evolution and devolution of a partially ordered state yield states of higher entropy with probability close to one. The Bayesian expectation is that the ancestor at $t = -k$ of N had more entropy than N, and this expectation is as well-founded as the expectation that N will evolve to a state of higher entropy at $t = +k$.

Moreover, the great majority of possible ancestor states, possessing col-lectively the bulk of the posterior probability in T^P, have little chance of evolving into states in N^V, being of higher entropy than N. Evolution of a high entropy state in T^P into a lower entropy state in N^V is possible but very improbable, as it implies an evolution running counter to the normal thermodynamic laws.

The posterior distribution over T^P summarizes all we can deduce about T from the present view N^V. Further deduction can only tell us things implicit in this distribution. As in the previous Section 8.6.3, with a uniform prior over T^P, Bayesian minimum loss deduction with a quadratic loss function is equivalent to posterior weighted averaging over T^P, which could be done by Monte Carlo sampling. As just shown, this will give the minimum loss value for the past state entropy as greater than the present entropy. The argument of Section 8.6.3 against the validity of more biased loss functions still applies.

As the minimum-quadratic-loss estimate of the entropy of T is so poor, we must doubt the quality of similar estimates of other properties of T (a doubt confirmed by many simulation results: see Figure 8.1 and Figure 8.2).

8.6.5 Alternative Priors

The counter-intuitive behaviour of Bayesian deduction noted above was ob-tained assuming a uniform prior $P_p(r)$ over the past states. Although there are arguments for the uniform prior, it is worth asking if a different prior could produce Bayesian posteriors more in accord with commonsense infer-ences of the past. For instance, one could consider a prior which placed more weight on past states of low entropy, in order to make the mean posterior entropy lower than in the observed present.

A sufficiently strong prior preference for low entropy past states might indeed bring the Bayesian deductions about T into line with common sense. However, it seems hard to provide a rational basis for such a "prior". For Bayes' theorem to be validly applied, it must be the case that the prior

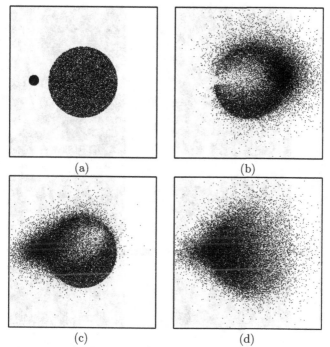

Fig. 8.1. (a) "Then" state of "Disc Collision" model; two-thirds scale; 31,623 atoms. (b) "Now" state of "Disc Collision" model; 35 time steps after "Then". (c) Reconstruction of Disc Collision "Then" from "Now" with one atom perturbed; deterministic devolution. (d) Reconstruction of Disc Collision "Then" from exact "Now" state; non-deterministic devolution.

probability distribution reflects the probabilities which might rationally be assessed on the basis of knowledge held *before* the present view is observed. In our case, this knowledge comprises only the microscopic laws of Physics, the system invariants, and the belief that the system has been isolated since before the time $t = -k$. That is, the prior knowledge is only what we have before we open the box to take the present view. There seems nothing in this knowledge to suggest we should believe that, had we opened the box at $t = -k$, the chances of our finding the system in some particular state q would depend on the entropy (or any other property) of q.

Further, adoption of a non-uniform prior $P_p(r)$ at time $-k$ leads to a non-uniform marginal distribution $P_m(s)$ for the present state, but $P_m(\cdot)$ is closer to uniformity than $P_p(\cdot)$, being smeared out by the variance of the evolution distribution $P_e(\cdot|\cdot)$. Thus, we would be supposing that the probability of finding the system in some particular state q depended on when we looked. Since we have no knowledge of when the system became isolated, or what its state was at that time, this supposition can have no basis.

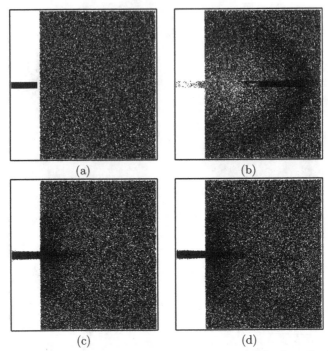

Fig. 8.2. (a) "Then" state of "Stick Collision" model; two-thirds scale. (b) "Now" state of "Stick Collision" model; 190 time steps later than "Then". (c) Reconstruction of "Then" state of "Stick" model from "Now" state with 100 atoms perturbed; deterministic laws. (d) Reconstruction of "Then" state of "Stick" model from exact "Now" state; non-deterministic laws.

The only prior which seems a defensible representation of our ignorance of T before observing the system is a prior which does not change with time. The uniform prior uniquely has this property. Price suggests that since we *know* the past to be more ordered, we can use this "fact" to modify the deduction of the past somehow to recover a lower entropy state from a present state. He does not say how, and in any case, we have so far seen no reason to accept his "fact".

Finally, it is not clear that a Bayesian prior is even applicable to the deduction of a probability distribution over possible "Then" states. We are happy to base predictions of future states on our knowledge (be it partial or exact) of the current state N, without reference to any "prior" beliefs about the future. The present state, by definition of a state, embodies all information relevant to the future behaviour of the system, so our calculations of the probabilities of possible future states need refer only to N, or our view N^V of N, unclouded by any distraction. Why should we lose our confidence in the sufficiency of the N state when computing the probabilities of possible past states? I see no compelling reason. If the sufficiency of N is accepted, we can make no better estimate of the probability of a possible past state T, given N, than that given by the Retrace Equality, and conclude

$$\Pr(T|N) = P_e(T'|N')$$

The result is of course equivalent to the assumption of a uniform Bayesian prior over past states.

8.6.6 A Tale of Two Clocks

A simple example will illustrate just how bizarre the results of attempting to deduce the past from the micro laws can be.

Suppose the closed system is an insulated box containing an ordinary mechanical clock with a mainspring coupled via gearing to the hands of the clock and to a conventional escapement governed by a balance wheel oscillating in resonance with a hairspring. Now suppose that at time "present" we inspect the clock and find that its mainspring is about half run down, and that the hands show 7 o'clock. We want to infer the state of the clock exactly one hour ago.

In such a clock, the movement is stationary most of the time, except for the balance wheel which oscillates, releasing the movement to advance by one tooth of the escapement wheel on each oscillation. For simplicity, assume that the present view is obtained at a time when the balance wheel is momentarily stationary at one extreme of its oscillation, and that the rest of the movement is also still. (This assumption is not crucial to the argument.) Now let us attempt to deduce its past state by applying our knowledge of the micro laws to backtracking from the present state. We can follow the process described in the previous sections.

First, choose randomly some state s consistent with the present view N^V, then time reverse it, giving s'. What will s' look like? Since there was no macroscopic movement taking place in s, there will be none in s'. Thermal noise vibrations on the micro-scale will be reversed in direction, but these have no significant macro effects. In fact, s' is consistent with N^V. So s' looks just like a clock with half-wound mainspring, balance wheel about to start a swing, and hands showing 7 o'clock. Now we let this state evolve for an hour in conformance with the micro laws. If s' looks like the state of a clock, it will evolve like a state of a clock. After an hour's evolution, corresponding to an hour's devolution of s, state s' will have done what a clock's gotta do, and will have changed to a state r' showing exactly 8 o'clock with a mainspring rather more than half run down. Finally, we time reverse r' to obtain the deduced past state r. As before, time reversal does nothing significant to the state except perhaps altering the phase of the balance-wheel oscillation by up to half a cycle, a jump of no more than half a second.

Our deduction, supported with extremely high probability by the fundamental laws of Physics, is that the clock, which now says 7.00, an hour ago said 8.00 and was more run down than it is now! (A grandfather clock would give similar results, but would need a bigger box.)

Since this clock seems as queer as a two-bob watch, we discard it and build our own. We build a disc free to turn on a vertical shaft. The disc carries two radial tracks on which two masses can slide freely from near the centre of the disc to its rim. Springs running from fixed points on the disc to the masses pull the masses towards the centre with a force exactly proportional to the distance of the mass from the centre. If the disc rotates with a certain speed, the inwards forces of the springs on the masses will exactly equal the outwards centrifugal force on them, no matter what the radial positions of the masses. To "wind" the clock, we spin the disc until it reaches this speed with the masses nearly at the rim, then let it spin by itself. It will continue to spin at the critical speed for a long time without slowing down, the masses gradually moving inward as kinetic and spring energy is released to overcome friction. Low-friction gearing couples the disc to the hands of our clock.

Now suppose our new clock becomes the closed system under study. Let our "present" view N^V be that we find the hands showing 7 o'clock as before and the masses are half way between centre and rim. Again, we attempt to deduce the state of the clock an hour ago by backtracking some state s in N^V. The time reversed state s' now has the disc spinning with the critical speed, but backwards, and otherwise looking just like s. To devolve the system by one hour, we deduce the state r' reached by evolving s' for an hour. If the friction is low enough, r' will still show the disc spinning with the constant critical speed, but the masses will have crept inwards. An hour's backward spin will have left the hands showing 6.00. Time reversing r' to get the inferred past state r, we find that an hour ago our new clock showed 6.00, not 8.00, but was less "wound-up" than at present. Great news! We now have a clock which not only keeps proper time when no-one is looking, but also winds itself!

If both clocks are put in the box and both now show 7.00, we have shown by impeccable reasoning that almost certainly, an hour ago one clock showed 6.00 and the other 8.00. The nature of the different behaviours of the clocks is related to the following observations.

- Clock 1 cannot be made to run backwards. Clock 2 can: one just "winds it up" by spinning the rotor in reverse until it reaches critical speed with the weights far out.
- Clock 1 (or any clock using an escapement) will not work properly without some slight friction and/or inelasticity in the escapement. It needs dissipative phenomena. Clock 2 does not: the less friction the better.
- Clock 1, once fully wound, will run for a time independent of the amount of dissipation in its works (within limits). Clock 2, once fully "wound", will run for a time inversely proportional to the amount of dissipation.

An hourglass, or a digital watch, will resemble clock 1. A sundial resembles clock 2. I hope this tale is sufficient to show that attempting to recover the past state of a closed system by statistical deduction based on the fundamental microscopic laws of Physics is both unusual and unwise.

8.7 Records and Memories

It may seem that by concentrating entirely on what can be deduced about the past from a single, instantaneous, present view, I am setting up a straw man. Even for some isolated closed system, we often have memories or records of the past which are informative of the system's past, and if we include these in our deductions, we should usually be able to deduce a lower entropy past state for the system. However, this argument fails because the system we think we are studying is then not really isolated. It must have interacted with the record-making machine or person which now holds the record or memory. By bringing a record of the system's past into the deduction, what we are really doing is enlarging our present view to include, not just the state of the nominal system but also the present state of pieces of paper and/or nerve cells. If these are included in the present view, our deductions of the past should be based on deductions of the past of the extended system which includes the recording apparatus. Suppose that we know nothing of the closed system save its energy, and that it is a box undisturbed for several hours. We open the box and find that it contains some physical system S and, separated by an insulating partition, a self-developing camera connected to a timer, whose clock now says 6.00, and which was clearly set to trigger the camera at 5.00. A picture has emerged from the camera showing the system S in a lower entropy state than it now has. Can we now deduce from this present view of S and picture that S really was in a low entropy state an hour ago? I agree with Boltzmann that we cannot.

As with any isolated system governed by reversible non-deterministic laws, devolution of the present state is far more likely to result in a past state of higher entropy than the present. The tale of two clocks suggests that devolution of the present view is likely to show the clock running backwards, and lead to a view at 4.59 in which the clock shows 7.01, the picture was already in its developed state (which has higher entropy than the metastable state of undeveloped film), and S was in a higher entropy state than now. During the period 4.59 to the present (6.00) the clock never reached an indicated time of 5.00, and the camera was not triggered. Despite its ridiculous nature, this "view" of 4.59 (or 5.00) has far more posterior probability given the present view than has the commonsense conclusion.

Human memory is of course a more flexible recorder than a camera, but I see no reason why it too should not fall victim to an argument similar to the above. In a deductive reconstruction of the past system state using a memory of the past, we must also reconstruct the past state of the observer. Very probably, the reconstruction will follow one of the individually wildly unlikely but collectively hugely numerous paths to a past along which the

Fig. 8.3. Two-chamber + Pipe model; original "Then" state; pipe flow left to right.

system-observer interaction now recollected never took place, or if it did, happened differently from what is remembered.

I am not arguing here for a rejection of the commonsense conclusion in favour of the high-posterior view, merely arguing that deduction, even using what appear to be records showing a past state in accord with common sense, does not lead to the commonsense conclusion.

8.8 Induction of the Past (*A la recherche du temps perdu*)

We are forced to the conclusion that the kind of deductive reasoning we habitually and successfully use in predicting the future of a system is inappropriate for inferring its past. I will argue that, whereas our predictions of the future are based on deduction from the present view to the possible and probable future states, our reasoning about the past is usually inductive. The term is used in its limited sense of reasoning from the particular to the general, or at least reasoning from a data statement to a statement more general and stronger than the data statement.

If we think about the kinds of conclusions we make about the past, these do not seem to have the form of "posterior distributions" or means thereof. Rather, we will often conclude by presenting a view T^P of a past situation, in the sense used here of an incomplete and imprecise specification of a state, often with an appended account of how that state might have evolved into the present state or view on which we base our conclusion. In other words, we tend to conclude with a view of the past which satisfactorily explains the present.

To return to a simple example, if at time zero we observe a gas in a box with two chambers connected by a narrow channel, and find in the present view that the left chamber has higher pressure than the right, and that the net flow of gas in the channel is from left to right, what might a reasonable person conclude? (Figure 8.3.) Given an assurance that the box was well insulated and that there had been no external interference in the last five seconds, the reasonable conclusion which most would vote for is that five seconds ago, the pressure in the left chamber was even higher than it now is, the pressure in the right chamber was less, there was then more gas in the left and less in the right than is now the case, and that gas has been flowing from left to right from then till now.

Why does this conclusion seem reasonable? The past view postulated is not extraordinary, if true it would lead with good probability to the present view, and the postulated process leading from the past view to the present is in accord with our knowledge and experience of macroscopic behaviour. In particular, the asserted past view has lower entropy than the present one in accordance with our knowledge of thermodynamics.

This last point is common to most of our conclusions about the past of well-isolated systems: we postulate a past view or state of lower entropy than the present. Since in a closed system there are far fewer low entropy states than higher entropy ones, the assertion of a low entropy past state is a stronger statement than can be deduced from the present view. In common with all non-empty inductive conclusions, it is a more general and stronger statement than the data statement on which it is based. To be an admissible hypothesis, it must not imply anything contrary to the present view. That is, the asserted past state or view must give high probability to the present view. It must have high likelihood. However, being a stronger statement than the present view, it will in general imply more about the present state N than can be deduced from the present view N^V. If these implications about the present state can be checked by more detailed observation of the present, they may be found to be false. Thus, an inductive conclusion about T is always in principle falsifiable by additional data about N (which may include observations at times later than zero.)

A principled inductive conclusion about T from the data N^V will be based on the likelihoods of possible past states, i.e., the probabilities these states have of evolving into N^V. Under reversible, non-deterministic micro laws, the evolution of any state will, with very high probability, conform to the normal macroscopic and thermodynamic laws (the "macro laws"), and in particular will not result in a decrease in entropy. Thus, the only states in T^P with a high likelihood given N^V are those states which can evolve into N^V in a manner consistent with the macro laws. Denote the set of such states as C (for "core"). Although C accounts for only a tiny fraction of the posterior probability distributed over T^P, it is among the states of C that we expect to find our inductive conclusion. It remains to consider what kind of inductive principle will best perform the task

8.8.1 Induction of the Past by Maximum Likelihood

A simple and well-known inductive principle is the "Maximum Likelihood" (ML) principle. It suggests the selection of that hypothesis which has the greatest likelihood of yielding the observed data. In our context, Maximum Likelihood suggests that we infer the state at $t = -k$ to be that state in T^P with the greatest probability of evolving into N^V. The ML inference T_{ml} will of course be some member of C. Unfortunately, T_{ml} will often be a poor choice, not conforming well with a "commonsense" inference.

Given a choice among a family of hypotheses of different complexity, ML notoriously is inclined to "overfit" the data. That is, ML will usually choose a rather complex hypothesis even when the complex details of the hypothesis are not justified by the data. In our problem, while we expect a partially ordered state to evolve most probably in accordance with the macro laws, we also expect its evolution to show minor deviations from these laws. As was found in Section 8.6.3 when considering loss functions favouring adherence to the macro laws, the likelihood of a past state is also highest when its evolution into some state in N^V closely follows the macro laws. Hence, the maximum-likelihood choice will in be expected to show in exaggerated form fine-scale macro structure in N^V which in fact arose from statistical fluctuations in the evolution from the true past state. Given a fairly precise view N^V, it is likely that the ML inference T_{ml} will show an exaggerated feature not present in T, and arising from some feature of N^V which arose through thermal noise since the time of T. Even if the system has been in equilibrium since before the time of T, a sufficiently detailed present view will show some deviations from the maximum entropy view, and T_{ml} will try to explain these as the vestiges of past order. That is, the ML inference will overfit the data by striving to fit noise in the data which is not really informative about T.

This defect of Maximum Likelihood inference is well known, and various techniques have been developed to overcome it. A useful but rather *ad hoc* technique is to apply standard tests of statistical "significance" to the various features asserted by the ML conclusion, and to delete from the ML inference those features found to be insignificantly supported by the data. This approach requires the essentially arbitrary specification of a "significance level", and has severe technical difficulties in determining appropriate significance tests to apply during the sequential deletion of several features. A more principled approach may be based on "penalizing" the likelihoods of the competing hypotheses by a penalty which reflects in some way their various complexities. A well-founded way of combining the likelihood of a hypothesis with a measure of its complexity is provided by the Minimum Message Length (MML) principle and similar methods such as Minimum Description Length (MDL) and Bayes Information Criterion (BIC).

8.8.2 Induction of the Past by MML

I do not here mean to argue the special merits of MML as an inductive principle, but merely choose it as an example to show that there is at least one respectable inductive principle which can be expected to lead to inferences about the past which are close to our commonsense conclusions. I have chosen MML because long work with and on the principle has given me some claim to a fair understanding of how it would perform in this context, but I expect any other effective inductive principle, such as a significance tested ML or Bayesian approach, would give similar results.

Given some data and a family of possible models for or hypotheses about the source of the data, MML considers the construction of a message which encodes the data in (say) the binary alphabet. The message must be intelligible to a receiver who knows the observational protocol used to collect the data, the family of possible hypotheses, the system invariants such as total energy, and some (not necessarily informative) prior probability distribution over hypotheses.

The message is required to have a particular form. It must begin with a first part which specifies a single hypothesis. This is followed by a second part which encodes the data using a code such as a Huffman or Arithmetic code which would be efficient were the stated hypothesis true. Such a message is called an *explanation* of the data.

Standard coding theory shows that the length of the second part is given by the negative log of the probability of the data given the hypothesis. That is, the length of the second part is the negative log likelihood of the hypothesis. The length of the first part, which states a hypothesis, is also supposed to use an efficient coding scheme. The optimal scheme will depend in part on the prior probability distribution (MML is inherently Bayesian), but in any reasonable scheme the amount of information (and hence message length) needed to specify a complex hypothesis will be greater than for a simpler hypothesis. Further, the more precisely any free parameters of the hypothesis are specified, the longer the statement becomes.

The MML principle asserts that the best inductive hypothesis is that which minimizes the total length (first and second parts) of the message encoding the data. Thus, there is a built-in compromise between the complexity of the hypothesis and its likelihood given the data. A complex hypothesis requires a long first part to specify it, but may give a high likelihood and hence a short second part. A simple hypothesis with severely rounded-off parameters can be stated in a short first part, but may have a lower likelihood requiring a long second part. The best compromise (shortest message) is found to be reached when the hypothesis has just the complexity and precision justified by the data.

In a context such as the recovery of the past, where the conceptually possible set of hypotheses is fixed and discrete (here, the set of states in the state space), the notion of precision of the inferred hypothesis needs some clarification. The two-part explanation will begin by asserting a single, precisely defined hypothesis (i.e., past state) and the second part will encode the data (N^V) using a code expected to be efficient were the assertion true. How then can we think of the assertion as being of limited precision? The answer is that the expected explanation length is minimized if the code used for nominating the hypothesis (which must be determined before the data is known) can only nominate one of a subset of the conceptually possible hypotheses. The smaller this subset, the shorter the first part becomes, but, when the data arrives, the sender must choose one of this subset as his inference. If the subset is too sparse, no member of the subset may have a high likelihood given the data, and so the second part will be long. The optimum choice of subset balances these conflicting effects.

The MML principle is supported by several theoretical arguments. For our purposes, the most relevant is that it can be shown that in the shortest message, the statement of the hypothesis contains almost all the information in the data which is relevant to the choice of hypothesis, and almost no irrelevant information. In the present context, an MML hypothesis about T will contain almost all the information in N^V which is about T, and almost none of the information in N^V which derives from thermal noise processes occurring in the period since T. The magnitude of the "almost" qualification is small. Although no exact general bounds are known, it appears to be less than the log of the number of free parameters whose values are stated in the hypothesis.

The exact construction of a code for hypotheses which will minimize the expected length of the two-part message is computationally infeasible except in the simplest problems, but good approximations exist for the message length which allow MML inference to be applied in problems of considerable complexity.

In the present problem of making an inductive inference about the past state T, an approximate treatment of MML described in Section 4.10 leads to the following construction, assuming all past states are equally likely a priori.

Given the present view N^V, construct a set Q of past states which minimizes the approximate message length

$$L_Q = -\log(|Q|/V) - (1/|Q|) \sum_{r \in Q} \log(\Pr(N^V|r))$$

where $|Q|$ is the number of states in Q, V is the number of states in the state space, and $\Pr(N^V|r)$ is the probability that past state r will evolve into a state in N^V.

The first term in L_Q is the negative log of the total prior probability of states in Q, and gives the length of the first part of the message. The second term is the expected length of the second part if a randomly chosen member of Q is asserted as T. It is easy to show that for the optimum choice of Q, the smallest log-likelihood given N^V of any state in Q is only one less than the average log-likelihood.

In words, MML inference finds a "view" Q of the past which maximizes the prior probability of the view times the geometric mean of the likelihoods of the states in the view.

Were the arithmetic average used instead, the quantity maximized would be the total Bayesian posterior probability of all states in Q, which would increase monotonically with the size of Q and be maximized only when $|Q| \geq |T^P|$, the set of all possible ancestors of N^V. By contrast, Q as defined can contain no state with very low likelihood. Q will basically comprise the "core" set C of ancestral states with a good likelihood of evolving into N^V. However, its total posterior probability may be quite small.

Another way of regarding an MML hypothesis is that, of all hypotheses with a good likelihood, it has the largest "margin for error". That is, it represents a large group of similar hypotheses all of which have good likelihood, so it need not be specified very precisely. This property is shared by the true, lower entropy "Then" state which led to the present view. Simulations show that when a low entropy T evolves to a higher entropy partially ordered N^V, the evolution and the final view is insensitive to substantial changes in the microscopic detail of T, and to the detailed random events occurring in non-deterministic evolution.

In the real world, an explanation of a present view conforming to MML ideas would not normally assert a specific past state T. Rather, we would assert a view of the past roughly equivalent to the set Q, that is, a view asserting just enough detail about the past to ensure that any state compatible with that view would have a high likelihood of evolving into the present view. In general, we might also have to include in our explanation of the present some account of incidents occurring during the evolution whose occurrence was not implicit in the asserted past view. These additional details would form the second part of the explanation, and while not implied by the past view, should have reasonable probability given the past view.

8.8.3 The Uses of Deduction

I am not asserting that deduction of the past from a view of the present has no role. In some situations, our view of the present is essentially exact, the relevant physical laws are well-known, and no appeal need be made to statistical thermodynamic reasoning. The classical example is the deduction of past positions of major solar system bodies such as the planets, their moons, and the major asteroids. Their "laws of motion" are well-known, the bodies are so large that indeterminacy is irrelevant, and present observations of their positions and velocities are very accurate. There are no "dissipative" effects of much consequence. Hence, we can make reliable deductions of their past behaviour some thousands of years into the past.

In some situations where dissipation is not negligible, bulk motions apparent in the present view may still provide a basis for deducing at least some properties of the past. For instance, if the present view shows a spherical shock wave in the atmosphere, even a deterministic back-tracking of some state conforming to the present view will fairly well retrace the recent history of the wave, showing it to have been expanding from its centre with about the correct speed (although it will probably suggest that the intensity of the wave was less in the past than at present, see Figure 8.2).

If we are prepared to accept the well-founded inductive inference from many data sources that the Second Law and its irreversible fellows held in the past, we may accept them as premises in making deductions about the past. The instabilities which can arise in using these "laws" backwards are not necessarily fatal over short periods. If we accept our inductive inference that the memories and records which are part of our present view of the world give fairly reliable information about the past, deduction from this information may reliably imply other details of the past. However it still seems to be the case that deductive reasoning without reliance on inductive inferences about the past has at best a very limited scope.

Finally, deduction plays a vital role in supporting any inductively produced assertion about the past. We should accept the assertion only if we can deduce from it that the present view is significantly more probable than were the assertion not to hold.

8.8.4 The Inexplicable

An MML "explanation" message is considered acceptable only if it is shorter than the raw statement of the data. An explanation of a present view is acceptable only if it is more concise than the description of the present view. If no such explanation can be found, the MML principle suggests that the data is simply random with respect to the kinds of hypothesis we can assert about T, and no such hypothesis is supported by the data. Presented with such a present view, MML will offer no explanation, leaving only the conclusion that the present state is random with respect to the state at time $-k$. That is, we must conclude that N^V no longer contains any information about T.

Inexplicable present views are not ruled out by the fundamental laws of Physics. A simple inexplicable view of a box containing an ideal gas which has been undisturbed for an hour would be a view which showed a strong standing wave with a natural period of, say, ten milliseconds. There is no conceivable state an hour ago which could be expected to evolve into this view with reasonable probability, although of course there is a multitude of past states which could possibly do so.

The "inexplicable" is not confined to the virtually impossible. In fact, most of our day-to-day views of the world are inexplicable in this technical sense. When we see an oddly shaped cumulus, or the disposition of the stars in the night sky, or a collection of people in a shopping mall, we can usually offer no explanation of the view which does not require a set of assumptions as great as or greater than the features which they "explain". These everyday views are not views of a complete closed system, and we accept that what we see is largely the result of "novelty" entering our view from a wider sphere beyond the current reach of our senses. Even when we inspect an effectively isolated system, say, a rock in a glass box, we do not expect to be able to explain the fine detail of what we see, and just accept that each particular fleck of quartz is where it is for reasons which are beyond rational discussion.

Even if we find in a closed system some mild but significant departure from equilibrium, say, a small temperature gradient along a metal bar, the inference that the system had a greater disequilibrium in the near past, whilst perhaps the best possible guess and perhaps even true, is not really an "acceptable explanation" of the present. It requires us to assume a proposition at least as lengthy as the one it is trying to explain, and leaves us no wiser as to *why* the gradient is present.

We may at times be forced to make inferences about the past which are not acceptable explanations of the present, but only the best explanations which can be found. In such cases, it seems plausible to argue that we should still follow the inductive principles which would have led us to an acceptable explanation were one possible. Acceptable explanations are most likely to be found when the present view comprises a number of features which are logically independent of one another (i.e., each could reasonably exist in the absence of the others). Then an inferred past describable by fewer features which implies a high probability of the co-occurrence of the observed features will almost certainly be acceptable. Each observed feature acts as corroborative evidence of the inference.

8.8.5 Induction of the Past with Deterministic Laws

A case has been made that inductive reasoning from present data using MML or some similar inductive principle can be expected to reach inferences about the past agreeing with the inferences we normally accept, whereas deductive reasoning, including probabilistic deduction, will not. The argument has so far assumed non-deterministic fundamental laws. In this section, I consider whether MML induction will behave similarly if the fundamental laws are known and deterministic. I assume some inexact partially ordered view N^V of the present, and suppose it to have evolved after k time steps from some unknown state T which common sense suggests will have lower entropy and greater order than the present.

Given an inexact view N^V of the present state, we can in principle proceed to enumerate the ancestors at $t = -k$ of all the states in N^V, e.g., by backtracking, and thus arrive at a set T^P of the possible states at time $-k$. We then know that the true state T is some member of T^P, but not which one. Since the laws are deterministic and reversible, the cardinalities of T^P and N^V are equal. This is an essentially deductive mode of reasoning, applying to the past the kind of reasoning used to predict the future. In the task now considered, it seems, while perfectly valid, to be of little help.

The vast majority of states in T^P bear little resemblance to T. In particular, the vast majority will, according to our argument and simulation results, have entropies greater than the entropy of the present state N, which itself has entropy greater than T. Standard statistical reasoning also does not advance us. All states in T^P have the same likelihood (namely one) of yielding the observed "data" N^V, so the maximum likelihood principle will not help us find in T^P any clue about T. A Bayesian approach (taking all states in the state space to have the same prior probability before the "data" are considered) leads to a uniform distribution of posterior probability over the members of T^P, since all have the same likelihood. Thus, no state in T^P can be preferred on the basis of posterior probability, and the mean entropy over the posterior distribution is a terrible estimate of the entropy of T.

MML also does not seem to work in this case. Since all states in T^P have the same likelihood, choosing Q to minimize L_Q will result in $Q = T^P$ (see Section 8.8.2), and a random selection of a past state from Q would almost certainly select a state of higher entropy than N.

An alternative approach trying to avoid the equality of likelihoods would be to aim to infer a past view rather than a past state. If a past view V is asserted, one could identify its prior probability with the total prior probability of past states compatible with it, which is proportional to the number of states in the view, and identify its likelihood with the fraction of states in the view which would evolve into a state in N^V. This does not work either, as maximization of the posterior probability of V would lead to a view including all of T^P.

There is one inductive scheme which would probably yield a "commonsense" inference. Let the state space be divided into "cells" each containing a large number of states, such that all states in the one cell differ only microscopically and would present virtually identical macroscopic views. Then we could apply MML or some similar inductive principle to the inference of a past cell, in effect replacing the set of states by the set of cells as the conceptual hypothesis space. The prior of each cell would be proportional to the number of states it contained (which might be equal for all cells), and its likelihood given N^V would be the fraction of its states which will evolve into a state in N^V, i.e., the fraction of its states which are in T^P. The approximate MML approach used above for non-deterministic laws would then be expected, given N^V, to generate a set Q of cells all having a high likelihood and all containing states of lower entropy than N. In effect, the "cell" scheme replaces states with zero or one likelihood by cells with a full range of likelihoods, and introduces a form of indeterminacy by the back door. The problem with this idea is that the results would depend somewhat on the choice of cell size and, more importantly, on the criteria of similarity used in determining which states can be put in the same cell. These criteria will depend on the nature of the observations and measurements used in defining a "view". There will be states which are indistinguishable by some measurements but revealed as different by others. Given the subtlety and precision of the measurements afforded by modern Science, it is hard to be confident that any similarity criteria could be defined which could not be violated by some form of view-forming measurement. This problem makes me unwilling to accept the scheme as general and well-founded.

I have been unable to think of any general, principled method of induction which can be expected to make a "commonsense" conclusion in the presence of deterministic fundamental laws. While my failure is no proof that no such method exists, I am at present inclined to think that principled inductive inference of the past will yield commonly accepted results if the fundamental laws are reversible, but only if they are also non-deterministic.

One could argue that in practice, the evolution of partially ordered states under deterministic laws seems, in simulations, to be virtually indistinguishable at macroscopic scales from evolution under non-deterministic laws. Hence, we might as a pragmatic concession pretend that we do not know the deterministic laws, and make the kind of inference, MML or whatever, which would be justified by non-determinism. While this might usually work, it scarcely seems principled.

8.9 Causal and Teleological Explanations

The nouns "cause" and "effect" presumably enter our language because they are useful in some areas of communication. They are human constructs which like "mass", "electron" and "energy", do not refer to things directly accessible to our senses, but which, if hypothesized to apply to certain phenomena, allow what we can observe of these phenomena to be encoded concisely and used in forming coherent mental models of phenomena having some predictive power. As with these other terms, their meaning is ultimately defined in the sets of sentences in which they appear and which can be "decoded" by others to convey valid accounts of observations or verifiable assertions about future observations. Their use is, of course, not limited to just these sentences, but it is these which finally must be appealed to if one wishes to claim some use of the terms to be incorrect or meaningless.

Some such terms, like "mass" and "electron", have proved to be enormously useful in sentences with a mathematical form. By establishing just which mathematical sentences do decode to match observations, it has been possible to define a "calculus" for their meaningful use, i.e., to frame well-defined rules governing which uses are "correct". This does not imply that their meanings are immutable or even uniformly understood by all those who use them. Our current uses of the term "mass" might surprise Newton, while embracing all the uses he might have made. Further, some terms, such as "attraction" (as in gravity or electrostatics) are now considered to lack an exact calculus, action at a distance having been divorced from our canon of natural law, but remain useful in slightly less formal discourse which is content to approximate reality.

In my opinion, "cause" and "effect" may suffer the same demotion as "attraction", and we may have little better hope of defining an exact calculus for their use than we have of defining calculi for "sin" or "beauty". That said, I will try to add my tuppenceworth.

In an account of system histories in which dissipation plays no significant role, there seems little need to invoke notions of cause and effect. The entire history is summarized by a sufficiently detailed description of its state at any chosen time in the history, and the earlier and later states are implicit in this description. Given the positions and velocities of the planets yesterday, asking for the cause of their positions today is a question which, no matter how answered, leaves us no wiser. This applies even if there is some indeterminism in the system.

Since a reductionist history of a system to the most fundamental level relies only on the reversible fundamental laws, dissipation is not a necessary part of the history. The elastic collision of a pair of atoms is not a dissipative event, although its outcome may be uncertain, and the result of a multitude of such events may be described as dissipative at the macro level. So, at the fundamental level, there is little role for cause-effect language. In fact, notions of cause and effect simply do not appear in the expression of fundamental physical laws (and very rarely even in macroscopic laws).

I am led to think that these notions have their use in discourse concerning the macroscopic behaviour of dissipative systems, and are most likely to be invoked when we have only a partial understanding of the processes involved. Consider an acceptable explanation of the current situation which hypothesizes a past situation and encodes the present in terms of its probability of evolving from the past. The system of interest need not be closed: part of the encoded evolution path may further hypothesize some external intervention. For the explanation to be acceptable, i.e., shorter than a plain statement of the present facts, the evolution from past to present must be highly probable, and hence, if dissipative, must follow macro thermodynamics. This fact allows the encoded history of the asserted evolution to be encoded using the macro laws to define the trajectory probabilities from state to state, and allows the evolving states to be described in purely macroscopic terms. We need not even know the underlying fundamental micro laws, since the macro behaviour depends little on the micro details. I will call such an acceptable explanation a Causal explanation, and suggest that any proposition asserted in the explanation may be regarded as a cause of the present situation.

If the explanation explains the extinction of the dinosaurs, the postulated past climate and modes of life of the dinosaurs are "causes" (of why dinosaurs but not crocodilians copped it), the arrival of a comet or asteroid is a cause, the resulting global winter was a cause, the postulated presence of sulphate rocks under the impact was a cause (of global acid rain) and so on.

Even if the explanation is a generalized explanation which asserts a new "law", that law may have the status of a cause. When a postulated universal gravitational attraction was first used to explain planetary motions, it would have been natural to think of the gravitational pull of the Sun as causing the planets to accelerate towards the Sun. If such a postulated law becomes generally believed, it perhaps can no longer be regarded as a cause: it is simply a statement of the way things are and always will be.

Returning to the main theme, it is not logically impossible for the present state of a closed system to admit an acceptable (i.e., short) explanation in terms, not of evolution from some more-ordered past state, but of devolution from some more-ordered future state. Such an explanation, which I will call Teleological, asserts that some state F (or view F^V) will hold at some time $t = f$ in the future, and then shows that with high probability, all trajectories leading to F will show at the present time states much like the present view N^V.

An acceptable Teleological explanation of the present has the same logical structure as an acceptable Causal explanation, and receives as much support from theories of inference such as MML as does a Causal explanation. However, few Teleological explanations of (parts of) the present seem to be adopted. It may be that the present state of the Universe is such that, while many of its parts admit of acceptable Causal explanations, few if any admit of acceptable Teleological ones. If this conjecture is true, it describes an empirical property of the present: an inherent asymmetry which we might call a "causal arrow of time".

There is a more direct reason to expect Teleological explanations to be shunned. Recall that, in an acceptable Causal explanation, the asserted "Then" state or view has a high probability of evolving to a state in the present view N^V, but, as we have seen, states in N^V have very low probability of devolving into T. This fact, while perhaps surprising, is not embarrassing. The present does not devolve, so we cannot observe the divergence of the devolving trajectory away from the trajectory leading to T, and so cannot use this divergence to disprove the asserted "Then". The corresponding position in an acceptable Teleological explanation is that the asserted future state F has a high probability of devolving into N^V, but states in N^V have very little probability of evolving into F. Thus, we need only watch N evolve for a little while to observe (with very high probability) that the evolution of N has diverged from the path to F. In other words, a Teleological explanation of the present, even if acceptable, must be expected to suffer empirical disproof.

8.10 Reasons for Asymmetry

My main case is that we use deduction to reason about the future, but usually use induction to reason about the past. If it is accepted, it poses the question of why we use different reasoning for past and future. A crude answer might be that we then avoid ever having to reason deductively in the backwards time direction, thereby never exposing our reasoning to Boltzmann's theorem. About the future, we deduce forwards in time from the present. About the past, we postulate a past by whatever means, then check the postulate by again deducing forwards in time from past to present to see if the postulated past implies something close to the present. Similarly, we avoid Teleological induction of the future because we would have to check the postulated future by deducing backwards in time from the postulated future to see if it accounts for the present.

More seriously, I suggest we reason differently in the two time directions because we have different interests in the past and future.

Our interest in the future lies in the possibility of adapting our behaviour to benefit from or at least cope with the future. If we can guess what is likely to occur, we can perhaps do something about it. If we have no possibility of effective action, we might as well be fatalistic about the future: whatever will be, will be, so why bother to guess? If indeed our interest in prediction comes from our possibility of choice of action, the deductive modes, and in particular probabilistic deduction, are appropriate. They lead to at least a rough probability distribution over future states, conditional on our chosen acts. If we can evaluate the possible future states on our own scale of values, we can choose the acts giving the greatest expected benefit.

Our interest in the past is quite different. The past is gone: there is nothing we can do to alter it and whether the possible past states were good or bad on our scale of values, the good or bad has been done. There is no immediate value in knowing the past as far as choosing our future acts is concerned. Rather, our interest in the past is in helping us to understand the present.

An inductive inference about the past must, when deductively evolved to the present time, show a view in good accord with the view we actually have of the present. If we find an acceptable explanation, we have not only made a good guess about the past but now have a more concise account of the present. But it can do more. It may imply details of the present state which we did not notice in our observations. Thus, induction of the past may improve our view of the present. Of course, since no inductive conclusion is certain, we would be well advised to check these implications by more observation of the present before accepting them, but at least the induction has suggested what to look for. If we find the implications to be verified, we gain confidence in the inductive conclusion; if we do not, we have a refined present view to use as data for a better inference of the past.

Our understanding of the present may be improved more radically. If we consider an inference of the past chosen because it gives a short MML-style explanation of the present, there is nothing to prevent our hypothesis (in MML the first part of the explanation) from asserting a hitherto unknown natural law, the assumption of which, together with a hypothesized past state, provides a shorter explanation of the present than can be found using previously known laws. For instance, in our ideal gas examples, we might assert a new, more detailed form of the non-deterministic reversible collision laws which imply some subtle change in evolution probabilities giving a more probable evolution to the present. This may be the route by which most scientific advances are made.

As to why we should want to understand the present, the motives are presumably based on the possibility of choosing future actions. The better we understand the present (and the laws governing the processes by which the present state was reached) the better we are able to predict the future.

In summary, our deductions discover the direction of the point of the arrow of time, and validly imply a high probability of increasing entropy in the future. Similar deductions would imply with equal validity a high probability that entropy was higher in the past: a second point at the other end of the arrow. But we have no real interest in *deducing* the past: this can tell us nothing more than is implied in the present. Instead, we seek in the past an explanation of the present, and must resort to induction to do so. The feathers on the arrow of time are our inductive inferences, unprovable but falsifiable.

Deductive and inductive inferences about the past state of a closed system can be very different. When they differ, which should we believe? I suppose my answer must be that it does not matter what we believe, since the past is truly lost, and all we can ever know of the past of a closed system is what we know of its present. It is what we believe about the present which matters. This is not to say that inductions based on the present view of a closed system are pointless. They may lead to the elucidation of features of the present otherwise unnoticed or unexplained, an understanding of the present as a partially coherent pattern rather than a collection of undigested data, and in some cases to the discovery of general laws. These benefits do not seem to arise from deductions about the past state.

Despite this, if presented somehow with a situation where I were required to estimate the entropy of a past state, and would suffer a loss quadratic in the error of my estimate, I am not sure I would reject the deduction based estimate that the entropy was higher than it is now. Note, however, that such a situation cannot arise if the system was genuinely isolated during the relevant time. For me to suffer a loss because of error, the actual past value must still be exerting an influence over future events otherwise than through the present state of the system. That is, the system cannot have been totally isolated, and its past state has interacted with my environment.

8.11 Summary: The Past Regained?

A case has been made for a distinction in the modes of reasoning normally employed regarding the past and future of closed systems. We use deductive modes for predicting the future, but seem to switch to inductive reasoning to reach conclusions about the past. Deductive reasoning may be employed in forming a view of the past, but I believe it is fair to say that this is usually done only in situations where dissipation is insignificant, or when the deduction uses as premises the assumption of irreversible laws whose validity in the past can only be supported by inductive reasoning.

If the fundamental laws are assumed to be reversible and non-deterministic, no loophole is evident in Boltzmann's theorem that both the past and the future of a partially ordered state are probably less ordered than its present. However, with just the same assumptions, a well-founded induction of the past is likely to conform with our commonsense expectation that the past was more ordered. Further, if a past-based explanation of the present is allowed to assert not only a view of the past but also a new or refined "law", inductive explanations may involve and justify the induction of new natural (or social, biological, economic, etc.) laws.

Under the same assumptions, Teleological explanations of the present, based on the induction of a future state, while logically possible, will almost certainly be disproved.

I have not been able to show that these conclusions follow if the fundamental laws are deterministic. On the other hand, I have not shown that the conclusions are incompatible with reversible deterministic laws.

Our different interests in past and future, based on the possibility of our effective action, are sufficient to motivate the switch of reasoning mode, and hence a subjective asymmetry in the Arrow of Time: deductive point at one end, inductive feathers at the other. Objective evidence may be found in an objective asymmetry in the present, if we find (as I believe to be the case) that the present admits acceptable explanations based on the inference of past states more often than it admits acceptable explanations based on the inference of the future.

8.12 Gas Simulations

Readers who are happy to accept the proposition that partially ordered states are most probably preceded and followed by less ordered states need pay little attention to the simulations, since they serve essentially only to establish the proposition in a very limited context.

The system imitates a two-dimensional ideal gas confined in an immovable rectangular box. In some experiments, internal divisions divide the box into two or more connected chambers. In order to implement exactly reversible "laws", all arithmetic in the simulation is done with integers. Floating-point arithmetic is in general irreversible because of rounding-off errors. Thus, the atoms of the gas have X and Y positions which are integers, so the space within the box is a rectangular grid of points. Similarly, atom velocity components U (in the X direction) and V (in the Y direction) are restricted to integer values. As assumed throughout, time is represented by an integer t.

In the simulation of a time step from t to $(t + 1)$, a two-phase process is used. First, at each point of space occupied by two or more atoms, collisions are simulated using a Collision Table which defines the collision physics and which is set up at the beginning of the simulation program. The several atoms at a point are collided in pairs, in order of their identifying indices. The first and second atoms collide, then the second and third, then the third and fourth, and so on. The collision between two atoms effects a rotation of their relative velocity vector, their centre-of-mass motion being unchanged. If (u_1, v_1) is the initial relative-velocity vector, the new vector

(u_2, v_2) is determined from the collision table. The collision table ensures that $(u_1^2 + v_1^2) = (u_2^2 + v_2^2)$ to conserve energy. The integer-velocity-component restriction also requires the collision table to preserve the parity (oddness or evenness) of u and v, which ensures that the atom velocity components remain integers. The table is also set up so that

$$\text{if } (u1, v1) \rightarrow (u2, v2) \text{ then } (-u2, -v2) \rightarrow (-u1, -v1)$$

For convenience, the table also obeys the further condition

$$\text{if } (u1, v1) \rightarrow (u2, v2) \text{ then } (u2, v2) \rightarrow (u1, v1)$$

Note that these conditions on atom collisions do not imply that if for some states r and q, $r = M(q)$ then $q = M(r)$. To limit the size of the collision table, a collision which would result in one of the atoms having an absolute velocity component exceeding a fixed limit is not permitted. The limit is high enough that it is rarely reached.

The second phase of the time step moves the atoms according to their velocities. In general, the (x, y) coordinates of an atom are replaced by $((x + u), (y + v))$, except that if the new position would lie outside the box, one or more elastic collisions with the walls are simulated to bring the new position within bounds.

To simulate internal compartments or other fixed features within the box, it is possible to place "obstacles" at grid points within the box. Collisions among atoms moved to an obstacle point are inhibited. Instead, during the collision phase, atoms at an obstacle have their velocities altered in a way depending on the type of obstacle. The most useful obstacle simply reverses the velocity of any atom landing on it, so the next move phase will return the atom whence it came. An internal wall within the box can be constructed with a thick slab of such obstacles. The slab acts as a "soft" wall: an atom with a velocity component perpendicular to the slab of, say, u, may penetrate into the slab by up to a distance $(u - 1)$, since in move phases, atoms move from place to place without regard to intervening points (except for the box walls). Thus, a soft wall of thickness W may be passed through by an atom of velocity component perpendicular to the wall exceeding W. In most uses, the slab is made thick enough to reflect virtually all atoms. Soft walls are "rough": they reverse both velocity components of impinging atoms.

"Hard" walls like the bounding walls of the box have been simulated in one set of experiments to set up two rectangular chambers connected by a long, thin pipe. Hard walls are "smooth": collision with a hard wall does not affect an atom's velocity component parallel to the wall.

Given the above description, it is clear that the simulated gas is governed by deterministic, reversible laws. Both forward and reverse state sequences through time can be followed. The only difference between a forward time step and a backward step is that in the latter, the move phase is done before the collide phase, and subtracts velocities from positions. The asymmetry

between the "collide-move" forward step and the "move-collide" backward
step does not represent a time asymmetry in the gas "physics". It arises
from the arbitrary decision to define the integer "time instants" as occurring
just before, rather than just after, a collision phase. The physical progression
simulated is

```
Earlier ..... C M C M C M C M C M ..... Later
Defined times ... T---T---T---T---T---T
```

It would of course have been possible, and exactly equivalent, to step back in
time by reversing all velocities, proceeding as for stepping forwards, and finally re-reversing velocities. The introduction of non-determinism is described
later.

8.12.1 Realism of the Simulation

Many tests have been done to see if this simple simulation resembles a thermodynamic system. The tests typically began with the gas in a partially
ordered state with individual atom positions and velocities chosen pseudo-
randomly in accordance with a specified macro view. The evolution of the
system was then followed for some hundreds or thousands of time steps while
tracking a physical quantity of interest.

In some experiments, the total gas energy was allowed to change by introducing a "heater" obstacle which in a collision phase, adds +1 to the V
component of any atom at the heater's position if the time is even, or −1
if the time is odd. This action is deterministic and exactly retraceable. The
effect of a heater is weakly to couple the gas to a heat source of infinite temperature, since the only velocity component distribution unchanged by the
heater is the Uniform distribution.

Pressure was measured by accumulating the momentum transfer between
one wall of the box and the atoms bouncing off it. A useful property of a gas
state is its velocity component distribution. This was formed in a histogram,
adding both U and V components to the same histogram since (at least near
equilibrium) the two distributions should be similar and nearly independent.
A "state entropy" was calculated from this distribution, which ignores the
entropy of the spatial distribution of atoms (which, near equilibrium, is expected to be uniform) but serves as an indicator of the true entropy.

A visual picture showing the positions of all the atoms could be displayed
at chosen intervals, and plot files written giving the history of chosen macro
quantities such as entropy, temperature, etc.

Experiments showed the following properties of the gas:

− Near equilibrium (i.e., after sufficient time steps to erase any initial order), the velocity component distribution matched the expected (discrete)
  Maxwell-Boltzmann distribution, and the state entropy approached its
  maximum possible value.

- When the collision laws were modified to allow collisions only when exactly two atoms occupied the same position, the effect is to make the laws independent of the identities of atoms, and to prevent any two atoms ever having the same position-velocity state at the same time. The atoms should then behave as fermions rather than classical particles. When this was done, the velocity distribution matched the Fermi-Dirac distribution. As the difference is only apparent at very low temperatures, and the modified simulation was slower to approach equilibrium, all further experiments used the Maxwell-Boltzmann form.
- Temperature was measured by a maximum-likelihood fit to the velocity distribution. This measurement is meaningful only near equilibrium.
- The gas obeyed the classical gas law

$$Pressure \text{ proportional to } Density \times Temperature$$

- The Specific Heat at Constant Volume had the expected value.
- The speed of sound, as measured by oscillations in a pipe, had the expected value

$$Speed = \sqrt{\frac{\gamma \times Pressure}{Density}}$$

where γ, the ratio of specific heat at constant pressure to specific heat at constant volume, has the value 2 for an ideal 2-dimensional gas.
- In an experiment where the gas is slowly heated by an obstacle at infinite temperature, the increase in entropy followed the classical law

$$\Delta Entropy = \Delta Heat/Temperature$$

- When the box was divided in two by a soft wall of thickness W and the initial state had all the gas on the left side of the wall with a mean speed much less than W, "hot" atoms slowly diffused to the right side, with a consequent initial cooling of the remaining gas on the left, but eventually the densities and temperatures of the gas on both sides became equal, with a temperature equal to the initial temperature. No quantitative analysis of the expected ideal gas diffusion behaviour has been done, but the simulation was at least qualitatively as expected.
  Similar qualitative realism was found when the box was divided into two chambers connected by a thin pipe, all with hard walls.
- Whatever the nature of the confinement or initial order, every simulation showed eventual convergence to an equilibrium view.

8.12.2 Backtracking to the Past

Whatever the details of the simulated situation, running for a long time showed convergence towards equilibrium. When the simulation was then reversed, the final state always returned exactly to the initial state, showing that the simulation indeed followed reversible deterministic laws. Further, if several runs were made with different initial states randomly chosen in accordance with the same initial macro view, the final states reached showed no significant macro differences. That is, the evolution of the macro properties of the gas is insensitive to initial differences on a micro scale.

If, however, the final state was perturbed by moving the position of a single randomly chosen atom by a single grid point, the reversed simulation invariably returned to an initial state with less order than the final state. In some experiments, the macro history running backwards would follow the true history fairly closely for 50 or so time steps before diverging towards a less ordered past, but divergence always occurred.

A simulation was done with a 10,000 atom system in a 200-by-50 box, where the initial "Then" state was in equilibrium with entropy close to the maximum possible, a uniform random distribution of atom positions, and a well-defined temperature of 2.0 (taking Boltzmann's constant as one.) Two "heater" obstacles were placed in the box, weakly coupling the gas to a high temperature. The simulation was run forward for 1000 steps to yield the "Now" state. As expected, the gas slowly heated up. At all times the gas had a well-defined temperature, which gradually increased from 2.0 in "Then" to 2.113 in "Now". Reverse simulation was then done from "Now" back to "Then". Starting with the exact "Now" state, deterministic devolution exactly recovered the "Then" state. Starting with the "Now" state with one atom shifted by one grid point, deterministic devolution followed the true temperature trajectory for about 50 steps, but thereafter showed an increase in temperature into the past, reaching 2.216 at the "Then" time. Non-deterministic devolution diverged from the true path almost immediately, reaching 2.229 at "Then". In both the latter cases, the backtracked "Then" had higher entropy than "Now", and of course higher than the true "Then".

8.12.3 Diatomic Molecules

An elaboration of the simulation imitates a gas such as hydrogen, in which pairs of atoms may combine with the release of energy to form diatomic molecules. In the simulation, the collision table has special entries which allow a pair of free atoms colliding with certain relative velocities to join, forming a molecule. The molecule then moves as a point unit according to the mean velocity of the atoms, and occupying a single grid position. The atoms retain individual velocities, their relative velocity cycling through a cycle of values under the control of another table, simulating a high-frequency internal vibration of the molecule involving a continual exchange between the kinetic energy of the atoms' relative motion and the potential energy of their bond. The total energy of a newly formed molecule exceeds the original total energy of the free atoms by a fixed "binding" energy. If nothing intervenes, the molecule will split into free atoms when its vibrational cycle returns to the state at which the molecule formed. However, collisions of its atoms with other free or bound atoms may alter its vibrational state, with the gain or loss of vibrational energy. If the molecule's vibrational energy falls below the binding energy, it cannot split.

A cool gas of free atoms may exist for a long time in a metastable state, since few of its collisions may reach the lowest relative velocity needed for combination. This threshold energy and the binding energy can be specified independently. If a small body of fast atoms enters the gas, triggering combinations, the local temperature may rise triggering more, and a simulated fire front or, in some cases, detonation shock wave appears.

The inclusion of this elementary "chemistry" into the simulated gas made no difference to the kinds of behaviour on which this study of the inference of the past are based. When the simulated laws are deterministic, the simulation remains exactly reversible, and a past state can be recovered exactly from an exactly known present state. Almost all partly ordered states evolve and devolve to less ordered states. When the present partly ordered state has evolved from a more ordered past state, exact retracing recovers the more ordered past state, but if retracing is attempted from an inexact view of the present, differing from the true present state only by a tiny perturbation, retracing almost always yields a past view less ordered than the present. With non-deterministic laws, virtually all attempts to recover the past state yield a less ordered view than the present.

8.12.4 The Past of a Computer Process

The reader may be forgiven for doubting the realism, and hence the relevance, of these simulation studies. A two-dimensional ideal gas in a quantized Newtonian space-time is a poor substitute for a real closed system. However, the simulation *is itself a real system with real conservation laws and real state transitions* even if it is only a pattern of data digits in a computer. It should be a very simple sort of system to reason about, since its state at any time step is very well defined and easily examined, its laws of behaviour are exactly known if one likes to inspect the program, and it is completely isolated.

The skeptical reader might ponder the following problem: given knowledge of the state of this real system after some thousand time steps, and given that the present state of the data arrays shows some order, what can he or she *deduce* about their state 500 steps ago? Given exact knowledge of the present data, there is no problem if the computer is not using an inaccessible random number source to simulate indeterminacy, but if it is, or if there are one or two tiny errors in the report of the present data, the results discussed here show that *deduced* past states (or probability distributions over past states) will almost certainly be misleading at least as to the degree of order in the past. If this real, but very simple, isolated and fully understood system presents deduction of the past with real trouble, can we expect to do better in the larger world?

8.13 Addendum: Why Entropy Will Increase (Additional Simulation Details)[1]

If we are presented with a closed system which appears to be in equilibrium, there is nothing which can be usefully predicted of its future, save that it will probably remain in equilibrium. We can hope to make interesting predictions about the future of a closed system only if it is now partially ordered, i.e., has entropy significantly below the maximum. I will now argue that we may confidently predict that its entropy will increase. Several cases will be treated.

(a) Deterministic Laws, Exact View.

Assume the Laws of Physics are deterministic, and that we have exact knowledge of the present state $s(0)$ at time $t = 0$. Suppose $s(0)$ is a state of low entropy H_0. The deterministic laws allow us to deduce the state $s(f)$ at some future time $f > 0$. The repeated application of the time-step mapping $s(t + 1) = M(s(t))$ will no doubt have changed the state and its entropy. Since the mapping is based on the microscopic laws, which make no reference to entropy and are reversible, there is nothing inherent in the mapping which can lead us to expect that the entropy of a state will be related in any obvious way to the entropy of its successor, although, since a single time step will usually produce only a small change in the macroscopic view of the system, we may expect the resulting change in entropy to be small. After the f steps leading to time f, therefore, we can expect the initial entropy to have changed by some amount, but otherwise to be almost a random selection from the set of possible entropy values. Recall that in a closed system, the number of states of entropy H is of order $\exp(H)$. There are therefore far more high entropy states than low entropy ones, so the chances are that $s(f)$ will very probably have an entropy greater than the initial state $s(0)$. In a deterministic closed system of constant energy, the sequence of states from $t = 0$ into the future is a well-defined trajectory through the adiabatic state space at that energy. The trajectory may eventually visit all states in the adiabatic space, or only some subset. Given that the state space is discrete, an adiabatic space can only contain a finite, albeit huge, number of states. The trajectory from state $s(0)$ thus must eventually return to state $s(0)$, and since the mapping is one-to-one, must form a simple cycle in which no state is visited more than once before the return to $s(0)$. Since it is cyclic and produced by time reversible laws, the trajectory must show just as many steps which decrease entropy as increase it. However, because the trajectory can be expected to visit far

---

[1] The editors included this section for completeness. It contains material from Section 8.4 together with material on simulations which appeared in previous versions of the chapter. Chris Wallace removed these details to avoid distracting readers from the main argument. However, the editors felt that his careful simulations were interesting in their own right.

more high entropy states than low entropy ones, if we start from some low entropy state and advance for some arbitrary number of steps, we must expect to arrive at a state of higher entropy. If this argument is accepted, the Second Law will in general apply even given deterministic, time reversible laws. No paradox is involved.

(b) Simulation of Deterministic Laws, Exact View.

The two-dimensional gas simulation has been used as a model system to check the above argument. Although it simulates a very simple system, and has no provision for important real-world processes such as gravity, chemical bonding, electromagnetic interactions, etc., it is not obvious that the argument is sensitive to such phenomena, so the simulation should give some useful indication of the validity of the argument.

Some hundreds of experiments have been conducted in which a partially ordered (sometimes very highly ordered) state was set up in the gas and allowed to evolve. As it is infeasible directly to compute the state entropy as defined above, the simulator computes a simple but useful surrogate. For each state entered, a histogram of atom velocity components is collected. For each atom, both the U and V velocity components are entered in the histogram, as these are expected to be statistically independent degrees of freedom.

The entropy of the velocity component distribution is then computed as

$$H_v = \sum_v -n[v] \log(n[v]/N)$$

where $n[v]$ is the histogram count for component velocity v and N is the total of all counts, which is of course twice the number of atoms.

H_v is a measure of one component of the state entropy, but does not include any spatial-distribution entropy. However, it suffices as a measure of disorder in the velocity distribution of the state. Note that the number of states having a particular value of H_v is approximately proportional to $\exp(H_v)$.

At each time step, the simulator records on a file the time t and the per-freedom entropy $G_v(t) = H_v/N$. This allows the G_v history to be displayed as a plot versus time. Note that G_v may display distinct oscillations as time goes by, superimposed on a general increase, as G_v shows only one component of the system entropy, and in some situations, e.g., an initial standing wave in the gas, there is a periodic exchange of entropy between velocity and spatial components. Such oscillations tend to decrease as the system approaches equilibrium.

Further, the simulator can display a picture of the spatial positions of the atoms at chosen times, allowing a qualitative assessment of the behaviour. Other properties of the gas can also be recorded, such as the gas pressure, total momentum, and an estimate of temperature. The temperature T is determined by choosing the value which gives the best fit between the observed velocity component histogram and the equilibrium Maxwell-Boltzmann distribution $\Pr(v)$, which is proportional to $\exp(-v^2/(2T))$ (our "atoms" have mass 1).

Many forms of partial order have been simulated, some in boxes containing barriers. Situations simulated include the passage of gas through a channel from a chamber of high initial pressure to one of low initial pressure, the oscillation of a standing sound wave, the impact of a ball of cold, dense gas on a body of cold, low density gas, the diffusion of fast atoms over an "energy barrier", the exposure of the gas to an external "hot spot", and many others. To summarize the results, in every case the entropy of the system increased, finally (perhaps after some tens of thousands of time steps) becoming close to the equilibrium value, but fluctuating slightly from step to step. Qualitatively, the pictures of the gas showed the gradual decay of the initially visible order.

Quantitative results were also encouraging. The final states showed velocity distributions conforming to Maxwell-Boltzmann statistics. The variation of pressure with density and mean atomic energy was as expected. In the sound wave simulation, the measurement of the standing wave frequency yielded an estimate of sound velocity. This allowed estimation of "gamma", the ratio of specific heat at constant pressure to the specific heat at constant volume.[2] The estimated value was 2.02, compared with the theoretical value of 2 for a two-dimensional ideal gas.

In one batch of experiments, the collision law was altered to forbid collisions at a point unless exactly two atoms occupied it. This restriction removes the effect of atom indices from the collision phase, with the result that the atoms are no longer identifiable, and makes it impossible for two atoms to have the same position and velocity. Thus, the atoms become fermions rather than classical. The equilibrium velocity distribution was then found to conform to Fermi-Dirac statistics rather than Maxwell-Boltzmann.

Overall, no simulation of the future of a partially ordered initial state showed any macroscopic behaviour at variance with thermodynamic predictions. In particular, the invariable increase in state entropy confirmed that the Second Law could apply to a deterministic system governed by time reversible laws. Turning to Boltzmann's theorem that velocity distributions tend towards high entropy, note that his assumption of no correlation between the initial velocities of colliding particles seems not to be essential. In many of the simulations, it can be shown that most of the collisions occur between atoms which have collided before. If collisions produce correlations in the resulting atom velocities, these correlations must certainly have been present. However, the simulations agree with the conclusions of the theorem: the velocity distributions did tend towards higher entropy, and did approach the Maxwell-Boltzmann equi-

---

[2] See Section 8.12.1.

librium distribution despite whatever correlations in atom velocities may have been induced in earlier collisions. (It is perhaps worth noting that, in a collision between two atoms whose initial velocities have been independently sampled from the Maxwell-Boltzmann distribution, the final velocities are not statistically correlated.)

(c) Deterministic Laws, Inexact View.

If, rather than knowing the precise initial partially ordered state $s(0)$, we have only an imprecise view of the initial state, the argument for expecting entropy to increase can only be strengthened. We may well feel more justified in regarding the entropy at future time f as being in some sense a random selection from a range of values, since we no longer have any certainty about the future state $s(f)$. Given this uncertainty, the larger number of high entropy than low entropy states becomes perhaps a sounder basis for expecting $H_f > H_0$. Because each time step has but a small effect on the macroscopic view, we also expect the macroscopic behaviour of the system to be insensitive to the fine detail of the initial state, and to be well predicted by a macroscopic, or at least inexact, initial view.

(d) Simulation of Deterministic Laws, Inexact View.

The above expectations were tested by simulations of two sorts. In the first sort, a "true" partially ordered initial state $s(0)$ was defined and its evolution followed to a future time f. Then several inexact copies of $s(0)$ were made, each resulting from a change of plus or minus one in a randomly chosen position coordinate of a randomly chosen subset of the atoms, the fraction of perturbed atoms being specified. The evolution of each copy was then followed to time f and compared with the results from the "true" $s(0)$. The copies showed macroscopic behaviour and entropy histories very close to those of the true state, but microscopic comparison of the final states showed great differences in the positions and velocities of the atoms.

In the second sort, the system was set up to have two chambers linked by a narrow channel. The initial view specified only the densities and mean atom energies in the two chambers. Several initial states were constructed consistent with this view, with the atoms in each chamber having random, high entropy spatial and velocity distributions meeting the different specifications of the two chambers (see Figure 8.3). The evolutions of the several initial states were then compared, and again showed no significant differences at the macroscopic level.

In these two-chamber simulations, the state entropy was calculated by forming separate velocity-component histograms for the two chambers and adding the entropies calculated from each histogram. This approach takes into account the partial order which is shown by different densities and/or temperatures in the two chambers, and gives a lower value than given by the overall velocity distribution if such partial order exists.

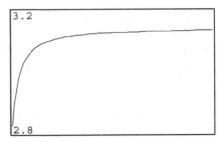

Fig. 8.4. Two-chamber + pipe model (from Figure 8.3); velocity entropy increase for 10,000 time steps from original "Then" state; non-deterministic laws.

(e) Nondeterministic Laws.

   If we suppose the fundamental microscopic laws to be nondeterministic (but still reversible) the state-to-state mapping $M()$ is replaced by a probabilistic law

$$\Pr(s(t+1) = r) = M(r|s(t))$$

   Given $s(0)$, we can at best compute a probability distribution over the possible states at time $f > t$, and are the more justified in expecting $s(f)$ to have higher entropy than $s(0)$.

(f) Simulation of Nondeterministic Laws.

   The simulation was modified to construct two different collision tables for effecting collisions. Both observed the constraints of time reversal, relative velocity component parity conservation, energy conservation, etc., described above. In collision phases, each pair-wise collision is effected by one or other table. The choice is determined by a conventional pseudo-random number generator of cycle length 2^{32}, and chooses each table with equal probability. The generator is a mixed congruential generator of 32-bit integers:

$$r[n+1] = ((69069 * r[n]) + 31415927) \text{ Modulo } 2^{32}$$

The inverse relation is

$$r[n] = (-1511872763 * r[n+1] + 670402221) \text{ Modulo } 2^{32}$$

which allows the simulation to be run backwards by retracing the sequence of choices. This scheme introduces uncertainty into the laws. Most collisions can have two equiprobable results. As expected, all simulations of the evolution of partially ordered states continued to show increasing entropy and macroscopic behaviour conforming to thermodynamic predictions. Figure 8.4 shows the long term increase in entropy in the two-chamber model.

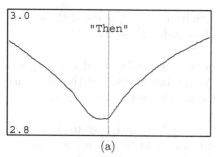
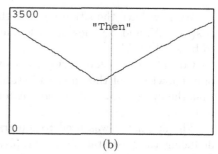

(a)   (b)

Fig. 8.5. Two-chamber + pipe model (from Figure 8.3). Deterministic laws. (a) State velocity entropy G_v evolution for + and − 250 time steps from original "Then" state; G_v range [2.8, 3.0]. (b) Atom count difference (Right chamber − Left) for + and − 250 time steps from original "Then" state; 30,000 total atoms.

8.13.1 Simulation of the Past

First, note that the correctness of the simulator has been well checked. In at least 100 experiments starting from partially or highly ordered states, the simulation has been run (usually but not always) until a near-equilibrium view was reached with a near-maximum G_v. The simulation has then been put into reverse and run backwards to time zero. In every case, the initial condition was exactly restored. In many cases, the simulation was then driven further back in time, to equally remote negative times in the past, then forwards again to time zero. Again, the initial state was always exactly restored. (When running backwards with nondeterministic laws, the pseudo-random number generator was run in reverse to reproduce the "random" choices of collision table at each collision.)

For all the initial states simulated, without exception, running the simulation backwards to a negative time b resulted in a state $s(b)$ with higher entropy than the initial state $s(0)$, provided only that b was earlier than about −10. Most runs began to show an increase in entropy after the first two or three backward steps from the "present" $t = 0$. Figures 8.5(a,b) illustrate the effect in the two-chamber-and-channel model.

To gain more insight into this phenomenon, another series of experiments was done with a different protocol. First, a partially ordered state was set up, say, T (for "Then"). This was run forwards in time for long enough to produce an unmistakable increase in entropy, giving a new state, say, N (for "Now"). The time was chosen so that N was still clearly ordered, with an entropy well below the equilibrium maximum H_m, but definitely greater than $H(T)$. Let the number of time steps from T to N be k. Then the simulator was reset to have N as initial state at time $t = 0$. Of course, running it backwards to time $t = -k$ recovered T, provided that, when "non-determinism" was enabled, the pseudo-random generator was put into reverse with an initial seed value equal to the final value reached in the generation of N from T.

We now have the situation that the "initial", $t = 0$, state of the system is $s(0) = N$, and we know its true ancestral state at the "past" time $t = -k$ to be $s(-k) = T$.

Consider the task of trying to recover the ancestral state at $t = -k$ given only knowledge of the "present" state N, the invariants, and the assurance that the system has not suffered external interference in the period $t = -k$ to $t = 0$.

The simulator was used to illustrate the effects of trying to do this by deducing the past state from the present state and the known microscopic laws. The simulation was run backwards from the final state N for k steps and the final state $\hat{T}$ at time $t = -k$ compared with T. The results can be briefly summarized. Running with deterministic laws and beginning with the exact state N, $\hat{T} = T$, i.e., the past is exactly recovered. This case exposes the loophole in Boltzmann's theorem. If N is the deterministic evolution of a lower entropy T, of course the collisions which produced the increased entropy will, when reversed, reduce the entropy of N back to that of T. The collisions in the evolution of T to N leave the final velocities in N correlated, but the correlation is not so much between the velocities of the last pairs of atoms to collide as in the entire pattern of atom positions and velocities.

More realistically, we cannot hope to have exact knowledge of N, so N was replaced by a state N_p obtained from N by randomly perturbing the positions of some of the atoms by plus or minus one grid point. The population of N_p states so generated can be thought of as a very precise but not quite exact view N_v of N. Then, running with deterministic laws and beginning with several perturbed versions of N, $\hat{T}$ was never equal to T and rarely looked close. In particular, $\hat{T}$ always had greater entropy than N, and so of course greater than T, even when the number of perturbed atoms was very small. It seems that the loophole in Boltzmann's theorem is very small, and the pattern in N required to slip back through the loophole to a lower entropy past state is very fragile.

Running with non-deterministic laws was done with an unreversed pseudo-random generator with various choices of seed, since we could have no direct knowledge of the random events which occurred leading to the present state N. In these runs, even starting with the exact state N, T was never recovered, and all instances of $\hat{T}$ had greater entropy than N. The non-determinism appears to erase any loophole finding correlation. Figures 8.6 and 8.7 show results on the two-chamber + pipe model. Figures 8.1(a-d) and Figures 8.2(a-d) show results from two simulations of collisions between two bodies of gas.

8.13.2 A Non-Adiabatic Experiment

All the argument and simulation to this point have assumed adiabatic (constant energy) systems. While this assumption follows from the assumption of a closed system, many of the conclusions seem to apply to systems which are subject to some external influences. A simulation was done with a 10,000 atom

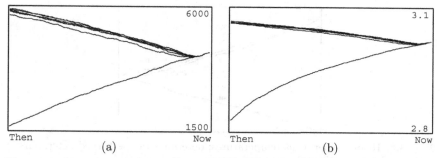

Fig. 8.6. Two-chamber + pipe model (from Figure 8.3). Deterministic laws. (a) Atom count difference devolved from "Now$_D$" to "Then"; traces show devolutions of "Now$_D$" with 0, 300, 100, 30, 10, 3 and 1 atoms perturbed by one grid point; only the unperturbed "Now$_D$" devolves to "Then". (b) State velocity entropy devolved from "Now$_D$" to "Then"; traces show devolutions of "Now$_D$" with 0, 300, 100, 30, 10, 3 and 1 atoms perturbed by one grid point; only the unperturbed "Now$_D$" devolves to "Then".

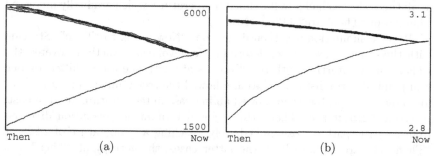

Fig. 8.7. Two-chamber + pipe model (from Figure 8.3). Non-deterministic laws. (a) Atom count difference devolved from "Now$_U$" to "Then"; traces show devolutions of "Now$_U$" with different random generator seeds; only the run with the seed equalling the final value reached in the evolution of "Now$_U$" devolves to "Then". (b) State velocity entropy devolved from "Now$_U$" to "Then"; traces show devolutions of "Now$_U$" with seven different random generator seeds; only the run with the seed equalling the final value reached in the evolution of "Now$_U$" devolves to "Then".

system in a 200-by-50 box, where the initial "Then" state was in equilibrium, with entropy close to the maximum possible, a uniform random distribution of atom positions, and a well-defined temperature of 2.0 (taking Boltzmann's constant as one.) Two special "heater" obstacles were placed in the box. In a collision phase, a "heater" adds +1 to the Y component of any atom at the heater's position if the time is even, or adds −1 if the time is odd. This action is deterministic and exactly retraceable. The effect of a heater is weakly to couple the gas to a heat source of infinite temperature, since the only velocity component distribution unchanged by the heater is the Uniform distribution.

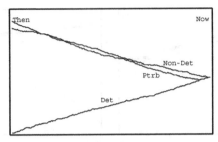

Fig. 8.8. Heater experiment temperatures; deterministic, perturbed deterministic and non-deterministic retraces; temperature range $[2.0, 2.25]$.

The simulation was run forward for 1000 steps to yield the "Now" state. As expected, the gas slowly heated up, so slowly that at all times the velocity distribution remained close to the equilibrium Maxwell-Boltzmann distribution, and G_v close to the maximum possible at the current total energy. Thus, at all times the gas had a well-defined temperature, which gradually increased from 2.0 in "Then" to 2.113 in "Now".

Reverse simulation was then done from "Now" back to "Then". Starting with the exact "Now" state, deterministic devolution exactly recovered the "Then" state. Starting with the "Now" state with one atom shifted by one grid point, deterministic devolution followed the true temperature trajectory for about 50 steps, but then showed an increase in temperature into the past, reaching 2.216 at the "Then" time. Non-deterministic devolution diverged from the true path almost immediately, reaching a temperature of 2.229 at "Then" (Figure 8.8). In both the latter cases, the entropy at "Then" was higher than at "Now", and of course higher than the true "Then" value. Although not shown in Figure 8.8, deterministic devolution of the true "Then" state to times earlier than "Then" also showed an increase in temperature into the past.

Clearly, this non-adiabatic situation presents the same problems for the deduction of past states as has been found in adiabatic systems, although now the increase in entropy occurring in forward evolution and backward devolution arises from an input of energy rather than the dissipation of order.

9. MML as a Descriptive Theory

Thus far, Minimum Message Length has been mainly presented as a normative principle for how statistical and inductive inferences should be assessed, with excursions into methods which, in limited contexts, can lead to inferences assessed as good according to this normative principle. If the normative principle is sound, one would hope to find that the history of scientific enquiry showed that inferences which MML assesses as good have been generally accepted, and conversely, that theories generally accepted in the light of the then-available evidence are not assessed as unacceptable (Section 1.5.2) by MML.

It would be unreasonable to demand of a descriptive hypothesis based on MML that it assert an exact correspondence at all times between generally accepted theories and theories which minimized the explanation length of the available evidence. First, a good theory can take many years to reach general acceptance simply because of the rather conservative nature of institutionalized Science. As Kuhn observed, sometimes it has to wait for all the old guard to die. Second, in most fields we cannot tell what theory exactly minimizes the explanation length of a body of data, because there is no algorithm for finding it if the set of possible theories is the set of computable functions. Third, for most past times, it would be hard to determine just what the relevant "body of available evidence" was. An MML descriptive theory must make a weaker assertion than that scientific enquiry has always adhered to the MML normative principle.

Instead, I propose the descriptive theory that scientific enquiry has shown a tendency for the generally accepted body of theory to evolve towards theories which provide concise explanations of widely known data, preferring those which give the most concise explanations, and rarely preserving theories which are unacceptable by MML assessment, i.e., which fail to provide explanations more concise than the raw statement of the data. Further, the descriptive theory asserts that the language of Science has shown a tendency to evolve in a way permitting the more concise statement of the theories to be widely accepted. Further still, it asserts that these tendencies have not been restricted to what would today be called Science, but predate it by centuries and have existed at least since the origins of human speech.

How should this theory be assessed? Since a case for MML as a normative principle for the assessment of theories has been the main focus of this work, the descriptive theory will be assessed to see if it is acceptable by MML's criterion. That is, does the descriptive theory provide a concise explanation for the historic data it professes to cover? (MML as a theory about the value of theories can encompass its own assessment without circularity. It is entirely possible that the descriptive theory about the history of accepted inferences might fail the normative criterion for acceptance.) In this chapter there will not be space for a comprehensive "explanation" of the history of scientific theory acceptance, nor have I the knowledge to frame one. At most, the chapter offers a faint outline which may allow the reader to assess the descriptive theory in the light of the reader's own knowledge of the history. In following this outline of an "explanation", remember that the normative principle does not require of an acceptable theory that it be universal, "true" in any absolute sense, exact or free from failure in specific cases. It certainly does not require that the theory give the best possible (shortest) explanation of the data, only that it give an explanation more concise than the bare facts. Of course, if the theory turns out to explain only some limited field of data, its assessment suffers because either the assertion must state the limitation or each case in the "detail" must carry a flag to indicate that it is an exception and is encoded in its raw-data form. Either way, the "explanation" message length is increased.

The "descriptive MML" theory of theory acceptance is obviously not the best possible theory about the development of Science. It has nothing to say about the social, economic, technological and personal factors which clearly have imposed their own patterns on this development. But in this omission it is no worse than Newton's theories about the orbital motions of the planets and moons, which say nothing about why the planets have the masses, orbital diameters and eccentricities and attendant moons which in fact they have. As a theory about the structure of the Solar system, it is limited to explaining just a few aspects of the data. Similarly, the descriptive theory proposed here is limited to just one aspect of history: which theories get accepted (and perhaps, which theories even get proposed). How, when and why theories are invented are outside its scope.

9.1 The Grand Theories

There are many examples of scientific theories which indubitably have great explanatory power and handsomely meet the MML criterion for acceptance. The theory of the roughly spherical shape of the Earth and its approximate size, the heliocentric theory, Newton's laws of motion and gravity, the element theory of chemistry, the periodic table and the atomic theory of the elements, the nucleus plus electron orbits theory of atomic structure, the wave theory of light, etc., all have quite concise assertions and enable the concise encoding of vast bodies of observations. Their cases are today so strong they seem to need no argument. It is less clear that they met the MML criterion when first proposed, and in some cases general acceptance was withheld long after the MML criterion was met. For instance, one reads that Columbus intended to sail to the East Indies, although knowledge of the size of the world which had been available for centuries showed the distance to be beyond the capacity and provisioning of his ships. Similarly Galileo and Wegener found little acceptance for their theories despite their immediate explanatory power.

In more recent times, acceptance has come more rapidly. Quantum mechanics and the Special Theory of Relativity gained general acceptance (at least among those sufficiently interested to consider the matter) within a few years rather than scores. It might even be said that acceptance of the present (2004) "Standard Model" of fundamental particles was granted before its explanatory power was really established. But there are notable exceptions. Darwin's theory of evolution is still rejected by many people, not all of them uninformed, yet no alternative explanation of the origin of species has demonstrated MML acceptability.

In summary, the Grand Theories of Science show plenty of examples meeting the MML criterion and there are few examples of theories which MML would assess as superior but which remain outside the canon. There do seem to be a few theories fairly widely accepted for which normative MML support is questionable. It is not obvious that Freudian psychology or the theory of free markets are able to offer explanations acceptable in MML terms. Overall, the acceptance histories of the Grand Theories lend some support for the descriptive MML theory, but only in a somewhat weakened form which allows for a variable and sometimes long period to elapse between the normative MML assessment and a corresponding general acceptance or rejection. In support of the descriptive theory, it is worth noting that few of the generally accepted grand theories would today be regarded as exactly true or complete, or even fully consistent with one another. This situation is not surprising if acceptance is based on ability concisely to explain data, but would not be expected if Science insisted its theories be consistent with every observation.

9.2 Primitive Inductive Inferences

The descriptive MML theory asserts that the history of pre-scientific inductive inferences shows rough compliance with normative MML principles, their acceptance allowing the concise encoding of observational data. In this context, we take "encoding" to mean both encoding in the neural machinery of our own minds and encoding in a natural language intelligible to our fellows.

Some inductive inferences we generally accept are so primitive that they may escape conscious attention. Our senses provide us with very indirect information about the external world, and to make much sense of them we must form and accept many hypotheses about the world which are not logically implied by our sensory input. We accept hypotheses that there are persistent objects, that some of these move and others do not, that their locations and movements inhabit a three-dimensional space and that what we see is to be explained by the rules of projective geometry (which we must learn), that one direction in the space is distinguished and dangerous, that some of the moving objects are flexible and autonomous, and some of these are in many ways like ourself, etc., etc. Each of these hypotheses enables us to recode our sense data into shorter forms within which many further patterns may be found leading to further hypotheses such as the existence of a distinguished other person and her likely responses to our cries, and, rather later, the useful classifications of other persons by size and sex, and the correlations between these classes and their behaviour. No doubt many of these hypotheses are largely wired-in at birth, having been "learnt" by the evolution of our ancestors, but how hypotheses are formed is not our present concern. We accept them, they enable shortened explanations of our sense data, and while ultimately we may form even better explanations, these will usually act to "explain" the already-compressed "details" resulting from the primitive explanations rather than directly explaining our raw sense data.

Whether these primitive inductions come close to minimizing the length of the explanations they provide us is not easily answered. I can only observe that, at the level at which they operate, I find it hard to postulate an alternative superior in MML terms.

9.3 The Hypotheses of Natural Languages

The most general variety of MML explanations which has been treated in any detail are those intended to be decoded by some Universal Turing Machine. Natural languages such as English and Hindi have at least the expressive power of the input language accepted by a UTM. Whether they have more has not been decided. They are at least as powerful because it is possible in such a language to define the structure and operation of a UTM and hence to give a verbal simulation of a UTM's computation. Natural languages also seem to share with the inputs to UTMs the property that there is no algorithm for determining in general whether an input (an utterance or writing intended to be intelligible to its receiver) is meaningful. They also share the property that they can be used to modify and extend themselves. Just as the input to a UTM can effectively modify the UTM to behave as a different UTM with a different input language, so statements in a natural language can be used to extend the language by defining new words or more radically by defining sub-languages such as mathematics. Often, the modification is intended to be temporary, applying only within a finite context, such as my subversion of the word "explanation" to serve a purpose limited to this work.

Natural languages differ from UTM languages in that the receiver who is to decode a natural language statement is not fully known to the transmitter, so the language must have sufficient redundancy to allow for the uncertain "priors" of the receiver. This difference also opens the possibility of a form of communication failure not evident in UTM inputs, namely the possibility of ambiguity. Not only may a message fail to make sense (which is possible for a UTM) but in a natural language it may be decodable in two or more ways with different meanings. These properties make it impossible to argue with any precision about whether the hypotheses embedded in a natural language are those which would minimize the expected length of statements in that language. However, some vague assessments may be possible.

9.3.1 The Efficiencies of Natural Languages

The basic hypotheses of a language are embodied in its words. A common noun is useful insofar as it names a class of phenomena whose members share so many features that in many statements it is unnecessary to deal with each member individually. A common verb is useful insofar as it names a class of actions whose members similarly share so many features that to announce a particular act by its class name is to convey all that need be conveyed about the act. In the statement "Jessie milked a cow", "Jessie" is a unique label for a unique object. (Actually, even this proper noun is a label for a class of phenomena: all those appearances within my senses of patterns which I "explain" by the inductive inference that there is a persistent object in the class of persons, and that one feature of these appearances is the common occurrence of the sound written as Jessie.) The common noun "cow" names a wider class, but for the purposes of the communication the particular instance involved is irrelevant, and perhaps unknown. Likewise, the verb "milked" serves as a sufficient specification of the act performed.

These verbal "hypotheses" may be held to be acceptable in MML terms just in case they allow a concise formulation of a message which might otherwise have to read even worse than "A visual image somewhat similar to those frequently accompanying the sound 'Jessie' approached another image which appeared to have four legs, two horns, an underslung bag with four smaller bags below ... and after some extreme parts of the first image became adjacent to the smaller bags a third visual element appeared intermittently of a generally white colour and unstable shape ...." The MML acceptability will depend on the frequency of the message which these verbal hypotheses help to encode.

By the kind of assessment outlined above, most of the shorter words of English can be considered to be good inductive inferences, especially if assessed in the context of the kind of message which might have been most common some hundreds of years ago. Today, the assessment might be different. We might well find that "fraudulent transaction" is a class label more in need in our newspapers than "milked". But no matter: the language adjusts itself and "scam" is invented. The fairly general agreement among many natural languages on which things, classes, concepts and acts deserve a word is some evidence that the different languages have all developed to convey similar messages. While this convergence could be explained by other criteria, brevity of expression is a plausible ingredient, along of course with common development from a small number of ancestral languages.

The development of specialized sub-languages and jargons for all sorts of specialized messages seems also driven in part by an aim for brevity. When a novel hypothesis or concept becomes accepted, a new or purloined word is usually adopted to name it. The words "mass", "force" and "energy" were adopted from common usage to have narrow but immensely useful meanings in Physics, two of them as names for quantities which were themselves hypotheses, being invisible to our senses. "Valence" and "bond" in Chemistry, "infection" in Medicine, "harmonic" in Music, "interest" in Commerce and so on, all suggest a language extension serving, if not consciously motivated by, brevity in stating the kinds of message frequent in the specialized discourse.

9.4 Some Inefficiencies of Natural Languages

Apart from the necessary redundancies of natural languages which cope with uncertain knowledge of the receiver(s), there seem to be inefficiencies which defy easy explanation by a descriptive MML theory. For instance, English in particular has many synonyms which have no distinctions in meaning. As far as most users of the language are concerned, substituting "large" for "big" would not alter the meaning of any statement, and many such pairs exist. There may be subtle differences in *usage* which might cause a sentence so substituted to sound a little odd, but this does not prove a difference in meaning. Do we really need to offer the readers of radio weather reports a choice among "now", "currently", "just now", "at the moment", "presently", "at present" and "right now" when announcing the temperature? (At this particular point in time, I cannot recall their ever choosing "now".)

Natural languages also contain words which do not seem to label any clear cluster of phenomena. Some, such as "god" and "unicorn", seem to correspond to hypotheses about the world which have little or no explanatory power except when the word is used as a label for the hypothesis itself rather than the hypothesized cluster.

Finally, natural languages are not used merely to encode real observations. The human appetites for fiction, myth, exaggeration and humour mean that a lot of the uses of language have no intended connection to reality. Also, personal, social and political motives sometimes drive the use of language to persuade, obfuscate and deceive. MML principles may explain some of the properties of natural languages, but far from all.

9.5 Scientific Languages

Scientific discourse uses a natural language with extensions, often but not always including the sub-language of Mathematics. The extensions used in a particular discipline do not appear to form self-contained universal languages in the Turing Machine sense of universality. That is, they do not seem in themselves to possess the means to talk about themselves, to modify and extend their own definitions and grammars. We seem rather to fall back on natural language, with its rich set of tenses, moods, argumentative devices and self-references, to define and modify these sub-languages and to glue together statements in the sub-languages.

The sub-languages of Science do appear to conform rather well to normative MML principles. Special terms used in them are adopted and dropped as the hypotheses they are used to express are accepted and rejected or superceded. Phogiston had a run during the popularity of a poor, but not ridiculous, theory about combustion, but is no longer in use. The concept of force in Physics was important for the expression of pre-relativistic theories of dynamics, but now has been relegated to metaphorical status in fundamental theory while retaining its importance in terrestrial-scale accounts of situations too complex for a fundamental treatment. The scientific usage of spatial and temporal terms has been modified to make them more useful in discourse accepting the theories of Relativity. The luminiferous aether, introduced as a necessary presumption for the wave theory of light and given a characterization in Maxwell's equations, has gone, not because it proved in conflict with observation but because it proved to be unobservable and so redundant. Special Relativity simplified the story and left it with no lines in the script. The meanings and grammars now governing the usage of "gene" are much modified from their originals. In general, it is fair to say that scientific languages are well adapted to allow the concise expression of explanations which are based on the currently accepted theories of the discipline. (I would not assert that this adaption is always apparent in the forms and lengths of the actual words employed. While Physics has stuck to fairly short words like "quark", "field", "charm", "charge", "mass", etc., despite fiddling with their meanings, some disciplines, such as Medicine, almost go out of their way to be polysyllabic.)

9.6 The Practice of MML Induction

The Minimum Message Length principle has been presented as a criterion for the acceptance, rejection and comparison of hypotheses or estimates based on bodies of data and prior expectation. It does not directly concern how "good" hypotheses or estimates might be found. In sufficiently simple cases, relatively standard optimization algorithms have been demonstrated to find inferences which at least approximate the ideal minimum message length, but these cases are much simpler and circumscribed than the problems which have led to notable scientific theories. For the general problems of scientific induction, we have, following Solomonoff, proposed a model in which the "explanation" of a body of data is framed as an input to a Universal Turing Machine which causes the UTM to output the original data. In Section 2.3.6 the form of the input was further required to satisfy conditions intended to ensure that the input begins by asserting a hypothesis. The discussion of Section 5.1.2 suggests that these conditions, as stated, are overly restrictive but the intent remains valid, namely to require the input to contain general assertions then used to compress the encoding of the specific data. The choice of UTM is supposed to be based on prior expectations. That is, the UTM (in the language applicable to present computers) is pre-programmed with routines implementing the kinds of logical and mathematical relations we expect to find in a good theory for the data domain, routines embodying previously accepted theory, routines for decoding the efficient representation of quantities for which we have prior distributions and so on. Ideally, we should try to choose a UTM which has at its fingertips the same kinds of knowledge and experience which a competent scientist might bring to the task. If we accept the thesis that human reasoning does not go beyond the power of a Universal Turing Machine, we may ask whether a UTM could in principle be programmed to find good theories within the unrestricted universe of computable functions.

The in-principle answer is yes, but. Suppose we have a suitably educated UTM M and a body of data represented by a binary string D. We want M to find a string I which, if M was presented with I as input, would cause M to output D. For the present we will ignore the conditions of Sections 2.2.2 and 2.3.6. A first attempt at a program to find I might look like this:

1   Read in D and store it in memory.
2   Initialize a routine E which, whenever called, will return the next binary string in some enumeration of strings in order of non-decreasing length.
3   Initialize a routine $S(J)$ which is a self-simulation of M. When called with string parameter J, it returns a string identical to the string which M would produce given input J.
4   Set X by calling E.
5   Set $Y = S(X)$
6   If Y does not equal D, go to Step 4.
7   Output X as the required input I.

This program is just an exhaustive search for the shortest I producing D. It will fail, because inevitably some trial input X will cause the self-simulation in Step 5 to run forever.

A more feasible program uses a device proposed by Levin [25]. See also [26] Section 7.5. Instead of running a single self-simulation, this scheme maintains a pool of partially completed simulated executions, each with its own input and output strings, and each with a record of how many simulated "clock cycles" have been performed and how many binary digits the simulation has read from its input. For some simulation, let Y be its binary output so far, L be the number of input digits it has consumed, and T be the number of simulated clock cycles completed. Define for that simulation the quantity $Q = L + C\log(T + 1)$ where C is some positive constant the same for all simulations. Also, define the value Z which is the length of some input which will cause M to produce D by some means, for instance by beginning with a "copy" program which simply copies the rest of the input to the output.

Initially, the pool will contain a single simulation with empty input and output strings and $Q = L = T = 0$. A global "limit" variable V is initialized to some small value. The various simulations are advanced by M in a series of rounds. In each round, each simulation in the pool is advanced by a burst of clock cycles until one of the following happens

- It "outputs" a digit to Y which causes Y not to be a prefix of D, in which case it is deleted from the pool.
- Its input length L exceeds Z. If so, the simulation is deleted from the pool.
- It attempts to read a new digit from its input. In this case, the simulation is replaced in the pool by two copies. In one copy the "input" is extended by a zero, in the other it is extended by a one.
- The simulation's value of Q reaches the global limit V. In this case, M records the state of the simulation and moves on to the next simulation in the pool.

At the end of each round, if no simulation has produced an "output" equal to D, the limit V is incremented and a new round begun. If some simulation has given D as output, its "input" is accepted as the desired input I and the program halts.

Note that no simulation whose output so far agrees with D and whose input length so far does not exceed Z is ever rejected from the pool. Also, every possible binary input meeting these conditions will be tried. As there exists some input of length Z which produces D, the program must eventually halt.

The input found by this "program" will not in general be the shortest explanation of the data. Rather, it will be one of the explanations which minimize the time-penalized quantity Q. The choice of the logarithmic time penalty imposes a fairly mild bias in favour of explanations which can be rapidly decoded, and has the useful property that the final result is independent of the computational efficiency of M, e.g., of the number of clock cycles it needs to perform a 20-digit division. The input found may not even be acceptable. Its length may exceed the length of D. Failure by the program to find an acceptable explanation does not imply that none exists. It merely shows that if one does exist, it must need a lot of computation to produce D. If one exists, its simulation would still be in the pool when the program halted with an unacceptable result, and a re-start of the program from this point with a reduced value of C and a suitably adjusted V might find it. Of course, it is never possible to know whether an acceptable input is waiting to be found or whether none exists.

As here presented, the program for M to find an explanation of D would be intolerably slow. It is almost a blind search. We have suggested that M be pre-programmed with "prior information" including routines corresponding to current theories in the data domain. If so, there is some hope that a new and better theory might be representable by a quite short input string which had the effect of modifying an existing theory, but as far as I am aware there is as yet no computer language which can efficiently represent any computable theory and which has the property that textually minor modifications to a program generally make minor changes to its behaviour and are unlikely to render the program meaningless. There has been progress in this direction in some declarative languages such as spreadsheets and logic programming but the devices which humans use such as analogy, looking for common formal structure, etc., are insufficiently understood as yet.

The potential failure arising from the undecidability of the "halting problem" may not really be a serious problem. Recall that the "assertion" part of an explanation is not quite a statement of the asserted theory as we might find it in a textbook. Rather, it is an algorithm for decoding data in a compressed form where the compression makes use of the theory. The scientific theories of which I have some knowledge seem to lend themselves to relatively simple "decode" algorithms which do not involve the kinds of inherently recursive computations, self-references or infinite regressions which are liable to result in unprovably unending simulations. Where a physical theory potentially involves an infinite set of interactions apparently leading to a silly result, as in some quantum accounts of empty space, scientists seem willing, if not happy, to "renormalize" the theory by discarding the silly bit provided the theory gives good explanations when not pushed to its limit. In short, the decode algorithms for current theories seem not to require the full computational power of a Universal Turing Machine. If so, there may be some hope of restricting the search for good theories to those which have decidable decoders. However, if the MML principle is close to a criterion for what constitutes a good scientific induction, good induction is never going to be easy.

The induction "program" has obvious scope for parallel computation. Rather than expecting the UTM M to do all the work, it could farm out the simulations in its pool to many similar machines all working at once. An interesting possibility is that the success of a parallel search for good explanations might be assisted if the different UTMs communicated their current best bets to one another. Communication certainly seems to assist human scientists.

9.6.1 Human Induction

The program M and its execution by a UTM is not claimed to be an accurate model of how human societies, in particular the scientific culture, actually behave in coming to form and accept theories. However, some features of the program do seem to have a relation to human induction. In program M, use is made of the ability of a UTM to simulate the behaviour of Turing Machines, including itself. Apart from the use of self-simulation in interleaving the examination of many potential explanations of the data, self-simulation protects against a fatal danger. Suppose a UTM, having by some search process decided that a certain string should be tried as a potential explanation, then just started executing that string directly instead of setting up a simulated execution. This process would suffer from two risks. The first is, of course, that the string would lead to an unending computation. The second is more subtle. It is possible that the execution of the string would indeed reproduce the data and be acceptably short, but leave the UTM in a state which was equivalent to a non-universal TM. No further exploration of alternative explanations would then be possible. Self-simulation, by which the UTM can in effect ask itself whether an explanation is good without committing itself to infinite indecision or irreversible acceptance, avoids these risks.

The human scientist is faced with at least the second risk, coupled with a more difficult environment. He or she, in assessing a potential explanation, may find it so convincing that he becomes unable ever fairly to assess a competitor. Further, in assessing an explanation, he must ask himself not only whether he can decode it to recover the data but also whether others will be able to do so. To assess an explanation without the risk of commitment, he must be able to "simulate" not only his own response to it but also to simulate the responses of an audience of peers with a range of prior background knowledge and expectations. It is therefore not surprising that a scientist's first exposition of a new theory for some data can often be longer than previous explanations of the same data. To accommodate the uncertain backgrounds of his audience, he must be prepared to spell out his theory in a redundant form, maybe even several times in different language or mathematical form, and to spell out the theory's implications about the data step by step rather than relying on the receiver to do all the deductive work. The result of this necessary prolixity is that the fundamental brevity of the explanation may be obscured, and become apparent to a receiver only after a good deal of thought.

The ability to simulate the "decoding" of an explanation by both oneself and (for a human if not for a UTM) a largely unknown audience appears an almost essential part of making an inductive inference in a communicable form. To be fully successful, such simulation may require the computational power of a Universal Turing Machine or something very like it. One might speculate that humans have an ability for theory formation far outstripping other animals because our mental abilities have crossed (or almost crossed) the threshold of universality. Obviously, the mental ability of a single person is less than universal: no-one has an infinitely capacious memory and no-one is immortal. However, like a UTM whose work tape has been truncated, we may have an information processing structure which would be universal were it not for limited time and memory. In any case, our ability to pass on information to others and from generation to generation affords our community a kind of universality.

9.6.2 Evolutionary Induction

We have previously suggested that natural selection has led to a kind of inductive inference by evolving organisms. Strictly speaking, what may have evolved are not necessarily inherited "theories" about the world. Natural selection mostly does not seem to favour an understanding of the natural laws and the organism's environment. Rather, it favours organisms whose actions, given the natural laws and their environment, promote the survival and reproduction of the organism. That is, the information processing ability which is favoured is one which selects an action as a function of environmental input. If I denotes the set of possible input histories and A the set of available actions, what is favoured is a good function $g(\cdot) : I \to A$. Logically, there is no need for this function to involve explicit use of a "theory" or even of a predictor function, except in the weak sense that we might see the function as embodying the theory "the world is such that given $x \in I$, my expected reproductive success is maximized by action $y = g(x)$". However, it is likely that the cost to the organism of implementing such a function is related to its complexity, in fact both to its algorithmic or structural complexity and to its computational complexity — the resources needed to compute it. Optimization of the function by mutation and selection might well be expected to behave rather like a Levin search, which similarly optimizes a mix of algorithmic and resource complexities.

If the world is indeed such that the success of an organism is probabilistically dependent on a computable function $f(x)$, $x \in I$ of its environmental input, the convergence theorems of Solomonoff prediction, MML and Levin search suggest that, as the resources which the organism commits are raised, the function will approach one which contains an approximation to $g(\cdot)$, and that this approximation will involve a computable model of relevant aspects of the world. If the organism can afford few resources, we would expect the action function $g(\cdot)$ to resemble more a simple decision tree rather than anything recognizable as a world model.

Since evolutionary induction is directed by reproductive success rather than by conciseness of explanation, the world model embodied in a well-resourced action function is not necessarily a good or even an acceptable explanatory theory for data of no great relevance for the reproductive success of the organism. In the minds of many people, an encounter with a snake triggers a violent aversive response leading to an aggressive or flight action, whereas a world view encompassing known statistics of threats to life suggests the snake usually presents no great danger. If we suppose that a human tribe or other reproductively semi-closed group has an overall reproductive success depending in part on cooperative behaviour within the group, an innate respect for authority, evolving from the respect for parental authority valuable for a species with a long immaturity, may enhance the success of the tribe. A world model with a dominant god who talks to the elders and punishes transgressions against their authority will help to direct individuals' action functions towards cooperation and promote the continuance of authority to the elders' successors. An afterlife theory lends some more weight towards social conformity by coupling cooperative actions directly to the individuals' survival and pain-avoidance instincts, and encourages youths to risk their own lives in battle with other tribes. The failure of the god theory to explain anything much is of little consequence, as it need not interfere with theories useful in growing food, hunting, raising children and other practical matters. A theory with no explanatory power, unacceptable as an inductive inference from real data, may still emerge by evolutionary induction as a component of a successful action function.

Evolutionary induction by mutation and selection has some theoretical advantages over Levin search or human inductive reasoning. Problems of undecidability, coding efficiency, uncertain audience, etc., become irrelevant. However, besides being slow and wasteful, it seems to suffer a serious disadvantage. A good but sub-optimal action function may be impossible to improve incrementally because all better functions lie beyond a lethal mountain range. To get to a better function by small steps may necessarily involve use, for at least one generation, of intermediate forms which are not conducive to survival.

9.6.3 Experiment

Inductive inference has so far been discussed in terms of inference from a defined, finite body of data. As an account of the practice of Science, this treatment may seem a gross over-simplification. Much progress in Science has resulted from experiments designed to create situations not occurring naturally and/or observations concentrated on carefully selected natural phenomena. We need not distinguish between these two practices. Both have the intent of providing new data which would not be provided by passive, undirected observation. The usual reasons for directed data acquisition are to get data in situations where current competing theories make different predictions, where current theory makes no prediction, or where the situation is so simplified that comparison with theory will be facilitated.

It may be thought that in neglecting directed intervention in the collection of data, we have badly compromised the discussion of induction. On the other hand, MML is intended as a criterion for assessing and comparing inductions. At any time when such a judgement is to be made, there will be some finite body of available relevant data, and it is surely on the basis of this data that the judgement should be made. The result of an MML comparison of two theories for the data may be inconclusive if their respective explanations have about the same length, in which case further data will be needed and its acquisition may well be directed to situations likely to resolve the comparison, but once the new data is found and added to the previous body, is the reason for its collection relevant to the assessment of theory? The answer is probably that it is not, but it will depend on exactly what is to be "explained" and on the causal independence of the old and new data.

Imagine that in a school of 1000 students, a test is given comprising ten rather difficult yes/no questions. The pupils' scores range from zero to ten, and their distribution resembles the distribution expected from a Binomial distribution with error probability 0.5. Of all students, only Jane Doe scores ten. This is the "old" body of data. Some new data is sought by asking the head teacher, who has seen the test results, for his comments. He remarks that he considered Ms. Doe to be his brightest student. This is the 'new' data. In this story, need we consider the history leading to these data in assessing the hypothesis that Jane is pretty clever? It seems clear that we should, because it is plausible that the teacher's opinion was affected by the test results. There is a possible causal link from the old to the new data. If the sequence of the data were reversed, with the teacher giving his opinion before the test, no plausible link from his opinion to the test scores exists, and the data could safely be treated as independent. There is still a possible link from the teacher's opinion to the existence of the "new" data (the test scores): the test may have been conducted to confirm or refute it. However, if the inductive inference process is not aiming to explain the history of the data but simply Jane's high score, this possible link is not relevant.

Consider a data string which is a historical record of data about the collisions between an electron and a proton. If an explanation of this data were sought by a Turing Machine or human with a good knowledge of science up to, say, 1890, the "scientist" might notice that the collision energies appearing in the data showed a strong tendency to increase from the earliest records to the later ones, and spend a good deal of effort devising a hypothesis about this tendency and its subtle details such as sudden jumps in energy followed by a stream of records all at about the same energy. A very good scientist might even arrive at the theory that the collisions have mostly been observed under artificial conditions, and that the energy increased in jumps as technical progress made higher energies achievable. This would be a fine inference if the history of the data collection was something to be explained, but not if our interest lay in the mechanics of the collisions.

When an MML inference is sought within a limited range of possible hypotheses, it is usually possible to define the range in such a way that undesired inferences can be excluded. Sometimes it may be more difficult. In the "binary addition" example of Section 7.1.7, MML was used to find a regular grammar for a set of strings each of which showed the argument and sum digits occurring in serial binary additions. The admittedly slow inference process used spent much time discovering that the strings all had lengths which were multiples of three, because the space of grammars it was searching had no such prior restriction. When the hypothesis space is very large, it is probably preferable to regard those parts or aspects of the data for which no explanation is sought as prior information already known to the intended receiver of the MML explanation, and so exclude the coding of this part or aspect from the message. In the ideal case where the receiver is a UTM, one may imagine the UTM to be provided with an extra read-only tape containing the "prior" data or a routine for checking for a "prior" data constraint such as the multiple-of-three condition. The UTM is thus relieved of the incentive or need to try to find hypotheses about this data or aspect.

10. Related Work

The proposal of an account of inductive inference similar to MML was probably inevitable given the success of Information Theory, including the theory and techniques of efficient coding, and the development of Turing's formal theory of computation. However, such an account spent a long time in gestation and then for many years remained almost unrecognised. As far as I know, the first published account to make a link between the lengths of UTM inputs and the probabilities of their resulting outputs is the work of Ray Solomonoff in 1964 [42]. It seems to have prompted the interest of the Russian statistician Kolmogorov but as a foundation for a definition of randomness rather than as a means of induction [23]. As it happened, Kolmogorov missed a trivial but vital point which made his effort lead nowhere. Very soon afterwards, in 1966, Gregory Chaitin independently proposed a definition of randomness which avoided the error [8] and later a paper showing a formal connection between probability and the lengths of UTM inputs [9]. His definition is similar in spirit to Solomonoff's but not identical. Chaitin's later work yielded remarkable results on the limits of mathematical reasoning, but he did not pursue the implications for induction. Again independently, David Boulton and I developed the MML concept as a way of attacking the problem of intrinsic classification, or mixture modeling [52]. This work seems to have been the first computer program to use the idea to make an induction. Yet again independently, and at about the same time, Jorma Rissanen developed his Minimum Description Length (MDL) principle.

This brief chapter will describe Solomonoff's and Rissanen's approaches and try to show why, although both resemble MML, they are aimed at different targets and neither quite does the same job. My descriptions are my interpretations of their approaches, which may not be quite accurate, and readers are advised to refer to the original works.

10.1 Solomonoff

Solomonoff's work is on prediction. Much data of interest comes to us as a potentially unending series from some real-world source. For some sources, each item in the series arrives without the need for experiment or any directed observation. An example is the daily rainfall in some city. We first consider such a source. We suppose that the data is encoded in a binary string. Solomonoff addresses the question of how, given the first n digits of the string, can we best predict future digits.

Before describing his solution, note that the problem addressed is one of *prediction* rather than the problem mainly addressed in this book, which is finding good *explanations*. The two problems are closely related. If, by MML or otherwise, one forms a theory about the data source which gives a good explanation of the data so far, the theory will almost certainly imply a (usually probabilistic) prediction about future data, and we have reason to expect the prediction to be quite good. However, since an inductively derived theory commits to a single theory and discards alternative explanations of the data, it will not in general give the best possible prediction. In simple Bayesian terms, if the source of the data is assumed to be described by some probability model $f(x|\theta)$ with unknown parameter θ drawn from a prior $h(\theta)$, an inductive theory $\hat{\theta}$ implies a probability distribution over future data y given by $f(y|\hat{\theta})$. However, the best predictive distribution for future data y is that derived from the posterior distribution of θ given the known data x, namely

$$\Pr(y|x) = \frac{\int f(y|\theta)h(\theta)f(x|\theta)\,d\theta}{\int h(\theta)f(x|\theta)\,d\theta}$$

which takes into account all possible sources of the data permitted by the assumptions.

Solomonoff has generalized this Bayesian predictor by admitting as possible models all computable probability distributions over finite binary strings and by taking as the "prior" over these models the distribution defined by some Universal Turing Machine M with unidirectional binary input and output tapes and a bi-directional work tape. Let x be a finite binary string of length $|x| = n$. Let $S_x = \{t_k : k = 1, 2, \ldots\}$ be the set of all binary input strings t_k such that when t_k is provided as input to M, M will output a string having x as prefix but will not do so when provided with any proper prefix of t_k. The set S_x is clearly a prefix set. Also, if x and y are two distinct strings of length n, the intersection of S_x and S_y is empty. Define for the chosen M and for all strings x of length n the quantity

$$P_n(x) = \sum_{t \in S_x} 2^{-|t|}$$

where $|t|$ is the length of the input t. $P_n(x)$ is just the probability that M will output a string beginning with x when provided with a completely random stream of digits as input. $P_n(\cdot)$ is a semi-measure over the set of all binary strings of length n because many inputs to M will cause the UTM to stop or compute forever without producing any output as long as n. Hence, in general

$$\sum_{x:\,|x|=n} P_n(x) < 1$$

Solomonoff defines a normalized probability measure $P'_n(\cdot)$ over finite strings of length n by the recurrence

$$P'_{n+1}(x0) = \frac{P_n(x0)}{P_n(x0) + P_n(x1)}$$

$$P'_{n+1}(x1) = \frac{P_n(x1)}{P_n(x0) + P_n(x1)} \qquad \forall x : |x| = n$$

with $P'_0(\Lambda) = 1$ for Λ the empty string. Then $\displaystyle\sum_{x:|x|=n} P'_n(x) = 1$ for all n. In fact the sum over any complete prefix set of strings is 1. The length subscript n can be dropped from $P'(\cdot)$, being implied by the length of its argument.

Solomonoff suggests that predictions of future data be based on the $P'(\cdot)$ measure. It predicts that the probability $\Pr(y|x)$ that a data string x will be followed by future data y is

$$\Pr(y|x) = P'(xy)/P'(x)$$

Of course, the halting problem prevents calculation of $P'(x)$, but it can be approximated if most of the shortest members of S_x are discovered. If the data stream is being generated according to a computable probability function $\Pr(c = 1\,|\,x) = \mu(x)$, where c is the next digit following the string x, it has been shown that the μ-expected sum of the squared errors in the true probabilities of successive digits and those predicted by $P'()$ is bounded independently of the length of x. Hence, the squared errors must in the limit decrease more rapidly than $1/|x|$. The value of the bound is approximately the length of the shortest program for M which can compute the function $\mu()$. This bound depends on the choice of the UTM M, so Solomonoff suggests that M be chosen to embody all prior information believed relevant to the prediction problem, in the expectation that the prior knowledge will enable M to accept a concise specification of $\mu()$. Solomonoff has also extended his predictor to handle data comprising an unordered collection of facts rather than a sequential stream [43]. In this form, it then predicts the probability that a specified potential "fact" is true of the source of the known facts. The techniques used are a fairly straightforward extension of those used for the sequential case. The journal in which this article appears, *The Computer Journal* 42(4), 1999, contains several articles on theoretical and practical aspects of Algorithmic Complexity.

This brief introduction to Solomonoff prediction is based on private communications and a paper by Marcus Hutter [20], which also proves a complementary convergence result for sequential prediction. If one knew the generating probability function $\mu()$, one could bet on the successive digits of a data string $x = x_1, x_2, x_3, \ldots$, betting that the next digit x_{n+1} would be one if $\mu(x_1, \ldots, x_n) > \frac{1}{2}$, and otherwise betting on zero. Hutter has shown that if the odds offered on each bet (assumed constant) and $\mu()$ are such that the μ-expected profit after $n \gg 1$ bets rises linearly with n, then betting on $x_{n+1} = 1$ if $P'(x1) > P'(x0)$ is also a winning strategy with an expected profit per bet which eventually approaches that using knowledge of $\mu()$. In summary, these results show that Solomonoff prediction can be very good provided the UTM M can accept a short specification of the unknown generator $\mu()$. These results do not explicitly cover situations where the generator of the data is not computable, but could be expected to apply with some weakening when $\mu()$ can be approximated by a computable function with a finite number of unknown and uncomputable parameters.

There is an obvious and close relationship between Solomonoff prediction and MML. The origins of the latter followed the former's by about four years, and the connection was not realised for about another five years, but the central idea in each seems the same. The weight given to some model of the data source depends not only on the fit of the data to the model but also on the complexity of the model with respect to prior knowledge, and the appropriate combination of these factors is found in the length of an encoding of the data which can be decoded by a receiver having the prior knowledge but no other knowledge of the model. In Solomonoff prediction, the receiver is a UTM primed with prior knowledge. In MML, the same receiver is contemplated, but for simplicity we usually consider a more limited receiver who can envisage only a limited range of models with the prior knowledge expressed as a Bayesian prior over this range.

The essential difference between us is in aims. Solomonoff aims to predict, MML aims to find a model which explains. Solomonoff finds the best predictions by a weighted average over all models, MML ideally commits *pro tem* to the single best model it can find. As models of scientific investigation, both have a respectable place. Scientific endeavour is certainly driven by a pragmatic desire to predict and society supports this aim fairly willingly. But the endeavour is also driven by a desire to understand and this aim is also supported, although perhaps more grudgingly. MML offers its single model as a basis for understanding what has been observed, with no real guarantee of predictive power. Thus, MML attempts to mimic the discovery of natural laws, whereas Solomonoff wants to predict what will happen next with no explicit concern for understanding why.

Actually, the difference is less clear-cut. MML inferences, although single models, are usually found to be good predictors. On the other hand, the convergence theorems and error bounds for Solomonoff prediction are based on the fact that if data is coming from a source well approximated by a computable function, the contribution to the $P'()$ measure will come to be dominated by one data-replicating input string, namely the one which tells the UTM how to compute this function. In other words, the proofs of the success of Solomonoff prediction are based on showing that if a good computable model of the data exists, MML inductive inference will find it!

10.1.1 Prediction with Generalized Scoring

Solomonoff has considered the efficacy of prediction as measured both by the accumulated squares of errors in predicted next-digit probabilities and by the accumulated number of digits predicted erroneously by higher computed probability. Vovk and Gammerman [50] have generalized the prediction problem to develop what they call the Complexity Approximation Principle. They show that, for several broad classes of loss function defining a loss incurred when a predicted datum is realized by a new observation, it is possible to define a "predictive complexity" of a finite data sequence z. This complexity is not computable, but is bounded above by the accumulated loss $L_S(z)$ suffered by some computable prediction strategy S plus a multiple of the algorithmic complexity of the strategy. Hence, they suggest that given z, future data should be predicted by finding and using a strategy giving the lowest known value of this bound. (The role of the multiplier applied to the complexity of the strategy is essentially to yield a value commensurate with the loss function.)

10.1.2 Is Prediction Inductive?

Some writers, including Solomonoff and Hutter, regard probabilistic prediction from known data as a form of induction. In fact, many seem to equate the terms. For instance, Solomonoff entitles one of his papers "Three Kinds of Probabilistic Induction", and Hutter begins the paper cited above with the assertion: "Induction is the process of predicting the future from the past or, more precisely, it is the process of finding rules in (past) data and using these to guess future data". (I cite these two authors, both very able, simply because I have their papers in front of me. Many others seem to have a similar view that induction equals prediction.) Hutter's assertion makes the confusion obvious. He starts by equating induction to data-based prediction, then qualifies this statement to require the prediction to be based on "rules found" in the data. The qualification suggests he realises that the traditional meaning of induction is the postulation of a general proposition based on a collection of specific facts. But having made the qualification, he then discusses a technique, Solomonoff prediction, which results in no such "rule" or general proposition. Such "rules" as the technique may find in the data are never explicitly apparent, and all "rules", whether or not supported by the data, are mixed together in forming the predictive distribution.

As mentioned at the beginning of this Section (10.1), Solomonoff prediction is a generalization of conventional Bayesian prediction. Both take some data, a range of possible generator functions (models) for the data, and some prior knowledge, and from these *deduce* a predictive distribution over possible future data. Both have the property that if the model range includes the true probabilistic generator of the data, their predictions will converge on the true probability distribution of future data. Both have the property that, given the model range, prior and data, their result (a predictive distribution) inevitably follows. Solomonoff's generalization is to expand the range of possible models to include all computable models, and to embody prior knowledge in the choice of the UTM, but this does not alter the essentially deductive nature of the process. In making the generalization, he has to address the undecidability inherent in UTMs, which he partly resolves by his normalization procedure. (There are other possible normalizations which could have been used, and to this extent the predictive distribution he gets is not quite an inevitable consequence of the givens. Solomonoff shows his choice of normalization leads to the most rapid convergence of the predictive distribution.)

I intend here no criticism of Solomonoff's work, which has been and is of great importance. It is certainly not to be criticised on the grounds that his ideal predictor can only be approximated because of undecidability. Just the same is true of MML inference with an unrestricted range of computable models. My objection is to the use of "induction" to apply to Solomonoff (or more restricted Bayesian) prediction, which appears to me to be a purely deductive process. To accept this misusage is to lose a hitherto useful and important distinction between deductive and inductive reasoning. The former starts with propositions, at least one being general, and by provably correct steps derives a more specific conclusion. The latter, for which I hope MML is a model, starts with specific propositions and arrives at a more general conclusion which is not a necessary consequence of the givens. Solomonoff himself, in a paper in 1996 ("Does Algorithmic Probability Solve the Problem of Induction?"), uses the term ALP (Algorithmic Probability) to refer to what we have called Solomonoff prediction. In Section 7 of the paper he writes: "ALP does not explicitly use induction. It goes directly from the data to the probability distribution for the future". He then goes on to point out that although ALP should be superior in practice to predictions based on "scientific laws", the latter serve other aims besides prediction. Clearly, he recognises the distinction between prediction and induction, but seems content to go along with others' confounding of the two.

The *Cambridge Dictionary of Philosophy* defines the narrow meaning of "induction" in the sense in which I have used the term, namely the inference of a general proposition from a collection of more specific ones. It also recognises a broader meaning, namely the inference of any conclusion which is stronger than any proposition which can be deduced from the premises, whether or not this conclusion has more generality. The quality common to both meanings is what the dictionary calls "ampliative": the conclusion, while compatible with the premises, cannot be deduced from them. Adopting the broader sense, it remains true that Solomonoff (or other) prediction which deduces a probability distribution over future data from defined premises of known data and prior information is not inductive. The predictive probability distribution is not ampliative. It is a necessary consequence of the premises, and hence no stronger than can be deduced from them. However, as a practical matter, this non-inductive predictive distribution may well lead to a genuinely inductive conclusion (in the broader sense) if the analyst, having derived the distribution, decided to accept *pro tem* that the most probable future data will indeed be observed.

A commonplace example may help. A follower of horse races may, using the results of previous races, breeding records and other information, *deduce* that Old Rowley has a probability of 0.02 of winning the Cup. If he then finds that a bookmaker is offering odds of 100 to 1, he may *deduce* that betting on Old Rowley is an action with positive expected return. These are not ampliative conclusions. They follow strictly from his premises. If he then concludes that Old Rowley *will* win the Cup (and maybe bets his house on it) this goes beyond the implications of his premises, is ampliative, and hence is an *inductive* conclusion in the broad sense. Thus, in the broader sense of induction, Solomonoff prediction per se is not inductive, but may lead someone to adopt an inductive conclusion.

10.1.3 A Final Quibble

The normalization used to go from the raw string probability $P(x)$ to the normalized form $P'(x)$ has the effect of disallowing all UTM input strings which would cause the UTM to produce only a finite output shorter than the data string x. This is of course proper when x is just the string of known data: no input which cannot produce x should contribute to the probability of x. It is less clearly proper when x is the known data string followed by a putative future extension whose probability we would like to predict. The normalization then has the effect of denying any probability to the event that the known data will not be extended by the real world.

I am not here envisaging trivial events such as the possibility of future observations being denied by the death of the observer, exhaustion of funding, etc. Rather, it may be that the real situation admits of only a finite string of the type of data being studied. Suppose, for instance, that the given data is a sequential string of records, each record giving a large number of physical, chemical and spectroscopic properties of a stable element isotope, and that the data has records for all stable isotopes in order of atomic weight from hydrogen to, say, silver. If the data has lots of information on each isotope and the chosen UTM M embodies some good knowledge of quantum mechanics, one might reasonably hope that the predictive process would predict high probabilities for future data matching the properties of the elements heavier than silver. But a really good predictor would flatly refuse to predict *any* data beyond a finite number of "future" records, because beyond a certain atomic number there are no stable isotopes. The shortest UTM input causing output which matched the records up to silver might have the form of a program which built models of nuclei with particular numbers of protons and neutrons, then calculated their binding energies and weeded out the nuclei which could decay. Keeping on going past the stable isotopes of silver, it would eventually reach lead-208 and thereafter loop endlessly, building ever-heavier models and rejecting them all.

The normalization used by Solomonoff would exclude such an input program, and instead insist on giving non-zero probabilities to records for non-existent or unstable isotopes by using less accurate programs.

This objection is no more than a quibble. The great majority of practical prediction problems deal with sources of unlimited data. However, it would be nice if the normalization could, without serious damage, be modified to permit the prediction: "That's all, folks!"

10.2 Rissanen, MDL and NML

More or less at the same time as the initial conception of MML, Jorma Rissanen independently conceived a rather similar idea which has come to be known as the Minimum Description Length (MDL) principle. A later development of the idea is called Normalized Maximum Likelihood (NML), but I will use the term MDL for his general approach and defer use of the term NML till the description of this version.

Unlike Solomonoff's work, Rissanen's is, like MML, concerned with the induction of a hypothesis about the source of data rather than prediction, and assumes a conventional framework of statistical models rather than being based on Universal Turing Machines. To use notation paralleling that used in describing MML, MDL deals with a set $X = \{x_i : i = 1, \ldots\}$ of possible data sets or vectors and a set Θ of possible probability models. MDL is particularly concerned with model sets comprising a number of *model classes*, where each model class has a number of real-valued parameters, the number usually being different for each class. It is hence convenient to write $\Theta = \{\Theta_k : k = 1, 2, \ldots\}$ and to write a parameterized probability model in Θ_k as $f_k(x \,|\, \theta_k)$ where $f_k(\cdot \,|\, \cdot)$ gives the general mathematical form of the models in class k and θ_k represents a vector of parameter values defining a specific model in the class. As in MML, the data set X is strictly speaking always discrete and countable, but providing the discretization of data values is sufficiently fine, it is possible and often convenient to treat X as a continuum and a specific data set x as a vector of real values.

MDL differs from MML in two important respects. First, Rissanen wishes to avoid use of any "prior information". His view is very much non-Bayesian. Second, given data x, MDL aims to infer the model class Θ_k which best explains the data, but does not aim to infer the parameter vector θ_k except (in some versions) as a step involved in inferring the class. MDL shares with MML the idea that the best inference from the data is that hypothesis which enables the data to be encoded most concisely, but in MDL the inferred hypothesis asserts only a model *class* and not (as in MML) a single fully specified model. The kind of message envisaged in MDL as an encoding of the data would therefore first assert a model class and then encode the data using a code which assumes the data comes from some unknown model within the class. As no informative prior is assumed over the set of classes, the assertion of a class would have the same length for all classes and so is ignored in choosing the class to minimize the description length of the data. The problem faced in MDL is how to compute the length $S_k(x)$ of an encoding of the data x assuming that it is drawn from some unknown member of a parameterized family of distributions $f_k(x|\theta_k)$. Rissanen terms the quantity $S_k(x)$ the *Stochastic Complexity* of data x with respect to the model class Θ_k. In discussing how MDL deals with this problem, we can drop the k suffix since every model class will be treated in the same way.

If a Bayesian prior density $h(\theta)$ were given over the unknown parameter θ of the model class distribution $f(x|\theta)$, the complexity of x with respect to the class could be calculated as

$$S(x) = -\log \int h(\theta) f(x|\theta) \, d\theta = -\log r(x)$$

and would equal what has been termed $I_0(x)$ in the discussion of MML. However, Rissanen seeks to avoid the need of any such prior. In early work, he proposed to approximate $S(x)$ by the length of a two-part message, the first part nominating a parameter estimate $\hat{\theta}$ and the second encoding x optimally for the distribution $f(x|\hat{\theta})$, much as in MML. His problem then became how best to devise a code for encoding $\hat{\theta}$ in the first part without assuming a prior density. The method employed chose a discrete set of codeable estimates similar to the MML Θ^* with the density of estimates in Θ being approximately $\sqrt{F(\theta)}$ where $F(\theta)$ is the (expected) Fisher Information determinant. Within an ignorable constant factor, this density of codeable estimates is very close to that given by an SMML code or the MML approximations of Chapter 5, and gives a near-optimal compromise between the cost of coding the estimate and the mismatch between the estimate and the data. The members of this Θ^* are then enumerated in some "natural" order. The first-part code need then only state the integer index of the chosen estimate $\hat{\theta}$ in this enumeration. If Θ^* has a finite number M of members, the length of the first part is just $\log M$, and if it has an infinite number, Rissanen proposed the index be coded using the universal log* code described in Section 2.1.16. For data x, the member of Θ^* of highest likelihood is chosen as $\hat{\theta}$.

In this scheme, the volume of the parameter space Θ included in the Voronoi region surrounding some codeable estimate $\hat{\theta} \in \Theta^*$ is approximately $1/\sqrt{F(\hat{\theta})}$. At least far from the origin of enumeration, the scheme encodes the estimates with lengths which vary very slowly if at all with the value of the estimate. Hence, the length of the first part of the message, if interpreted as the negative log of a discrete "prior probability" associated with $\hat{\theta}$, would imply a "prior probability density" $h(\theta)$ in Θ approximately proportional to the local density of codeable estimates, i.e., proportional to $\sqrt{F(\theta)}$. A "prior density" of this form is known as a "Jeffreys prior" (Section 1.15.3). Of course, in MDL, no such interpretation as "prior information" is intended.

A further development of this method of computing $S(x)$ was made by modifying the coding of the second part of the message. Instead of coding x using the model distribution $f(x|\hat{\theta})$, the model distribution was renormalized to sum to one over just that subset of data vectors which would result in the estimate $\hat{\theta}$. This "complete coding" results in a somewhat higher probability, and hence shorter second part, for every data vector in the subset. It removes the redundancy in the original scheme which would have permitted a data vector x to be encoded using any $\theta \in \Theta^*$ for which $f(x|\theta) > 0$.

At this point of its development, MDL could still have been open to the charge that, at least when Θ^* is an infinite set, the calculation of $S(x)$ depended on the order in which its members were enumerated, and that one person's "natural" order might be another's arbitrary choice. A recent advance has largely removed this arbitrariness.

10.2.1 Normalized Maximum Likelihood

This discussion is based on a 1999 paper by Rissanen entitled "Hypothesis Selection and Testing by the MDL Principle" [36], which gives a good presentation of Normalized Maximum Likelihood (NML). Given some data x, the shortest possible encoding of it using a model of the form $f(x|\theta)$ is obtained by setting $\theta = \theta_M(x)$, the maximum likelihood estimate. The length of the encoded data would then be $L_M(x) = -\log m(x)$ where $m(x) = f(x|\theta_M(x))$. Of course, such an encoding is useless for conveying the data to a receiver who does not already know $\theta_M(x)$. Instead, one must choose some other coding scheme. Suppose some non-redundant code is devised which will be intelligible to a receiver who knows only the set X of possible data, the model class, the range Θ of possible parameter vectors and hence the conditional probability function $f(x|\theta)$. One such code would be the complete-coding MDL two-part code described above, but let us consider others. (Note that an SMML or MML code will not do, since these are inherently redundant.) Let the chosen code be named q, and let it encode x with length $L_q(x)$. It therefore assigns to x a coding probability $q(x) = 2^{-L_q(x)}$.

If a prior distribution over θ were available, and hence a marginal distribution over x, one could choose the q to minimize the expected length $EL_q(x)$ and hence the expected excess length $E[L_q(x) - L_M(x)]$. Avoiding priors, Rissanen chooses q to minimize the *worst case* excess length, that is the excess for the worst case data vector in X. Since the excess for data x is

$$L_q(x) - L_M(x) = -\log \frac{q(x)}{m(x)}$$

minimizing the largest excess is done by maximizing the smallest ratio $\dfrac{q(x)}{m(x)}$

subject to $\sum_X q(x) = 1$, $q(x) \geq 0$ for all x.

This is achieved when all ratios are equal, for suppose the contrary. Then some x_l has the minimum ratio $R = q(x_l)/m(x_l)$, and some other vector x_h has a larger ratio $q(x_h)/m(x_h) = R + \varepsilon$. Writing $q(x_l)$ as q_l, etc., consider the modified code which has

$$q'_l = q_l + \varepsilon \frac{m_l m_h}{m_l + m_h}, \quad q'_h = q_h - \varepsilon \frac{m_l m_h}{m_l + m_h}$$

and $q'(x) = q(x)$ for all other x. Then

$$\frac{q'_l}{m_l} = \frac{q'_h}{m_h} = R + \varepsilon \frac{m_h}{m_l + m_h} > R, \quad \sum_X q'(x) = 1, \quad q'(x) \geq 0 \text{ all } x$$

Thus, the modified code has one fewer members of X with the low ratio R. Repeating this modification will result in a code with no x having ratio R, and hence with a higher minimum ratio, leading eventually to a code with all ratios equal.

The final NML result is a code for data vectors in X with coding proba-bilities proportional to their maximum likelihoods:

$$q(x) = \frac{m(x)}{\sum_{y \in X} m(y)} = \frac{f(x|\theta_M(x))}{\sum_{y \in X} f(y|\theta_M(y))}$$

and code lengths

$$L_q(x) = -\log f(x|\theta_M(x)) + \log C \quad \text{where } C = \sum_{y \in X} m(y)$$

For any sufficiently regular model class with k free parameters and x an i.i.d. sample of size n, Rissanen has shown that

$$L_q(x) = -\log f(x|\theta_M(x)) + \frac{k}{2} \log \frac{n}{2\pi} + \log \int_\Theta \sqrt{F'(\theta)}\, d\theta + o(1)$$

where $F'(\theta)$ is the per-case Fisher Information. Converting to our convention of writing the Fisher Information for the given sample size as $F(\theta)$, we have $F'(\theta) = F(\theta)/n^k$ and so

$$L_q(x) = -\log f(x|\theta_M(x)) + \frac{k}{2} \log \frac{1}{2\pi} + \log \int_\Theta \sqrt{F(\theta)}\, d\theta + o(1)$$

Compare this expression with the MML I1B formula of Section 5.2.12

$$I_1(x) \approx -\log \frac{h(\theta')}{\sqrt{F(\theta')}} - \log f(x|\theta') - (D/2) \log 2\pi + \frac{1}{2} \log(\pi D) - 1$$

with the number of scalar parameters D set to k and the prior density $h(\theta)$ set to the Jeffreys prior proportional to the square root of the Fisher Information (Section 1.15.3) $h(\theta) = \dfrac{\sqrt{F(\theta)}}{\int_\Theta \sqrt{F(\phi)}\,d\phi}$ giving

$$
\begin{aligned}
I_1(x) &\approx -\log \frac{\sqrt{F(\theta')}}{\sqrt{F(\theta')}\int_\Theta \sqrt{F(\phi)}\,d\phi} - \log f(x|\theta') - \frac{k}{2}\log 2\pi \\
&\quad + \frac{1}{2}\log(k\pi) - 1 \\
&\approx -\log f(x|\theta') + \frac{k}{2}\log\frac{1}{2\pi} + \log\int_\Theta \sqrt{F(\phi)}d\phi \\
&\quad + \frac{1}{2}\log(k\pi) - 1
\end{aligned}
$$

With the Jeffreys prior, $I_1(x)$ is minimized by the Maximum Likelihood estimate $\theta_M(x)$, so the difference between the values of $L_q(x)$ and $I_1(x)$ reduces to

$$
I_1(x) - L_q(x) \approx \frac{1}{2}\log(k\pi) + o(1)
$$

The greater length of $I_1(x)$ represents the message length cost of the small redundancy inherent in MML explanations, because the two-part code employed does not insist on using the minimizing parameter estimate. However, except for this small difference, it is clear that the Stochastic Complexity of the data with respect to the model class is very similar to the MML explanation length assuming the same model class and a Jeffreys prior.

For model classes sufficiently regular for the $L_q(x)$ expression to hold, and where the Jeffreys prior is a tolerable representation of the state of prior belief about θ, there will be little difference between MML and NML assessments of the class as a possible source of the data, and little difference between their comparison of different model classes in the light of the data.

There is a potential problem with the Jeffreys prior, and more generally with the normalization of the maximum likelihood. The normalization constant $C = \sum_{y \in X} f(y|\theta_M(y))$ diverges for some model classes, and the Jeffreys prior may not be normalizable either. For such classes, Rissanen suggests that the class be curtailed by restricting the allowed range of the parameter θ to a portion Ω of its original range Θ, and further, that for NML the set X of possible data vectors be restricted to the subset X'

$$
X' = \{x : x \in X,\ \theta_M(x) \in \Omega\}
$$

where Ω is chosen to ensure that the maximum likelihood is normalizable over X'. It is of course necessary also to ensure that X' includes the observed data vector x. The problem is analogous to that encountered in MML (or Bayesian analysis generally) when an attempt is made to choose a prior expressing a lack of any prior knowledge of the parameter. For some types of parameter, the preferred "uninformative" prior is improper, so either its range must be restricted or another prior chosen, either way importing some sort of prior knowledge. However, in the MML or Bayesian analysis, it is never necessary to disregard possible data values. The restriction of X' to data whose maximum likelihood estimate lies in Ω may be troublesome in model classes where the maximum likelihood estimate is strongly biased or inconsistent, as in the Neyman-Scott problem of Section 4.2. In such cases, the observed data may have a maximum likelihood estimate outside what would otherwise be a plausible choice of Ω.

In the cited paper, Rissanen suggests that the model class giving the lowest Stochastic Complexity or highest normalized maximum likelihood be selected as the preferred model of the data. This is the essential inductive step of the MDL/NML approach, suggesting the acceptance pro tem of a specific model class for the data. Note that no estimate of the parameter(s) accompanies this selection. The induction does not name a preferred model within the model class, and although the NML criterion $L_q(x)$ involves use of the maximum likelihood estimate, this estimate is not endorsed.

In a later section, Rissanen considers the use of NML in a Neyman-Pearson style of hypothesis test, and there makes his choice of X' in order to maximize the power of the test. This leads in a couple of Normal examples to a *very* restricted choice of Ω centred on the observed sample mean and shrinking with increasing sample size. The approach seems inconsistent with his criterion for model selection. However, my instinct is to regard this inconsistency, if it indeed exists, as an argument against Neyman-Pearson testing rather than as a criticism of NML.

10.2.2 Has NML Any Advantage over MML?

MDL and NML are inductive methods for selecting a model class among a (usually nested) family of classes. Although each model class may have free parameters, the methods do not inductively choose a specific model within the selected class, and eschew the use of any prior probability density over the parameters of a class. MML, however, makes an inductive choice of a fully specified single model for the data. In its purest SMML form, MML does not even pay much attention to which class the chosen model belongs to, although in practical application the MML explanation message would typically begin by nominating the class before stating the parameter estimates and then the data.

Perhaps the principal argument for MDL over MML is that it avoids priors and hence may appeal to a natural desire to be "objective" or at least more so than Bayesians. Of course, the objectivity has its limits. One must make a choice on some grounds or other (dare I say) prior experience of the family of model classes to consider, and at least for some families, must be prepared for a fairly arbitrary restriction of the set of possible data vectors in order to normalize. But given these limits, MDL/NML does require less non-data input than does MML. However, the price paid is not trivial.

- MDL offers no route to improved estimation of model parameters, leaving maximum likelihood as the default. MML does give an estimator which in some cases is less biased than maximum likelihood, and is consistent in model classes where maximum likelihood is not.
- MDL as presented does not seem to contemplate the conflation of models in different classes when the data is such as cannot be expected reliably to distinguish between the models. MML can do so, and for instance, in the causal net learning of Section 7.4, this ability was shown to have a significant and beneficial effect on the inferences.
- The rationale presented for NML seems less transparent than that of MML. NML defines the complexity of a data vector with respect to a model class as the length of the string encoding the vector in a particular code q. The code is chosen to minimize the worst-case excess of the code length $L_q(x)$ over the "ideal length" $L_M(x) = -\log f(x|\theta_M)$, the worst case being over all possible data vectors in X. It is not clear why $L_M(\cdot)$ should be chosen as the gold standard to which q should aspire, since $L_M(\cdot)$ is not a code length in any code usable in the absence of knowledge about the data. Further, it is not obvious why the construction of q should concentrate on the worst-case data vector, which may for some model classes be one which most people would think very unlikely to be observed. By using the worst-case criterion, NML makes the "complexity" of the observed data depend on the properties of a data vector which, in general, was not observed and which, having an extreme property, is atypical of data generated by the model class.

  By contrast, MML designs a code which is optimized in the usual Shannon sense for the *joint* probability distribution of data and model. In this, MML seems closer to the notion of the Algorithmic Complexity of the data. Although the usual definitions of Algorithmic Complexity (AC) do not explicitly involve the joint distribution, almost all the theorems demonstrating useful properties of AC in fact rely on bounding the AC of the data from above by a string length which encodes both a generating model and the data vector, and whose length therefore may be taken as the negative log of the joint probability of model and data under some universal distribution. See for instance V'yugin [51].
- NML has normalization difficulties which are reflected in possibly serious violation of the likelihood principle (Section 5.8). For the binomial problem where the data is the number of successes s in a fixed number N of trials, the NML formula gives an acceptable value for the complexity of s. However, for the Negative Binomial problem where the data is the number of trials N needed to get a fixed number s of successes (Section 5.7), the Jeffreys prior diverges and the maximum likelihood of N approaches a constant for large N, so the maximum likelihood cannot be normalized

and some arbitrary limit on N must be applied to get a complexity. The value of this limit will affect the complexity assigned to the observed number of trials. The different behaviour of NML in the two problems violates the likelihood principle. As shown in Section 5.8, while MML also technically violates the likelihood principle in treating these two problems, the violation is insignificant.

– MDL aims to infer the model class which best models the data out of a defined family of model classes. While this is a natural and useful aim, it is not always obvious how a set of possible models should be classified into classes. The classification of models envisaged in MDL is always such that all models in a class have the same number and set of free parameters, but even with this natural restriction much choice remains in classifying a large set of models. Consider the set of polynomial models which might be used to model the dependence of one data variable on another. Most obviously, the polynomials can be classed according to their order, but in some contexts different classifications may be of interest. If the two variables are physical quantities with natural origins of measurement, models with few non-zero coefficients or only coefficients of even or odd order are scientifically interesting. Of course, an MDL classification could be tailored to such cases, but it seems simpler to use an MML analysis which, as in Section 7.4, can be arranged to reject insignificant coefficients where the data permits.

– With some natural classifications of models, it can occur that the "best" fully specified model is not in the "best" model class. This situation has been found in multivariate mixture models for some data sets. If models are classified according to the number of component simple distributions in the mixture, for some data one finds that the MDL best class has k components, and contains several quite distinct but fairly good models, whereas there exists a single model in the $k + 1$ class which gives a much better fit.

This list is of course not impartial, and an advocate of MDL might offer an equal list of disadvantages in MML. When used intelligently, MDL is capable of yielding very good inferences from complex data, and several impressive applications have been published. Of those I have read, it seems that an MML analysis would have reached very similar inferences.

Bibliography

1. H. Akaike. Information theory and an extension of the maximum likelihood principle. In B.N. Petrov and F. Csaki, editors, *Proceedings of the 2nd International Symposium on Information Theory*, pages 267–281, 1973.
2. A. R. Barron and T. M. Cover. Minimum complexity density estimation. *IEEE Transactions on Information Theory*, 37:1034–1054, 1991.
3. R. A. Baxter and D. L. Dowe. Model selection in linear regression using the MML criterion. Technical report 96/276, Department of Computer Science, Monash University, Clayton, Victoria 3800, Australia, 1996.
4. J.M. Bernardo and A.F.M. Smith. *Bayesian Theory*. Wiley, New York, 1994.
5. R.J. Bhansali and D.Y. Downham. Some properties of the order of an autoregressive model selected by a generalization of Akaike's FPE criterion. *Biometrika*, 64:547–551, 1977.
6. D. M. Boulton and C. S. Wallace. The information content of a multistate distribution. *Journal of Theoretical Biology*, 23:269–278, 1969.
7. R. Carnap. *The Logical Structure of the World & Pseudoproblems in Philosophy*. Berkeley and Los Angeles: University of California Press, 1967. Translated by Rolf A. George.
8. G. J. Chaitin. On the length of programs for computing finite sequences. *Journal of the Association for Computing Machinery*, 13:547–569, 1966.
9. G. J. Chaitin. A theory of program size formally identical to information theory. *Journal of the Association for Computing Machinery*, 22:329–340, 1975.
10. D. M. Chickering. A tranformational characterization of equivalent Bayesian network structures. In P. Besnard and S. Hanks, editors, *Proceedings of the Eleventh Conference on Uncertainty in Artificial Intelligence (UAI95)*, pages 87–98, Montreal, Quebec, 1995.
11. J. H. Conway and N. J. A. Sloane. On the Voronoi regions of certain lattices. *SIAM Journal on Algebraic and Discrete Methods*, 5:294–305, 1984.
12. D. L. Dowe, J. J. Oliver, and C. S. Wallace. MML estimation of the parameters of the spherical Fisher distribution. In *Proceedings of the 7th International Workshop on Algorithmic Learning Theory (ALT'96), Lecture Notes in Artificial Intelligence (LNAI) 1160*, pages 213–227, Sydney, Australia, 1996. Springer Verlag.
13. D. L. Dowe and C. S. Wallace. Resolving the Neyman-Scott problem by Minimum Message Length. In *Computing Science and Statistics – Proceedings of the 28th Symposium on the Interface*, volume 28, pages 614–618, Sydney, Australia, 1997.
14. R.M. Fano. The transmission of information. Technical report 65, Research Laboratory of Electronics, MIT, Cambridge, MA, 1949.
15. G. E. Farr and C. S. Wallace. The complexity of strict minimum message length inference. *The Computer Journal*, 45(3):285–292, 2002.

16. L. Fitzgibbon, D.L. Dowe, and L. Allison. Univariate polynomial inference by Monte Carlo message length approximation. In *Proceedings of the 19th International Conference on Machine Learning (ICML-2002)*, pages 147–154, Sydney, Australia, 2002. Morgan Kaufmann.

17. B.R. Gaines. Behaviour structure transformations under uncertainity. *International Journal of Man-Machine Studies*, 8:337–365, 1976.

18. P. D. Grünwald. *The Minimum Description Length Principle and Reasoning under Uncertainty*. Institute for Logic, Language and Computation, Amsterdam, 1998.

19. D.A. Huffman. A method for the construction of minimum-redundacy codes. *Proceedings of the Institute of Electrical and Radio Engineers*, 40(9):1098–1101, 1952.

20. M. Hutter. New error bounds for Solomonoff prediction. *Journal of Computer and System Sciences*, 62:653–667, 2001.

21. K. G. Joreskog. Some contributions to maximum likelihood factor analysis. *Psychometrica*, 32:443–482, 1967.

22. M. Kearns, Y. Mansour, A. Y. Ng, and D. Ron. An experimental and theoretical comparison of model selection methods. *Machine Learning Journal*, 27:7–50, 1997.

23. A. N. Kolmogorov. Three approaches to the quantitative definition of information. *Problems of Information Transmission*, 1:4–7, 1965.

24. T.C.M. Lee. Tree-based wavelet regression for correlated data using the Minimum Description Length principle. *Australian and New Zealand Journal of Statistics*, 44(1):23–39, 2002.

25. L. A. Levin. Universal search problems. *Problems of Information Transmission*, 9:265–266, 1973.

26. Ming Li and P.M.B. Vitanyi. *An Introduction to Kolmogorov Complexity and its Applications*. Springer-Verlag, New York, 2nd edition, 1997.

27. K. V. Mardia, J. T. Kent, and J. M. Bibby. *Multivariate Analysis*. Academic Press, London, 1979.

28. J. J. Oliver. Decision graphs – an extension of decision trees. In *Proceedings of the Fourth International Workshop on Artificial Intelligence and Statistics*, pages 343–350, 1993. Extended version available as TR 173, Department of Computer Science, Monash University, Clayton, Victoria 3800, Australia.

29. J.D. Patrick. *An Information Measure Comparative Analysis of Megalithic Geometries*. PhD thesis, Department of Computer Science, Monash University, 1978.

30. H. Price. *Time's Arrow and Archimedes' Point*. Oxford University Press, New York, 1996.

31. J. R. Quinlan and R. L. Rivest. Inferring decision trees using the Minimum Description Length principle. *Information and Computation*, 80(3):227–248, 1989.

32. J.R. Quinlan. *C4.5: Programs for Machine Learning*. Morgan Kaufmann, San Mateo, CA, 1993.

33. A. Raman and J. Patrick. Inference of stochastic automata using the Sk-strings method. Tr IS/2/95, Information Systems Department, Massey University, Palmerston North, New Zealand, 1995.

34. J. J. Rissanen. A universal prior for integers and estimation by Minimum Description Length. *Annals of Statistics*, 11(2):416–431, 1983.

35. J. J. Rissanen. Stochastic complexity. *Journal of the Royal Statistical Society (Series B)*, 49:260–269, 1987.

36. J. J. Rissanen. Hypothesis selection and testing by the MDL principle. *The Computer Journal*, 42:223–239, 1999.

37. J. J. Rissanen and G.G. Langdon. Arithmetic coding. *IBM Journal of Research and Development*, 23(2):149–162, 1979.

38. G. W. Rumantir and C. S. Wallace. Minimum Message Length criterion for second-order polynomial model selection applied to tropical cyclone intensity forecasting. In M. R. Berthold, H. J. Lenz, E. Bradley, R. Kruse, and C. Borgelt, editors, *Advances in Intelligent Data Analysis V, LNCS2810*, pages 486–496. Springer-Verlag, Berlin, 2003.

39. G. Schou. Estimation of the concentration parameter in von Mises-Fisher distributions. *Biometrica*, 65:369–377, 1978.

40. G. Schwarz. Estimating dimension of a model. *Annals of Statistics*, 6:461–464, 1978.

41. C. E. Shannon. A mathematical theory of communication. *Bell System Technical Journal*, 27:379–423, 623–656, 1948.

42. R. J. Solomonoff. A formal theory of inductive inference. *Information and Control*, 7:1–22,224–254, 1964.

43. R. J. Solomonoff. Two kinds of probabilistic induction. *The Computer Journal*, 42(4):256–259, 1999.

44. A. Thom. A statistical examination of the megalithic sites in britain. *Journal of the Royal Statistical Society (Series A)*, 118(3):275–295, 1955.

45. A. Thom. The egg shaped standing stones rings in Britain. *Archives Internationales D'Histoire Des Science*, 14:291–303, 1961.

46. A. Thom. *Megalithic Sites in Britain*. London: Oxford University Press, 1967.

47. V.N. Vapnik. *Estimation of Dependencies Based on Empirical Data*. Nauka, Moscow (in Russian), 1979. (English translation: V. Vapnik, *Estimation of Dependencies Based on Empirical Data*, Springer, New York, 1982.).

48. M. Viswanathan and C. S. Wallace. A note on the comparison of polynomial selection methods. In *Proceedings of Uncertainty 99: the 7th International Workshop on Artificial Intelligence and Statistics*, pages 169–177, Fort Lauderdale, FL, 1999. Morgan Kaufmann.

49. M. Viswanathan, C. S. Wallace, D. L. Dowe, and K. B. Korb. Finding cutpoints in noisy binary sequences – a revised empirical evaluation. In *Proceedings of the 12th Australian Joint Conference on Artificial Intelligence, Lecture Notes in Artificial Intelligence (LNAI) 1747*, pages 405–416, Sydney, Australia, 1999.

50. V. Vovk and A. Gammerman. Complexity approximation principle. *The Computer Journal*, 42(4):318–322, 1999. Special issue on Kolmogorov Complexity.

51. V. V. V'yugin. Algorithmic complexity and properties of finite binary sequences. *The Computer Journal*, 42(4):294–317, 1999.

52. C. S. Wallace and D. M. Boulton. An information measure for classification. *The Computer Journal*, 11:185–194, 1968.

53. C. S. Wallace and D. L. Dowe. MML estimation of the von Mises concentration parameter. Technical report 93/193, Department of Computer Science, Monash University, Clayton, Victoria 3800, Australia, 1993.

54. C. S. Wallace and D. L. Dowe. MML clustering of multi-state, Poisson, von Mises circular and Gaussian distributions. *Statistics and Computing*, 10(1):73–83, 2000.

55. C. S. Wallace and P. R. Freeman. Estimation and inference by compact coding. *Journal of the Royal Statistical Society (Series B)*, 49:240–252, 1987.

56. C. S. Wallace and P. R. Freeman. Single factor analysis by MML estimation. *Journal of the Royal Statistical Society (Series B)*, 54(1):195–209, 1992.

57. C. S. Wallace and K. B. Korb. Learning linear causal models by MML sampling. In A. Gammerman, editor, *Causal Models and Intelligent Data Management*, pages 89–111. Springer Verlag, 1999.

58. C. S. Wallace and J. D. Patrick. Coding decision trees. *Machine Learning*, 11:7–22, 1993.
59. I.H. Witten, R. Neal, and J.G. Cleary. Arithmetic coding for data compression. *Communications of the Association for Computing Machinery*, 30(6):520–540, 1987.
60. M. Woodroofe. On model selection and the arc sine laws. *Annals of Statistics*, 10:1182–1194, 1982.
61. P. Zador. Asymptotic quantization error of continuous signals and the quantization dimension. *IEEE Transactions on Information Theory*, 28:139–149, 1982.

Index

The Cross-Entropy Method

R.Y. Rubinstein and D.P. Kroese

The cross-entropy (CE) method is one of the most significant developments in randomized optimization and simulation in recent years. This book explains in detail how and why the CE method works. The CE method involves an iterative procedure where each iteration can be broken down into two phases: (a) generate a random data sample (trajectories, vectors, etc.) according to a specified mechanism; (b) update the parameters of the random mechanism based on this data in order to produce a ``better" sample in the next iteration. The simplicity and versatility of the method is illustrated via a diverse collection of optimization and estimation problems. The book is aimed at a broad audience of engineers, computer scientists, mathematicians, statisticians and in general anyone, theorist and practitioner, who is interested in smart simulation, fast optimization, learning algorithms, and image processing.

2004. 300 p. (Information Science and Statistics) Hardcover
ISBN 0-387-21240-X

Probabilistic Conditional Independence Structures

M. Studeny

Probabilistic Conditional Independence Structures provides the mathematical description of probabilistic conditional independence structures; the author uses non-graphical methods of their description, and takes an algebraic approach. The monograph presents the methods of structural imsets and supermodular functions, and deals with independence implication and equivalence of structural imsets. Motivation, mathematical foundations and areas of application are included, and a rough overview of graphical methods is also given. In particular, the author has been careful to use suitable terminology, and presents the work so that it will be understood by both statisticians, and by researchers in artificial intelligence. The necessary elementary mathematical notions are recalled in an appendix.

2005. 285 p. (Information Science and Statistics) Hardcover
ISBN 1-85233-891-1